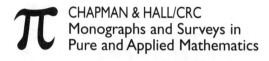

CHAPMAN & HALL/CRC
Monographs and Surveys in
Pure and Applied Mathematics    **124**

# INTERFACIAL

# PHENOMENA

# AND CONVECTION

**π** CHAPMAN & HALL/CRC
Monographs and Surveys in
Pure and Applied Mathematics **124**

# INTERFACIAL

# PHENOMENA

# AND CONVECTION

## ALEXANDER A. NEPOMNYASHCHY
## MANUEL G. VELARDE
## PIERRE COLINET

## CHAPMAN & HALL/CRC

A CRC Press Company
Boca Raton  London  New York  Washington, D.C.

## Library of Congress Cataloging-in-Publication Data

Nepomniashchii, A. A. (Aleksandr Abovich)
    Interfacial phenomena and convection / Alexander A. Nepomnyashchy, Manuel G. Velarde, and Pierre Colinet.
        p. cm. — (Monographs and surveys in pure and applied mathematics ; 124)
    Includes bibliographical references and index.
    ISBN 1-58488-256-5
        1. Heat—Convection. 2. Marangoni effect. 3. Surface tension. 4. Interfaces (Physical science) I. Velarde, Manuel G. (Manuel García) II. Colinet, P. (Pierre) III. Title. IV. Chapman & Hall/CRC monographs and surveys in pure and applied mathematics ; 124.

QC327 .N46 2001
536′.25—dc21

2001047301

### Visit the CRC Press Web site at www.crcpress.com

© 2002 by Chapman & Hall/CRC

No claim to original U.S. Government works
International Standard Book Number 1-58488-256-5
Library of Congress Card Number 2001047301
Printed in the United States of America 1 2 3 4 5 6 7 8 9 0
Printed on acid-free paper

*Dedicated to Professors Grigory I. Barenblatt and Robert S. Schechter and to the memory of Professors Grigory Z. Gershuni, Francisco Moran, Jose M. Saviron, and Efim M. Zhukhovitsky*

# Contents

# Preface

Interfacial phenomena (adsorption, desorption, evaporation, boiling, wetting, spreading, drop and bubble formation and migration, rippling, etc.) are essential for life and for many natural and artificial technological processes. In particular, interfacial convection is ubiquitous in nature; it appears all around us and inside us (small scale flows and microhydrodynamics when body forces and inertia, generally, have negligible influence). Its role in engineering processes (metallurgy, electrochemistry, welding, painting, drying, etc.) has been emphasized by numerous writers (see e.g. Levich and Krylov, 1969, Szekely, 1979). Recently, it has become obvious that in various processes in the (free fall) microgravity or variable effective-gravity environment of space laboratories, and the International Space Station, due to the practical absence of buoyancy, interfacial convection is the basic mechanism of fluid motion even in large scale processes (see, e.g. Ostrach, 1982; Walter, 1987; Ratke *et al.*, 1989).

At an open surface, or at the interface between two liquids, the surface or interfacial tension accounts for the jump in normal stresses proportional to the surface curvature across the interface (Laplace hypothesis), and hence affects the surface shape and its stability. Gravity competes with the Laplace (overpressure) force in accommodating equipotential levels with curvature (minimizing the corresponding free energy). This balance permits, for instance, the stable equilibrium of the (practically) spherical shape of drops and bubbles as we see them around us.

When surface tension varies with position along an interface, its change takes care of the jump in the tangential stresses. Hence its gradient acts like a shear stress applied by the interface on the adjoining bulk phases and thereby generates flow or alters an existing one (Marangoni effect). Surface tension gradient-driven (Marangoni-driven) flows are known to affect the evolution of growing fronts, and measurements of heat and other diffusion coefficients.

Gradients may be due to heat or mass transfer processes occurring along or across an interface (hence leading to thermocapillarity or solutocapillarity as forms of the Marangoni effect) or, indirectly, may originate in buoyancy-driven convection or in any other form of flow. Electric and magnetic fields can influence surface tension and hence flow at an interface. Electrocapillarity and magnetocapillarity can also, by extension, be considered as forms of the Marangoni effect. An interface, clearly, provides a nontrivial coupling mechanism between two bulk phases, surface and body forces, flow, and transport processes. From such coupling occurs a large variety of phenomena. Noticeable is that although no Marangoni effect may originally occur in a liquid-liquid system, interfacial (Marangoni-driven) convection could develop due to temperature variations resulting from positive or negative heats of solution.

Needless to say, concentration and/or temperature gradients may help in reducing or producing convection. Take the growth of crystals from a melt, which is a process where the elimination of convection in the melt is desirable because flow produces non-uniform growth conditions and, thereby, an increase in the generally uncontrollable number of dislocations and other defects in the crystal (see, e.g. Ostrach, 1982, 1983; Regel, 1987; Walter, 1987; Ratke *et al.*, 1989). The evaporative purification of a levitated melt is a process in which the opposite result is desired: convection is important here because it increases the rate of purification by replenishing the impurity concentration at the surface and because it tends to maintain a uniform composition throughout the melt. Furthermore, interfaces often contain traces of surface active substances (surfactants) that alter, significantly, the surface tension. In general, surface tension-lowering solutes adsorb preferentially in the interface (Gibbs adsorption hypothesis). The longer the characteristic time for a solute to redistribute itself between the interface and the bulk of the liquid, the more surface active the solute is. If the interface expands locally, these surfactants are swept outward with the movement, creating a gradient in their concentration. This concentration gradient implies a surface tension gradient which acts opposite to the movement. This effect was well explained by Levich (1962), particularly when dealing with drop migration in the presence of thermal gradients (see also Bakker *et al.*, 1966).

Progress in the theory, numerics, and experiments with interfacial phenomena has been made in the past decades, leading to our understanding of the conditions for interfacial instability and/or interfacial convection to occur. Some examples are the study of cellular (Benard) convection and its evolution, rippling and the generation of (nonlinear) waves and solitons, drop and bubble migration in the presence of thermal gradients, and three-dimensional surface tension gradient-driven (Marangoni-driven) flows and related spatio-temporal problems. Interfacial convection has, however, been much less studied than buoyancy-driven, natural convection, and other flows driven by pressure gradients and body forces. In fact, the study of liquid flow along an interface from places with low surface tension to places with a higher surface tension started long ago when the Italian scientist Marangoni studied the conditions for spreading of one liquid on another in, among other places, the largest basin of the Tuileries gardens in Paris. He stated that a liquid A spreads on a liquid B when the sum of the interfacial tension and the surface tension of A is lower than the surface tension of B. He reported on this phenomenon in an 1865 brochure, and made his research more widely available six years later (1871a,b), because of his fear that publications by the Belgian scientist, and Plateau's son-in-law, Van der Mensbrugghe (1870, 1873) and the German scientist Lüdtge (1869), the latter partially in error, would render his priority on this subject unacknowledged. Plateau, in a book published in 1873, gives proper credit to the work of Marangoni. He also describes the work of Dupre de Rennes (1869) with findings identical to those of Marangoni. Young (1805) and later Maxwell (1871, 1878) correctly stated the spreading laws that, finally, Harkins and Feldman (1922) established on sound thermodynamic ground.

James Thomson (1855, 1881), the older brother of Lord Kelvin, also established, albeit qualitatively only and not with fully correct understanding of the phenomena, though he distinguished buoyancy from surface tension effects, that when gradients in surface tension arise due to concentration differences within one fluid, flow arises as well (1855). He explained the tears found in a glass of wine or any other strong alcoholic liquor in terms of surface tension gradients, and for this reason the Marangoni effect has on occasion been referred to as the Thomson effect. Worth mentioning also is the work of Weber (1854), who described convective motions occurring at the surface of bubbles placed in alcohol solutions. Earlier, Varley (1836) had described curious motions in evaporating drops observed under a microscope.

In observing but not explaining some of the above-described phenomena, both Marangoni and Thomson were far outdated by other authors. In 1686, Heyde observed dancing camphor on olive oil. Much earlier, Plutarch and Pliny the Elder (Levich, 1962) reported the calming of the sea as a result of sailors spreading olive oil over the sea surface, a phenomenon also studied by Benjamin Franklin (1774; Tanford, 1989; see also Tomlison, 1864, 1869 and Scriven and

Sternling, 1960).

In the present book, we provide a succinct account of results concerning interfacial phenomena and convection in various systems, albeit in each case limited to the simplest possible but significant model-problem from which we have extracted universal features. In the Introduction, we give a brief description of the hydrodynamics needed to understand the remainder of the book in a relatively selfconsistent way. We recall the Navier-Stokes, Fourier, and Fick equations as well as their corresponding boundary conditions, as interfaces are treated as boundaries rather than genuine, autonomous phases. We also discuss phenomena related to chemical reaction, heat and mass transfer along or across interfaces, and features of adsorption and desorption phenomena. We delineate the role of the Marangoni effect that flow at an interface or an open surface occurs whatever there is variation of surface tension, but whether or not the initial flow disturbance is sustained and/or penetrates in the adjoining bulk phases depends on the strength of such gradient relative to viscous damping and, eventually, on the ratio of their kinematic viscosities and heat or mass diffusivities.

We then proceed with the consideration of the relatively simple but important case where convection is caused by imposed concentration or temperature gradients along the interface (Chapter 2). The following chapters are devoted to the development of linear and nonlinear theory of surface tension gradient-driven flows which appear spontaneously as a result of instability of the interface, and eventually, flow generated by growing disturbances induced by the Marangoni effect.

In Chapter 3 we concentrate on the phenomena of drop (bubble) migration and features of drop spreading due to the Marangoni effect. In particular, we show how a surface tension gradient-driven instability may either augment drag or, the opposite, may help overcoming hydrodynamic drag, hence leading to the spontaneous selfpropulsion or autonomous motion of a drop whose surface is affected by a (surface) chemical reaction or by internal heating.

In Chapter 4 we provide a succinct description of salient features of interfacial convection in the form of patterned, cellular flows. As in other cases, our study is limited to the simplest albeit significant model-problems, with particular attention paid to the consequences, hence the solutions, of an equation proposed by Knobloch, together with significant generalizations to account for the salient features of patterns in surface tension gradient–driven (Benard-Marangoni) convection.

In Chapter 5 we deal with oscillations and waves of different physical nature though all excited by the Marangoni effect. A detailed analysis is provided of two types of interfacial waves. One is the capillary-gravity wave which corresponds to membrane-like transverse oscillations. The other, first studied by J. Lucassen, corresponds to an elastic, longitudinal expansion-compression (sound-like) motion of a membrane that can only occur when there are tangential stresses and the Marangoni effect. The latter, as well as viscosity, indeed affect capillary-gravity waves, but as secondary factors. For the longitudinal (also called dilational) waves, the Marangoni effect and viscosity are key for their onset and evolution and hence dilational waves are strictly dissipative waves. We also provide a succinct account of recent findings about mode mixing and resonance of interfacial wave modes. Furthermore, we recall major (theoretical, numerical, and experimental) results about solitonic aspects of interfacial waves, whose observation was made three decades ago by H. Linde before the soliton concept was coined. In particular, as the simplest significant model we discuss the generalization of the Boussinesq-Korteweg-de Vries equation for waves in shallow liquid layers when the Marangoni effect is added.

In Chapter 6 discussion is provided first about the stability of flows generated by a longitudinal surface tension gradient (including a short account of the combined action of thermocapillarity and buoyancy) already introduced in Chapter 2, and subsequently, about film flows with transverse thermal gradients. In the latter case, the model-problem chosen is the Kapitza-Shkadov falling film case with the Marangoni effect added. This chapter also contains a study of flows in two-layer systems and the corresponding stability analysis.

Finally, Chapter 7 is devoted to speculative comments about a few problems and topics that we feel have great interest, both basic and application-oriented.

When writing the book we had in mind graduate students and researchers from applied mathematics, the nonlinear sciences, and various engineering branches. We have done our best to make the material reasonably self-contained and accessible to the reader. Throughout the book we offer not just a description of phenomena but as much as possible heuristic argumentation and significant portions of methodologies whose utility exists well beyond the domain of problems discussed here.

In the Bibliography we offer the reader reference to original publications as well as somewhat redundant references, thus offering alternative and diverse reading material, particularly when citing books and review articles.

This monograph is part of the training effort done by the authors in the framework of the Interfacial Convection and Phase Change (ICOPAC) Network sponsored by the European Union.

# Acknowledgments

Over the years, the authors have learned and benefited from collaboration with many colleagues. This list is alphabetically ordered and therefore does not discriminate the degree of gratitude and rank. With apologies to those unhappy about this format, we take pleasure in manifesting our appreciation to the following individuals: D. Bar, P. Berge (deceased), M. Bestehorn, R. V. Birikh, J. Bragard, L. Braverman, V. A. Briskman (deceased), J. W. Bush, M. Castans, A. Castellanos, P. Cerisier, C. I. Christov, X.-L. Chu, P. Dauby, M. Dubois, K. Eckert, A. Hari, M. Hennenberg, A. N. Garazo, A. Golovin, H. Gouin, E. Guyon, I. B. Ivanov, D. Johnson, Y. Kliakhandler, E. L. Koschmeider, S. Kosvintsev, V. Kurdyumov, G. Lebon, J. Cl. Legros, A. Linan, H. Linde, A. Mendes-Tatsis, K. C. Mills, S. W. Morris, R. Narayanan, V. I. Nekorkin, B. Nicolaenko, K. Neuffer, A. Paterson, J. Pantaloni, F. Petrelis, L. Pismen, Y. Pomeau, J. Pontes, A. Porubov, D. Quere, A. Ye. Rednikov, Yu. S. Ryazantsev, A. de Ryck, J. Salan, A. Sanfeld, M. Santiago-Rosanne, G. S. R. Sarma, M. F. Schatz, D. Schwabe, V. Shkadov, V. Shkadova, I. Simanovski, B. L. Smorodin, V. D. Sobolev, V. M. Starov, A. Steinchen-Sanfeld, A. Thess, U. Thiele, S. Van Vaerenberg, J. VanHook, M. Vignes-Adler, P. D. Weidman, J. E. Wesfreid, A. Wierschem, and R. Kh. Zeytounian.

For the preparation of the compuscript, Dr. Valeri Makarov and Ms. Iouliia Makarova have been most invaluable. Our sincere thanks.

Economic support for this research came from the Ministerio de Educacion y Cultura (Spain), the Ministerio de Ciencia y Tecnologia (Spain), Fonds National de la Recherche Scientifique (Belgium), the European Union, the BBV Foundation (Spain), the Minerva Foundation (Germany), the InterUniversity Poles of Attraction Programme (Belgian Federal Office for Scientific, Technical and Cultural Affairs and the European Space Agency (ESA) thanks to its Microgravity Office and the enthusiasm of G. Seibert, H. U. Walter and O. Minster.

# Chapter 1

# Introduction

In this chapter we shall recall the concept of surface or interfacial tension, the Navier–Stokes, Fourier, and Fick hydrodynamic equations, and we shall discuss various heat and mass transfer phenomena occuring at an interface, deformable or not. The Marangoni effect is also discussed.

## 1.1 The interface as a physical system

### 1.1.1 Interfacial tension

The transition layer (Fig. 1.1) between two immiscible fluids (as well as between a fluid and a solid body) has a thickness of *microscopic* size (usually few molecular diameters) and is considered as a *two-dimensional surface* from the point of view of *macroscopic* theories (thermodynamics, fluid mechanics, etc.).

In the thermodynamic approach, the interfacial region should be described as a specific two-dimensional medium which posesses its own interfacial internal energy, $U^S$ (physically, it is the excess energy caused by the molecular interactions in the transition layer), interfacial entropy, $S^S$ (caused by a new possibility of randomness relative to that of bulk fluids), and interfacial numbers of molecules of various species, $n_j^S$. Let us note that $n_j^S$ may be either positive (*adsorption*) or negative (*desorption*). The first law of thermodynamics in reference to the interface may be written as (see, e.g. Defay, Prigogine, Bellemans and Everett, 1966; Landau and Lifshitz, 1980; Adamson, 1982)

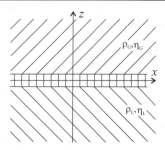

Figure 1.1: Geometry of an interface separating two liquids or
a liquid (L) and ambient air, gas (G), or vapor. $\rho_G$ ($\rho_L$) and
$\eta_G$ ($\eta_L$) denote density and viscosity, respectively. In the simplest
case, one neglects the internal structure and when the fluctuations
of the interface out of the $z = 0$ plane can be considered small, the
boundary conditions (b.c.) for each bulk phase can be expressed
in terms of the fluid displacements evaluated at the geometrical
level $z = 0$. Otherwise if one neglects the structure and transverse
size of the interface but not its deformability, b.c. are taken at
the moving interface surface, $z(t)$.

$$dU^S = TdS^S + \sum_j \mu_j dn_j^S + \sigma dA, \qquad (1.1)$$

where $T$ is temperature, $\mu_j$ are interfacial chemical potentials, $A$ is
*area* of interface which replaces the volume characteristic for three-
dimensional media. The parameter $\sigma$ which describes the work which
is necessary for changing the area of the interface is called *interfacial
tension*. In the case where a liquid is in contact, equilibrium, with its
own saturated vapour, this parameter is called *surface tension* of the
liquid. Certainly, the interfacial internal energy may depend also on
some additional thermodynamic parameters like density of electric
charge, etc.

For a stable interface, the interfacial tension should be positive
(otherwise, the interface would be destroyed by fluctuations). The
dependence of the interfacial tension on the thermodynamic parameters (temperature, interfacial concentration of species, etc.) is similar
to the equation of state for the three-dimensional medium. Generally, this dependence is not obtained from first principles and is taken
from experiment. Typically, the interfacial tension decreases when
the temperature is increased. This phenomenon is called *normal*

*thermocapillarity*. However, several systems are known that display an opposite kind of behavior (*anomalous thermocapillarity*).

The dependence of the interfacial tension on the interfacial concentration of species deserves a special discussion. In a thermodynamic equilibrium state where all the *intensive* variables (temperature, interfacial tension, etc.) are constant, the *extensive* variables $U^S$, $S^S$, and $n_i^S$ are all proportional to the area of the interface $A$. Thus, Eq. (1.1) may be written as

$$U^S = TS^S + \sum_j \mu_j n_j^S + \sigma A. \tag{1.2}$$

Combining Eqs. (1.1) and (1.2), one can find *the Gibbs adsorption equation*

$$-S^S dT + \sum_j n_j^S d\mu_j + A d\sigma = 0. \tag{1.3}$$

For sake of simplicity, let us consider the case of a single relevant species (e.g., the interface between a liquid and a gas soluble in the liquid) and omit the subscripts for $n_1$ and $\mu_1$. For the *interfacial concentration*

$$\Gamma \equiv \frac{n^S}{A} \tag{1.4}$$

we find:

$$\Gamma = -\left(\frac{\partial\sigma}{\partial\mu}\right)_T. \tag{1.5}$$

At the same time, the following thermodynamic inequality can be established

$$\left(\frac{\partial\mu}{\partial\Gamma}\right)_T > 0. \tag{1.6}$$

Thus, the interfacial concentration

$$\Gamma = -\left(\frac{\partial\sigma}{\partial\Gamma}\right)_T \left(\frac{\partial\Gamma}{\partial\mu}\right)_T$$

is positive (positive adsorption) if and only if

$$\left(\frac{\partial\sigma}{\partial\Gamma}\right)_T < 0, \tag{1.7}$$

i.e., the interfacial tension decreases as the interfacial concentration increases. In other words, the interface tends to be enriched in the species which diminishes the surface tension.

The equilibrium interfacial concentration $\Gamma$ depends on the *volume concentration*, $C$, of the species in the bulk fluid. For dilute solutions, they are proportional. For higher volume concentrations, the interfacial concentration may tend to a limit value corresponding to a *monomolecular layer* created by molecules adsorbed on the interface which are called surface-active or surfactant molecules. Typical examples of surfactants generating interfacial monolayers even for quite small volume concentrations are hydrocarbons (and other substances with molecules containing hydrocarbon chains) on the interfaces water/air or water/oil.

In some cases, the thermodynamic equilibrium between the volume and interfacial concentrations of surfactant is established during a rather long time. Therefore, the *interfacial kinetics* should be taken into account. The flux $j$ characterized the mass exchange between the bulk solution and the interface is taken in the form

$$j = k_a C - k_d \Gamma, \tag{1.8}$$

where $k_a$ and $k_d$ are the adsorption and desorption rate constants. The adsorption-desorption kinetics is discussed in more detail in subsection 1.2.2.

### 1.1.2   Hydrodynamic properties of the interface

Let us discuss now the hydrodynamic phenomena caused by the presence of an interface.

First of all, the deformation of the interface changing its area changes its energy (1.1). This circumstance leads to additional force connected with the interface. Then the pressures $p_1$ and $p_2$ in two contacting media are not equal, and the difference (Laplace surface pressure or overpressure) is

$$p_1 - p_2 = \sigma \left( \frac{1}{R_1} + \frac{1}{R_2} \right), \tag{1.9}$$

where $R_1$ and $R_2$ are the principal radii of curvature at a given point of the interface. The pressure is higher in the medium whose surface is convex, hence it is higher inside a drop or a bubble relative to the surrounding fluid. The difference of pressures may be considered as a *normal force* per unit area

$$\mathbf{f}_n = \sigma \left( \frac{1}{R_1} + \frac{1}{R_2} \right) \mathbf{n}$$

directed along the normal **n**.

Besides the normal force, there may be a force *tangential* to the interface if inhomogeneity of the interfacial tension exists (generated e.g. by the inhomogeneities of temperature or surfactant concentration on the interface). The tangential force per unit area is equal to

$$\mathbf{f}_t = \nabla \sigma$$

(because $\sigma$ is defined only on the interface, $\nabla$ denotes the surface gradient and its corresponding stress is named after Marangoni). This force produces flow or alters an existing one and hence one has a surface tension gradient-driven convection, thermocapillary convection or *Marangoni–driven convection*. Generation of motion by interfacial tension inhomogeneities has also been called the *Marangoni effect* (Block, 1956; Scriven and Sternling, 1960). Near the interface, the motion is directed towards the region with larger surface tension; because of the incompressibility of fluids, a bulk motion in the opposite direction supported by a pressure gradient will arise (Fig. 1.2).

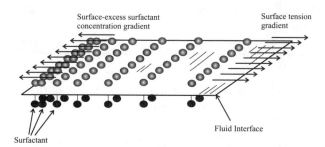

Figure 1.2: The Marangoni effect. Since surfactant adsorption at a fluid interface generally lowers the interfacial tension, an inhomogeneous distribution of surfactant within the interface results in the creation of local interfacial tension gradients. The highest tension occurs in those regions where the surfactant density is lowest; thus, the tensile restoring force acts in an opposite direction to the surfactant density gradient. The ensuing fluid motion arising from this interfacial tension gradient is named after Marangoni (reproduced by courtesy of Edwards, Brenner, and Wasan, 1991).

Thus a system with an interface between two fluids may be in a *mechanical equilibrium state* only if the surface tension is constant on the whole interface. Otherwise, convection arises, however small are the inhomogeneities of the surface tension. However, in the case where the temperature or volume concentration gradients are *normal* to the interface, they do not produce inhomogeneity of the surface tension, and hence they do not destroy the mechanical equilibrium of the system. Thus the interfacial or surface tension (up to the sign) is to the surface what pressure is to the bulks.

It should be emphasized, however, that such a mechanical equilibrium state may be *unstable*, and one observes convection even though the external conditions may be homogeneous. As a rule, such an instability occurs only if the applied normal gradients are strong enough.

Various flows and instabilities leading to flows past a given threshold generated by interfacial tension inhomogeneities and the Marangoni effect are the subject of this book.

**Table 1.1.  Surface tension of various fluids**

| Liquid | $\sigma$(mN/m) | in contact with | T($^\circ$C) |
|--------|-----------|-----------------|--------|
| Benzene | 28.9 | air | 20 |
| Ethylalcohol | 22.3 | air | 20 |
| $n$-Hexane | 18.4 | air | 20 |
| Mercury | 470 | air | 20 |
| Water | 74 | air | 20 |
| Helium | 0.24 | vapor | -270 |

**Table 1.2.  Order of magnitude of bulk properties of various fluids**

| Liquid | $\rho$(kg/m$^3$) $\times 10^3$ (g/cm$^3$) | $\eta$(kg/m s) $\times 10^3$ | $\nu$(m$^2$/s) $\times 10^6$ (cm$^2$/s) | $\lambda$ (W/m K) | $\kappa$(m$^2$/s) $\times 10^6$ (cm$^2$/s) | $\alpha$ (1/K) $\times 10^3$ | $Pr$ |
|--------|--------|--------|--------|--------|--------|--------|--------|
| Mercury and molten metals | $10^4$ 10 | 1 | $10^{-3}$ cm$^2$/s | 10 | $10^{-1} - 10^{-2}$ cm$^2$/s | | $\geq 10^-$ |
| Helium | | | | $10^{-2}$ | | | |
| Water (room temperature) | $10^3$ | 1 | 1 $10^{-2}$ cm$^2$/s | $10^{-1}$ | $\geq 10^{-1}$ $10^{-3}$ cm$^2$/s | 0.32 | (5.85) $\leq 10$ |
| Silicone oils | $\leq 10^3$ | 10 | $\geq 10$ $\geq 1$cm$^2$/s | $10^{-1}$ | $\geq 10^{-2}$ $10^{-3}$ cm$^2$/s | 1 | $\geq 10^2$ |
| Olive oil | $10^3$ | | | | | | |
| Glycerine | | $10^3 - 10^7$ | $\geq 10$cm$^2$/s | | $10^{-3}$ | | $10^2$-10 |
| Fluorinert | $2 \times 10^3$ | $\geq 10^2$ | $\geq 10$ | $\geq 10^{-1}$ | $10^{-2}$ | 1 | $\geq 10^2$ |
| Ethanol | $\leq 10^3$ | 0.5 - 40 | 1.5 | | | | |
| Methanol | $\leq 10^3$ | $\geq 10^{-1}$ | $\geq 10^{-1}$ | $\geq 10^{-1}$ | $\geq 10^{-1}$ | 1 | 10 |
| $n$-octane | $\leq 10^3$ | $\geq 10^{-1}$ | $\geq 10^{-1}$ | $\geq 10^{-1}$ | $\geq 10^{-1}$ | 1 | 10 |
| Air and standard gases | 1 | $\geq 10^{-2}$ | $\geq 10$ $10^{-1}$cm$^2$/s | 2.2 | 0.02 $10^{-1}$cm$^2$/s | 3.67 | (0.71) $\leq 1$ |

## 1.2   Mathematical formulation

In the present section we shall describe some basic mathematical models corresponding to the above-mentioned phenomena.

## 1.2.1  Heat transfer in a system with an interface

### i. Equations and boundary conditions

Let us consider two fluids situated in regions $L_1$ and $L_2$ and separated by an interface. In a spatially inhomogeneous temperature field, convective motion of fluids arises for two main reasons. First, the temperature inhomogeneity produces a spatial inhomogeneity of the interfacial tension that leads to *thermocapillary forces* generating *thermocapillary convection* (Marangoni effect). Also, because of thermal expansion, the densities of fluids are spatially inhomogeneous, which may cause bulk *buoyancy forces* generating *buoyancy-driven convection* (Rayleigh, 1916).

For description of the bulk motions of both fluids we shall use the Boussinesq approximation (Oberbeck, 1879; Boussinesq, 1901, 1903; Mihaljan, 1962; Perez-Cordon and Velarde, 1975; de Boer, 1984, 1986; Joseph, 1976; Velarde and Perez-Cordon, 1976; Gershuni and Zhukhovitsky, 1976; Velarde *et al.*, 2000). This approximation is valid in the case were the density variations caused by temperature inhomogeneities are relatively small, while the characteristic temperature gradient is much larger than the adiabatic temperature gradient, $g\beta T/c_p$, where $g$ is gravity acceleration, $\beta$ is the thermal expansion coefficient, $T$ is absolute temperature, and $c_p$ is the specific heat at constant pressure. Within this approximation, the dependence of the density, $\rho$, on the pressure is ignored. Flow motions are limited to velocities much lower than the speed of sound in the liquid and hence the "incompressibility" assumed. Also, the dependence of the density on the temperature is ignored in the continuity equation, while in the equation of motion it is taken into account. The equations of motion are identical to those of an incompressible fluid with addition of a bulk *buoyancy force*, $-\mathbf{g}\beta T$. In the equation of heat transfer, the compressibility effects and the viscous heat generation are neglected. Also, the values of (dynamic) shear viscosity, kinematic viscosity, heat conductivity and heat diffusivity $\eta$, $\nu$, $\lambda$, and $\kappa$ are assumed to be constant; $\eta = \rho\nu$. A denotes the temperature gradient.

Using the subscript $m$ for the quantities corresponding to $m$-th fluid ($m = 1$, 2), we can write the equations in the following form (Lamb, 1945; Moran, 1960; Chandrasekhar, 1961; Aris, 1962; Levich, 1962; Batchelor, 1967; Segel, 1972; Lin and Segel 1974; Gershuni and Zhukovitsky, 1976; Joseph, 1976; Normand et al, 1977; Segel, 1977;

Velarde and Castillo, 1982; Miller and Neogi, 1985; Rosner, 1986; Landau and Lifshitz, 1987; Acheson, 1990; Zeytounian, 1998; Guyon *et al.*, 2001):

$$\frac{\partial \mathbf{v}_m}{\partial t} + (\mathbf{v}_m \cdot \nabla)\mathbf{v}_m = -\frac{1}{\rho_m}\nabla p_m + \nu_m \nabla^2 \mathbf{v}_m + g\beta_m T_m \mathbf{e}, \quad (1.10)$$

$$\frac{\partial T_m}{\partial t} + (\mathbf{v}_m \cdot \nabla)T_m = \kappa_m \nabla^2 T_m, \quad (1.11)$$

$$\nabla \cdot \mathbf{v}_m = 0. \quad (1.12)$$

Here $\mathbf{e}$ is a unit vector directed upward, $p_m$ is the *hydrostatic* pressure; the total pressure is equal to $p_m - \rho_m gz$, where $z$ is the vertical coordinate.

Now, we have to write down the boundary conditions on the interface between the fluids. For simplicity, we shall assume that the interface is described by the equation

$$z = h(x, y, t). \quad (1.13)$$

Taking into account both Laplace surface pressure and the thermocapillary stresses, we can write the following boundary condition describing the balance of stresses on the interface:

$$\left[ (p_1 - \rho_1 gh) - (p_2 - \rho_2 gh) - \sigma \left( \frac{1}{R_1} + \frac{1}{R_2} \right) \right] n_i = \quad (1.14)$$

$$(\tau'_{1,ik} - \tau'_{2,ik})n_k + \frac{\partial \sigma}{\partial x_i},$$

where

$$\tau'_{m,ik} = \eta_m \left( \frac{\partial v_{m,i}}{\partial x_k} + \frac{\partial v_{m,k}}{\partial x_i} \right)$$

is the viscous stress tensor for the $m$-th fluid, $n_i$ is the normal vector directed into the 1st fluid; because $\sigma$ is defined only on the interface, the expression $\frac{\partial \sigma}{\partial x_i}$ denotes a surface gradient. Introducing also orthogonal tangential vectors $\tau^{(1)}$ and $\tau^{(2)}$, we can rewrite separately the balance conditions for normal and tangential stresses:

$$(p_1 - \rho_1 gh) - (p_2 - \rho_2 gh) - \sigma \left( \frac{1}{R_1} + \frac{1}{R_2} \right) = (\tau'_{1,ik} - \tau'_{2,ik})n_i n_k, \quad (1.15)$$

$$(\tau'_{1,ik} - \tau'_{2,ik})\tau_i^{(l)} n_k + \frac{\partial \sigma}{\partial x_i}\tau_i^{(l)} = 0, \, l = 1, 2. \quad (1.16)$$

Eq. (1.16) defines the Marangoni stress.

Also, at the interface the velocities of both fluids are equal:

$$\mathbf{v}_1 = \mathbf{v}_2. \tag{1.17}$$

The motion of the interface itself is governed by the following kinematic condition:

$$\frac{\partial h}{\partial t} + v_{1,x}\frac{\partial h}{\partial x} + v_{1,y}\frac{\partial h}{\partial y} = v_{1,z}. \tag{1.18}$$

Finally, the conditions for temperatures and heat fluxes should be written. The temperature field is continuous:

$$T_1 = T_2, \tag{1.19}$$

and the normal components of heat fluxes are equal:

$$\left(\lambda_1\frac{\partial T_1}{\partial x_i} - \lambda_2\frac{\partial T_2}{\partial x_i}\right)n_i = 0. \tag{1.20}$$

Also, some boundary conditions should be fixed on the external boundaries of domains $D_1$ and $D_2$, depending on the physical nature of these boundaries.

If the energy spent by the interface deformation is taken into account, the following equation is obtained (Napolitano, 1978)

$$\left(\lambda_1\frac{\partial T_1}{\partial x_i} - \lambda_2\frac{\partial T_2}{\partial x_i}\right)n_i = \alpha T_1\nabla_s\cdot\mathbf{v}_1 + \gamma\left(\frac{\partial T_1}{\partial t} + \mathbf{v}_1\cdot\nabla T_1\right),$$

where $\nabla_s = \nabla - (\mathbf{n}\cdot\nabla)\mathbf{n}$, $\alpha = -d\sigma/dT$, $\gamma = d(\sigma + \alpha T)/dT$. However, the correction is small under realistic conditions (Pukhnachev, 1987), and we shall neglect it.

Let us note also that the deformation and expansion of the interface may generate additional dissipation processes known as the "surface viscosity" (Boussinesq, 1913a-c; Scriven, 1960; Aris, 1962; Goodrich, 1981; Edwards et al., 1991). Experimental data concerning this phenomenon are scarce. We shall not include the surface viscosity into our description.

## ii. One-fluid approach

In the case of a liquid/gas interface, a simplified mathematical model may be used. Because of relatively low (dynamic) shear viscosity of the gas, we can expect that the influence of the gas motion on the motion in the liquid is negligible and hence the terminology "free surface". Also, we can ignore the temperature field in the gas using some empirical boundary conditions for temperature. In this case, we can consider the processes that take place *only* in the liquid phase. Such a "one-fluid" approach may be justified only under some conditions (see, e.g., Golovin *et al.*, 1995), and in some cases makes it difficult to permit a comparison with experiments in the framework of such an approach (because of unmeasurable empirical coefficients used). Nevertheless, this approach may be useful for the qualitative description of the physical phenomena.

Considering the motion of just the liquid, we drop the subscripts in equations (1.10)-(1.12):

$$\frac{\partial \mathbf{v}}{\partial t} + (\mathbf{v} \cdot \nabla)\mathbf{v} = -\frac{1}{\rho}\nabla p + \nu \nabla^2 \mathbf{v} + g\beta T\mathbf{e}, \qquad (1.21)$$

$$\frac{\partial T}{\partial t} + (\mathbf{v} \cdot \nabla)T = \kappa \nabla^2 T, \qquad (1.22)$$

$$\nabla \cdot \mathbf{v} = 0. \qquad (1.23)$$

In the boundary conditions for stresses we neglect the stresses caused by the gas (m=2) and omit the subscript 1 corresponding to the liquid:

$$(p - \rho gh) - \sigma\left(\frac{1}{R_1} + \frac{1}{R_2}\right) = \tau'_{ik}n_i n_k, \qquad (1.24)$$

$$\tau'_{ik}\tau_i^{(l)}n_k + \frac{\partial \sigma}{\partial x_i}\tau_i^{(l)} = 0, \, l = 1, 2. \qquad (1.25)$$

The condition (1.17) is cancelled, while the kinematic condition (1.18) is rewritten as

$$\frac{\partial h}{\partial t} + v_x\frac{\partial h}{\partial x} + v_y\frac{\partial h}{\partial y} = v_z. \qquad (1.26)$$

Instead of exact conditions for temperatures and heat fluxes we shall use some empirical condition, say,

$$\lambda\frac{\partial T}{\partial x_i}n_i = K(T - T_g), \qquad (1.27)$$

where $K$ is a heat exchange coefficient which may depend on local curvature of the surface, local temperature etc, $T_g$ is some characteristic temperature of the ambient gas (recall that vector $\mathbf{n}$ is directed into the liquid).

### iii. Non-dimensional parameters or dimensionless groups

Let us rewrite the system (1.21) - (1.27) in a non-dimensional form. Let us assume that $a$ is a characteristic spatial size of the region filled by the fluid, and $\theta$ is a characteristic temperature difference across this region. We shall choose $a$, $a^2/\kappa$, $\kappa/a$, $\eta\kappa/a^2$ and $\theta$ as units for length, time, velocity, pressure and temperature, respectively. The system of equations (1.21) – (1.27) takes the following form (expecting no confusion by the reader, we use the same letters denoting non-dimensional variables):

$$\frac{1}{P}\left[\frac{\partial \mathbf{v}}{\partial t} + (\mathbf{v}\cdot\nabla)\mathbf{v}\right] = -\nabla p + \nabla^2\mathbf{v} + RT\mathbf{e}, \qquad (1.28)$$

$$\frac{\partial T}{\partial t} + (\mathbf{v}\cdot\nabla)T = \nabla^2 T, \qquad (1.29)$$

$$\nabla\cdot\mathbf{v} = 0, \qquad (1.30)$$

where $P = \nu/\kappa$ is the *Prandtl number* which is the ratio of the typical heat diffusion time, $\tau_\kappa = a^2/\kappa$, and the viscous momentum transfer time, $\tau_\nu = a^2/\nu$, $R$ is the *Rayleigh number*, $R = \beta g a^4 A/\nu\kappa$, which governs buoyancy–driven, natural convection. There are cases where the *Grashof number*, $Gr = R/P$, is used. A thorough discussion of scaling and adimensionalization can be found in the books by Palacios (1964), Lin and Segel (1974), Barenblatt (1996) and Guyon *et al.* (2001) and the reviews by Segel (1972), Ostrach (1977) and Castans (1991).

Let us assume that the dependence of the surface tension $\sigma$ on the temperature is linear: $\sigma = \sigma_0 - \alpha T$. The boundary condition on the interface $z = h$ (recall that we use the same letters denoting non-dimensional variables) are:

$$p - Gah - Ca^{-1}(1 - \delta_\alpha T)\left(\frac{1}{R_1} + \frac{1}{R_2}\right) = \tau'_{ik}n_i n_k, \qquad (1.31)$$

$$\tau'_{ik}\tau_i^{(l)}n_k - M\frac{\partial T}{\partial x_i}\tau_i^{(l)} = 0,\ l = 1,2, \qquad (1.32)$$

$$\frac{\partial h}{\partial t} + v_x\frac{\partial h}{\partial x} + v_y\frac{\partial h}{\partial y} = v_z, \qquad (1.33)$$

$$\frac{\partial T}{\partial x_i}n_i = -Bi(T - \bar{T}_g), \qquad (1.34)$$

where $Ga = ga^3/\nu\kappa$ is the *(modified) Galileo number* (the Galileo number is, traditionally, defined as $G = ga^3/\nu^2$ and later we shall use it like this. Then $G$ is a ratio of length scales that recalls the role of gravity and hydrostatic pressure variations relative to viscous flow), $Ca = \eta\kappa/\sigma_0 a$ is the *capillary or crispation number* (other authors use its inverse), $M = \alpha a^2 A/\eta\kappa$ is the *Marangoni number*, $Bi = Ka/\lambda$ is the *Biot number*, $\delta_\alpha = \alpha\theta/\sigma_0 = M/Ca$, and $\bar{T}_g = T_g/\theta$. Let us mention here also the (static) *Bond number* $Bo = \rho ga^2/\sigma_0 = GaCa$ and the dynamic Bond number $Bd = \rho ga^2/\alpha = R/M$. The physical meaning of each parameter will be discussed later. On the other hand, note that, for simplicity, here we denoted with $R$, $M$, and $P$ the Rayleigh, Marangoni, and Prandtl numbers, respectively. Later we shall be using $Ra$, $Ma$, and $Pr$ whenever we feel it might help clarity in the notation. For the Galileo number we shall be using $Ga$ or $G$ according to the context. The same would be the case with the capillary number, $C$ or $Ca$.

### iv. Influence of convection on the deformation of the interface

It is necessary to emphasize that the formulated boundary problem is *incorrect* from the physical point of view if both the Rayleigh number, $R$, and the Galileo number, $Ga$, are taken of the same order. The Boussinesq approximation is based on the assumption of small relative deviations of density : $\delta_\beta = \beta\theta = R/Ga \ll 1$. If the latter condition is violated, the consideration of non-Boussinesq corrections in the equation of motion and in the continuity equation are mandatory (Velarde *et al.*, 2000). Also, such corrections are necessary for a self-consistent description of effects caused by small but nonzero $\delta_\beta$. Otherwise, some spurious results may be produced by using the system (1.28) – (1.34).

Thus, when considering buoyancy–driven convection ($R = O(1)$) we should assume that $Ga \gg 1$, to be within the Boussinesq approximation. Typically, on Earth $Ga$ is large and $C$ is small, while the Bond number is of order of unity. In the limit $Ga \to \infty$, the terms in the right-hand side of the boundary condition (1.31) may be omitted in the leading order, because they are $O(1)$. Also, the quantity $p$ describing the difference of pressures on both sides of the interface may be considered as a constant, because the gradient of pressure is $O(1)$. Thus, as earlier noted, the shape of the interface is determined by the balance of normal stresses which are of hydrostatic and capillary origin:

$$h + Bo^{-1}(1 - \delta_\alpha T)\left(\frac{1}{R_1} + \frac{1}{R_2}\right) = const. \qquad (1.35)$$

In the case where the motion takes place in a closed cavity, the equation (1.35) is solved with a boundary condition on a lateral boundary corresponding to a certain contact angle. The influence of the lateral boundary vanishes on distances large compared with $Bo^{-1/2}$. For an "infinite" layer ($L \gg Bo^{-1/2}$, $L$ is the dimensionless horizontal size of the system) one can assume $h = 0$. In the opposite limit (for instance, under reduced effective gravity) the second term in the left hand side of (1.35) prevails over the first term. If $\delta_\alpha$ is small, the interface has a constant curvature.

The equation (1.35) may be considered as a zeroth approximation for the full problem in the limit of small $\delta_\beta$. The fields of variables (velocity, pressure, temperature) calculated in the region with the shape governed by the equation (1.35) may be used for calculations of the next order corrections to the interface shape. For instance, in the case of an infinite layer the shape of the interface $h = O(Ga^{-1})$ in the presence of buoyancy–driven convection and with no temperature dependence of the surface tension may be calculated from the equation

$$h - Bo^{-1}\left(\frac{\partial^2 h}{\partial x^2} + \frac{\partial^2 h}{\partial y^2}\right) = Ga^{-1}\left[p - 2\frac{\partial v_z}{\partial z}\right]\Bigg|_{z=0}. \qquad (1.36)$$

The next step (calculation of the influence of the surface deformation on the convective motion) cannot be done within the Boussinesq approximation, because this influence is of the same order in $\delta_\beta$ as some terms omitted in this approximation.

Let us discuss now the case of finite values of (modified) Galileo number. Within the Boussinesq approximation, $\delta_\beta \to 0$ and thus $R \to 0$, so only thermocapillary convection may appear. This situation may take place in thin layers or under reduced effective gravity conditions. If the capillary number $Ca$ is large enough, while $\delta_\alpha$ is small, the shape of the surface is not essentially influenced by the thermocapillary motion. The zeroth order solution may be constructed for a fixed shape of the surface, then the corrections to the surface shape may be calculated, etc. However, in the case of long-wave convection on the background of a flat surface, the capillary term in the boundary condition for normal stresses does not prevail, and the full problem should be considered.

### 1.2.2   Mass transfer in a system with an interface

#### i. Mass transfer in the absence of a surface adsorption

Let us consider a binary mixture characterized by the volume concentration $C$. Let us assume that the interfacial kinetics is so fast that the interfacial concentration $\Gamma$ is determined by the volume concentration $C$ in a unique way. In this case the equations and boundary conditions governing the problem are actually identical to those formulated in the previous subsection for the case of heat transfer (1.21) – (1.27). It is sufficient to introduce the concentration $C$ instead of the temperature $T$, and replace the coefficients $\beta$ and $\alpha$ by

$$\beta_c = -\frac{1}{\rho}\left(\frac{\partial \rho}{\partial C}\right)$$

and

$$\alpha_c = -\frac{\partial \sigma}{\partial C}$$

Besides, the coefficient $\kappa$ (and $\lambda$) should be replaced by the corresponding diffusion coefficient $D$. The same changes are done in the definitions of the nondimensional parameters. Thus, both models are isomorphic.

A point to be noted is that there are systems for which $\alpha$ or $\alpha_c$ may be negative or even vanish. Schwarz (1970) studied two-phase systems like Cyclohexanol/Water with diffusion substances Methanol, n-Propanol, n-Butanol, n-Amylol and n-Hexanol in concentrations from 2 to 8 %. These systems are characterized by $d\sigma/dC < 0$ (the first three) and by $d\sigma/dC > 0$ (the last two). Other cases showing positive growth in the surface tension or a minimum in the surface tension which can be considered as anomalous behavior, $d\sigma/dC > 0$ or $d\sigma/dT > 0$, relative to that of pure water have been described by several authors (Vochten and Petre, 1973; Petre and Azouni, 1984; Petre *et al.*, 1993; Azouni and Petre, 1998).

In the case of a two-phase system, where the component characterized by the volume concentration $C$ diffuses through the interface, similar substitutions should be done for each fluid. The equation are written in the following form:

$$\frac{\partial \mathbf{v}_m}{\partial t} + (\mathbf{v}_m \cdot \nabla)\mathbf{v}_m = -\frac{1}{\rho_m}\nabla p_m + \nu_m \nabla^2 \mathbf{v}_m + g\beta_{c,m}C_m\mathbf{e}, \quad (1.37)$$

$$\frac{\partial C_m}{\partial t} + (\mathbf{v}_m \cdot \nabla)C_m = D_m \nabla^2 C_m, \quad (1.38)$$

$$\nabla \cdot \mathbf{v}_m = 0. \quad (1.39)$$

The boundary conditions (1.14), (1.17), (1.18) are not changed, the equation for normal components of fluxes is

$$\left(D_1 \frac{\partial C_1}{\partial x_i} - D_2 \frac{\partial C_2}{\partial x_i}\right)n_i = 0. \quad (1.40)$$

The essential difference between the cases of heat transfer and mass transfer is that the limit values of the concentration on the interface from both fluids are not necessarily equal but proportional:

$$C_2 = kC_1. \quad (1.41)$$

This difference may be removed by a suitable transformation of variables.

## ii. Mass transfer with surfactant adsorption-desorption

Expanding on what was said in subsection 1.1.1, let us consider now a general case where the "surface gas phase" of the surfactant is taken into account. The interfacial concentration of the surfactant should be considered as an additional variable. If there is no chemical reaction on the interface, the evolution of the interfacial concentration is governed by the following equation (Levich, 1962; Rosner, 1986; Sadhal and Johnson, 1983; Edwards $et\ al.$, 1991; Myers, 1999):

$$\frac{\partial \Gamma}{\partial t} + \nabla_s \cdot (\Gamma \mathbf{v_s} - D_s \nabla_s \Gamma) = j, \quad (1.42)$$

where $\mathbf{v}$ denotes velocity along the surface, $D_s$ is the coefficient of the surfactant surface diffusion, $\nabla_s$ is the gradient along the surface, $j$ is the flux of the substance from the bulk to the interface.

Because the interface unit area possesses a mass $m\Gamma$, where $m$ is the mass of the surfactant molecule, the boundary condition (1.14) should be rewritten in the form

$$m\frac{d}{dt}(\Gamma v_{1,i}) = -\left[(p_1 - \rho_1 gh) - (p_2 - \rho_2 gh) - \sigma\left(\frac{1}{R_1} + \frac{1}{R_2}\right)\right]n_i$$

$$+(\tau'_{1,ik} - \tau'_{2,ik})n_k + \frac{\partial \sigma}{\partial x_i}, \tag{1.43}$$

where $d/dt$ is the Lagrangian derivative with respect to time, $\sigma$ is a function of $\Gamma$, which is the surface excess surfactant concentration. Usually, the mass of the surfactant layer is relatively small or processes are quasisteady, and the term in the left-hand side of (1.43) may be omitted.

The flux $j$ characterizes the mass exchange between the bulk and the interface. If the surfactant is insoluble, $j = 0$. If the surfactant is soluble in the bulk liquid, the exchange is determined by its diffusion in the bulk fluid and the interfacial kinetics. From one side, the flux of the surfactant from the bulk to the surface is determined by the relation

$$j = -D\mathbf{n} \cdot \nabla C. \tag{1.44}$$

From another side, it is determined by the adsorption-desorption kinetics (1.8), recalled here:

$$j = k_a C - k_d \Gamma. \tag{1.45}$$

The adsorption-desorption kinetics may be fast or slow. For fast enough kinetics a local adsorption-desorption equilibrium could be introduced which corresponds to $j = 0$. That is local equilibrium of the surface excess solute with the solute in the adjacent subphase. As it follows from (1.45), the (local) equilibrium values $C$ and $\Gamma$ are connected by the relation

$$\Gamma = k_a k_d^{-1} C \tag{1.46}$$

which, in fact, corresponds to the ideal *Gibbs adsorption isotherm* for very dilute solutions.

Deviations from the equilibrium of the system with surfactant may be due to various reasons, namely, initial nonequilibrium conditions, desorption and adsorption of surfactant, chemical reaction at the interface, etc. If every term in the left hand side (l.h.s) of (1.45) exceeds greatly the diffusion flux $j$, then the mass transfer process could be treated as locally in equilibrium (then $j$ is a small difference between two large values in the l.h.s.). In this case, at local equilibrium the relation (1.46) between $C$ and $\Gamma$, as a function of space and time, is fulfilled and the adsorption and desorption processes are controlled by bulk diffusion (1.44).

For moderately dense surfactant solutions, the departure from (1.46) is governed by the *Langmuir adsorption isotherm*

$$\Gamma = \Gamma^\infty \frac{C}{a + C}, \tag{1.47}$$

and the surface tension obeys the *Szyszkovsky state equation*

$$\sigma - \sigma_0 = -RT\Gamma^\infty \ln\left(1 + \frac{C}{a}\right), \tag{1.48}$$

that provides the surface pressure, where $\Gamma^\infty$ corresponds to a complete coverage or the surface excess saturation or maximum realizable excess concentration, and $a$ is constant for a given surfactant (adsorption coefficient). From (1.48) follows the relation

$$\Gamma = \left(-\frac{1}{RT}\frac{d\sigma}{dC}\right)C. \tag{1.49}$$

For small deviations from an initial equilibrium state

$$-\frac{1}{RT}\frac{d\sigma}{dC} \equiv L = \text{const.}$$

Then (1.49) takes the form of the Gibbs equation (1.46),

$$\Gamma = LC. \tag{1.50}$$

For more general adsorption-desorption kinetics one needs to apply the non-equilibrium relation (1.45) with e.g. the equilibrium Langmuir adsorption isotherm (1.47). Otherwise one could use nonlinear kinetics as

$$k_a\left(1 - \frac{\Gamma}{\Gamma^\infty}\right)C - k_d\Gamma = j. \tag{1.51}$$

This and more general forms of nonlinear relations have been used in the literature (Bojadjiev and Beshkov, 1984; Ravera *et al.*, 1993, 1994; Stebe and Barthes-Biesel, 1995; Liggieri *et al.*, 1996).

If the adsorption and desorption processes are very slow, $k_a$ and $k_d$ are practically zero, and again $j$ vanishes. The mass transfer in the sublayer near the interface is kinetically frozen, so the surface excess concentration, $\Gamma$, is effectively unchanged.

# Chapter 2

# Interfacial flows

As a straightforward follow-up of Chapter 1, in this second chapter we shall illustrate simple flows due to the Marangoni effect. However, we shall not consider the problem of *stability* of interfacial flows which will be the subject of subsequent chapters.

## 2.1 Flows generated by a longitudinal surface tension gradient

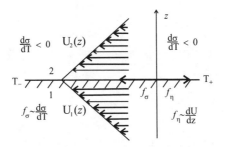

Figure 2.1: Interface between two liquids and the Marangoni effect. The actual flow fields $(U)$ in fluids 1 and 2, with motion from hotter to cooler regions due to the (tangential) Marangoni stress along the interface, depend on the values of the corresponding (dynamic) viscosity $(\eta)$.

Typical flows generated by the Marangoni effect appear in Figs. 2.1–2.3. In all cases non-uniform or unequal heating leads to inhomogeneity of the interfacial or surface tension and hence flow from points

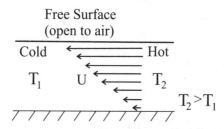

Figure 2.2: Thin liquid layer placed on a solid support and the Marangoni effect. The liquid layer is open to ambient air. Due to unequal heating $(T_2 > T_1)$ along the horizontal, flow $(U)$ driven by the Marangoni stress along the free surface brings liquid from the hotter to the cooler regions.

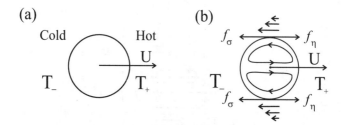

Figure 2.3: Drop or bubble subjected to unequal heating at two opposite poles and the Marangoni effect. (a) A drop or a bubble tends to move towards the hotter region as a reaction to the (outer) surface flow due to Marangoni stresses created by the unequal heating $(T_+ > T_-)$; (b) For a drop, flows outside and inside are schematically shown for moderate temperature differences at, say, front and rear poles.

of lower to higher tension values (see, e.g., Terada, 1928; Hershey, 1939; Levich, 1981). For the particular case of a liquid layer placed on a horizontal solid surface (Fig. 2.2) if the free surface of the layer is heated inhomogeneously, for instance, with a temperature $T_0$ of the free surface growing linearly in the direction $x$: $T_0 = Ax$ then this inhomogeneity of temperature generates two kinds of convection: buoyancy–driven convection and surface tension gradient–driven (Marangoni) convection or thermocapillary convection.

In order to get analytical results, we shall assume that the system is infinite in both horizontal directions: $-\infty < x < \infty$, $-\infty < y < \infty$. When considering the flow in an infinite region, we should realize that this mathematical problem may be a limit of some different physical problems. If the liquid is situated in *a box* bounded by some very distant lateral walls inpenetrable for liquid, the physical solutions are characterized by a zero mean flux of the liquid. A horizontal pressure gradient balancing the action of interfacial stresses will appear in such a system. If the liquid flows in an open *channel* between two vessels where the liquid is under the same pressure, the pressure gradient is absent while the flux is non-zero.

Let us solve the problem in both cases.

### 2.1.1  Return flow

We shall use the one-layer approach described in the previous chapter. We choose the mean depth of the layer $a$ as a unit of length, and $\theta = Aa$ as a unit of temperature. Other units are taken as in the previous chapter. In the case of zero flux, a pressure gradient appears in the system. Because of that, the surface of the fluid is not exactly horizontal (1.36). However, we shall neglect this phenomenon assuming $Ga \to \infty$. In this limit the depth of the fluid layer is equal to $a$ everywhere. Also, the Boussinesq approximation can be used to account for buoyancy–driven convection.

The problem is governed by the system of equations (1.28) - (1.30). Because we neglect the deformation of boundaries, we omit the boundary conditions (1.31) and (1.33). At the surface $z = 1$ we take the boundary conditions (1.32) and assume that the temperature of the free surface is fixed:

$$T = x. \tag{2.1}$$

On a rigid surface $z = 0$ we should take the boundary condition

$$\mathbf{v} = 0 \tag{2.2}$$

and a certain boundary condition for the temperature, for instance, the condition of vanishing transverse heat flux:

$$\frac{\partial T}{\partial z} = 0 \tag{2.3}$$

or the condition of a zero transverse temperature difference across the layer:

$$T = x. \tag{2.4}$$

We shall find the simplest solution of the described problem corresponding to a stationary parallel flow:

$$\mathbf{v} = (u(z), 0, 0). \tag{2.5}$$

The convective heat transfer, which is inhomogeneous with respect to the vertical coordinate $z$, generates some vertical temperature profile:

$$T(x, z) = x + \tau(z). \tag{2.6}$$

The pressure does not depend on the variable $y$:

$$p = p(x, z). \tag{2.7}$$

Substituting (2.5) - (2.7) into equations (1.28) - (1.30), we find:

$$0 = -\frac{\partial p}{\partial z} + R[x + \tau(z)], \tag{2.8}$$

$$0 = -\frac{\partial p}{\partial x} + \frac{d^2 u}{dz^2}, \tag{2.9}$$

$$u = \frac{d^2 \tau}{dz^2}. \tag{2.10}$$

In addition to the boundary conditions which give

$$z = 1 : \frac{du}{dz} + M = 0 \tag{2.11}$$

and

$$z = 0 : u = 0, \tag{2.12}$$

we use the condition for the zero horizontal flux of fluid:

$$\int_0^1 u(z)dz = 0. \tag{2.13}$$

Because the temperature distribution on the free surface is fixed,

$$z = 1 : \tau = 0. \tag{2.14}$$

On the rigid boundary

$$z = 0 : \frac{d\tau}{dz} = 0 \tag{2.15}$$

if it is a really poor heat conductor where disturbances in heat flux are not allowed (on occasion called an insulating or adiabatic boundary, but one has to be careful about use of terms), or

$$z = 0 : \tau = 0 \tag{2.16}$$

if the rigid surface of the layer is kept at a fixed temperature distribution.

Solving equation (2.8), we find

$$p = Rxz + \int_0^z \tau(\zeta)d\zeta + p_0(x), \tag{2.17}$$

where $p_0(x)$ is a still unknown function of $x$ only. Substituting the formula (2.17) into (2.9), we find

$$\frac{d^2u}{dz^2} = Rz + \frac{dp_0}{dx}. \tag{2.18}$$

Because $u$ depends only on $z$, $dp_0/dx$ should be constant:

$$\frac{dp_0}{dx} = C. \tag{2.19}$$

Thus,

$$p = Rxz + \int_0^z \tau(\zeta)d\zeta + Cx + C_1, \tag{2.20}$$

where $C_1$ is constant. Solving (2.18) with boundary conditions (2.11) and (2.12), we find the following expression for the velocity profile:

$$u = R\left(\frac{z^3}{6} - \frac{z}{2}\right) - Mz + C\left(\frac{z^2}{2} - z\right). \tag{2.21}$$

In the case of a closed region the solution should satisfy the zero flux condition (2.13). We find that

$$C = - \left( \frac{5R}{8} + \frac{3M}{2} \right). \tag{2.22}$$

Finally, the velocity profile is (Birikh, 1966)

$$u = R \left( \frac{z^3}{6} - \frac{5z^2}{16} + \frac{z}{8} \right) + M \left( -\frac{3z^2}{4} + \frac{z}{2} \right). \tag{2.23}$$

Following Davis (1987), we shall call this flow the *return flow*. The first term in the expression describes the advective flow caused by buoyancy, while the second term is generated by thermocapillarity.

The field of temperature may be found by solving equation (2.10) with boundary conditions (2.14) and (2.15) or (2.14) and (2.16). Let us note that because of zero mean flux of the fluid (2.13) one can find by integrating (2.10) that

$$\left. \frac{d\tau}{dz} \right|_{z=1} = \left. \frac{d\tau}{dz} \right|_{z=0}.$$

In the case of fixed flux at the rigid boundary (2.15) the temperature profile is

$$\tau = R \left( \frac{z^5}{120} - \frac{5z^4}{192} + \frac{z^3}{48} - \frac{1}{320} \right) - M \left( \frac{z^4}{16} - \frac{z^3}{12} + \frac{1}{48} \right); \tag{2.24}$$

in the case of a fixed boundary temperature (boundary condition (2.16) the temperature profile is

$$\tau = R \left( \frac{z^5}{120} - \frac{5z^4}{192} + \frac{z^3}{48} - \frac{z}{320} \right) - M \left( \frac{z^4}{16} - \frac{z^3}{12} + \frac{z}{48} \right). \tag{2.25}$$

In a cavity of a finite horizontal size, the obtained solution should be corrected near the lateral vertical walls. For large values of $R$ and $M$, such parallel flow becomes *unstable* which leads to the appearance of motions with more complicated spatial and temporal structure (Chapter 6).

## 2.1.2   Unidirectional flow

The case of an *open channel* where the mean flux of fluid is nonzero but the pressure on the interface is constant may be considered in a similar way. The only difference is that the condition (2.13) should be replaced by the condition

$$z = 1 : p = 0. \tag{2.26}$$

Substituting the expression (2.20) into (2.26), we find $C = -R$, thus

$$u = R\left(\frac{z^3}{6} - \frac{z^2}{2} + \frac{z}{2}\right) - Mz. \tag{2.27}$$

This flow is an *unidirectional flow*. In the case of purely thermo-capillary convection ($R = 0$), it is called *linear flow* (Davis, 1987).

The corresponding temperature profile is

$$\tau = R\left(\frac{z^5}{120} - \frac{z^4}{24} + \frac{z^3}{12} - \frac{1}{120}\right) + M\left(-\frac{z^3}{6} + \frac{1}{6}\right) \tag{2.28}$$

for boundary condition (2.15) and

$$\tau = R\left(\frac{z^5}{120} - \frac{z^4}{24} + \frac{z^3}{12} - \frac{z}{120}\right) + M\left(-\frac{z^3}{6} + \frac{z}{6}\right) \tag{2.29}$$

in the case where the temperature field is fixed on the rigid boundary (2.16). Let us consider also the case where the temperature field $T = x$ is kept on the rigid boundary, while the free boundary is insulated. The corresponding temperature field is

$$\tau = R\left(\frac{z^5}{120} - \frac{z^4}{24} + \frac{z^3}{12} - \frac{z}{8}\right) + M\left(-\frac{z^3}{6} + \frac{z}{2}\right). \tag{2.30}$$

## 2.1.3   Multilayer flows

A similar analysis may be performed in the case where the system contains several fluid layers separated by horizontal interfaces. Such is the case when there is competition of thermocapillary flows caused by different interfaces that may effectively reduce the intensity of the motion. This idea is used in the liquid encapsulation technique of crystal growth in reduced effective-gravity laboratories.

The problem was considered for a two-layer system with a free surface and a liquid-liquid interface, both subject to the thermocapillarity, by Doi and Koster (1993). Buoyancy–driven and thermocapillary–driven convection in a three-layer system caused by a temperature gradient along the interfaces was studied by Prakash and Koster (1993). The thermocapillary–driven convection in a liquid cylinder surrounded by a coaxial layer of another fluid, in the presence of thermocapillarity at both free surface and interface, was investigated by Briskman *et al.* (1991). In all cases, in correspondence with the physical origin of the flow, return flows were considered.

As the basic example of a parallel flow in a two-layer system, let us consider the more general problem with an arbitrary orientation of the imposed thermal gradient.

Let the space between two parallel rigid plates $z = -a_1$ and $z = a_2$ be filled by two immiscible viscous fluids. The temperature on these plates is fixed in the following way: $T(x, y, -a_1) = Ax + \Theta$, $T(x, y, a_2) = Ax$. Thus, a constant temperature gradient $A$ is imposed in the direction of the $x$ axis. For fixed values of $x$ and $y$, the difference between the temperatures of the lower plate and the upper plate is equal to $\Theta$. It is assumed that the interfacial tension coefficient $\sigma$ decreases linearly with temperature: $\sigma = \sigma_0 - \sigma_1 T$ and buoyancy is neglected. For convinience here we use $\sigma_1$ rather than $\alpha$.

In the present section the interface is assumed to be flat at $z = 0$. Strictly speaking, the interface can be perfectly flat only if horizontal pressure gradients appearing in both fluids are equal. Generally, these gradients are not equal, so that the interface is deformed in such a way that the pressure difference generated by the thermocapillary motion is balanced by the hydrostatic pressure and the interfacial tension. However, in some cases the deformation is negligible. The relevant parameters characterizing the interface deformation are $\epsilon_\sigma = Al_x \sigma_1 / \sigma_0$ (Pshenichnikov and Tokmenina, 1983) and $\epsilon_g = Al_x \sigma_1 / (\rho_2 - \rho_1) g a_2$ (Tan *et al.*, 1990); as above $g$ is the acceleration of gravity, $l_x$ is the characteristic scale of the region in $x$-direction (see also subsection 2.2). Here we assume that these parameters have small enough values, and hence the interface deformation caused by the difference of horizontal pressure gradients can be neglected.

For length, time, velocity, pressure and temperature, we use the following units: $a_1$, $a_1^2/\kappa_1$, $\kappa_1/a_1$, $\eta_1\kappa_1/a_1^2$ and $Aa_1$, respectively. Besides, $\rho = \rho_1/\rho_2$, $\nu = \nu_1/\nu_2$, $\eta = \eta_1/\eta_2$, $\lambda = \lambda_1/\lambda_2$, $\kappa = \kappa_1/\kappa_2$, and $a = a_2/a_1$.

The complete system of nonlinear equations can be written in the following dimensionless form:

$$\frac{1}{P}\left[\frac{\partial \mathbf{v}_1}{\partial t} + (\mathbf{v}_1 \cdot \nabla)\mathbf{v}_1\right] = -\nabla p_1 + \nabla^2 \mathbf{v}_1, \qquad (2.31)$$

$$\frac{\partial T_1}{\partial t} + \mathbf{v}_1 \cdot \nabla T_1 = \nabla^2 T_1, \ \nabla \cdot \mathbf{v}_1 = 0;$$

$$\frac{1}{P}\left[\frac{\partial \mathbf{v}_2}{\partial t} + (\mathbf{v}_2 \cdot \nabla)\mathbf{v}_2\right] = -\rho\nabla p_2 + \frac{1}{\nu}\nabla^2 \mathbf{v}_2, \qquad (2.32)$$

$$\frac{\partial T_2}{\partial t} + \mathbf{v}_2 \cdot \nabla T_2 = \frac{1}{\kappa}\nabla^2 T_2, \ \nabla \cdot \mathbf{v}_2 = 0,$$

where $P = \nu_1/\kappa_1$ is the Prandtl number of the lower fluid.

On the rigid horizontal plates, the following boundary conditions are used:

$$z = -1 : \mathbf{v}_1 = 0, \ T_1 = x; \qquad (2.33)$$

$$z = a : \mathbf{v}_2 = 0, \ T_2 = x - b, \qquad (2.34)$$

where the parameter $b = \Theta/Aa_1$ describes the relation between the characteristic vertical and horizontal temperature differences. At the interface, the normal components of the velocity vanish:

$$z = 0 : v_{z1} = v_{z2} = 0; \qquad (2.35)$$

and the continuity conditions for the tangential components of the velocity

$$z = 0 : v_{x1} = v_{x2}, \ v_{y1} = v_{y2}, \qquad (2.36)$$

for the tangential stresses

$$z = 0 : \eta\frac{\partial v_{x1}}{\partial z} = \frac{\partial v_{x2}}{\partial z} - M\eta\frac{\partial T_1}{\partial x}, \ \eta\frac{\partial v_{y1}}{\partial z} = \frac{\partial v_{y2}}{\partial z} - M\eta\frac{\partial T_1}{\partial y}, \ (2.37)$$

for the temperature

$$T_1 = T_2, \qquad (2.38)$$

and for the heat fluxes

$$\lambda\frac{\partial T_1}{\partial z} = \frac{\partial T_2}{\partial z} \qquad (2.39)$$

are fulfilled. Here $M = \alpha A a_1^2/\eta_1\kappa_1$ is the Marangoni number.

In the limit of an infinitely extended layer, it is necessary to impose some additional conditions determining the pressure gradients in the system. If the flow occurs in a channel that connects two vessels kept under the same pressure, the mean longitudinal pressure gradient in the system is zero. The corresponding thermocapillary flow is the above mentioned "linear flow" (Davis, 1987). In the case of a closed cavity, the mean longitudinal flux of fluid is zero, so that there is "return flow" characterized by a non-zero longitudinal pressure gradient. In the latter case one has:

$$\int_{-1}^{0} dz\, U_1^{(0)}(z) = 0, \ \int_{0}^{a} dz\, U_2^{(0)}(z) = 0. \tag{2.40}$$

The boundary value problem (2.31) - (2.39) has an exact solution corresponding to a parallel flow in the direction opposite to the direction of the temperature gradient:

$$\mathbf{v}_i = U_i^{(0)}(z)\mathbf{e}_x, \ p_i = B_i^{(0)}x, \ T_i = x + \Theta_i^{(0)}(z), \ i = 1, 2,$$

where $\mathbf{e}_x$ is the unit vector of the axis $x$ (Doi and Koster, 1993).

In the case of the zero longitudinal pressure gradient (linear flow),

$$U_1^{(0)}(z) = -\frac{a\eta M}{1 + a\eta}(z + 1); \tag{2.41}$$

$$U_2^{(0)}(z) = \frac{\eta M}{1 + a\eta}(z - a); \tag{2.42}$$

$$\Theta_1^{(0)}(z) = -\frac{a\eta M}{6(1 + a\eta)}\left[(z + 1)^3 - \frac{1 + 2\kappa a^2 + 3\lambda a}{1 + \lambda a}(z + 1)\right] \tag{2.43}$$

$$-\frac{b(z + 1)}{1 + a\lambda};$$

$$\Theta_2^{(0)}(z) = \frac{\eta M}{6(1 + a\eta)}\left[-\kappa(-z + a)^3 + \frac{a(\kappa\lambda a^2 + 2\lambda + 3\kappa a)}{1 + \lambda a}(-z + a)\right] \tag{2.44}$$

$$-\frac{b(\lambda z + 1)}{1 + a\lambda};$$

$$B_1^{(0)} = B_2^{(0)} = 0. \tag{2.45}$$

If $b \geq 0$ then $\Theta_i(z) \leq 0$.

For zero longitudinal fluxes of fluids (return flow),

$$U_1^{(0)}(z) = -\frac{a\eta M\left(1 + 4z + 3z^2\right)}{4(1 + a\eta)}; \tag{2.46}$$

$$U_2^{(0)'}(z) = -\frac{\eta M\left(a^2 - 4az + 3z^2\right)}{4a(1 + a\eta)}; \tag{2.47}$$

$$\Theta_1^{(0)}(z) = \frac{a\eta M(1 + z)\left[a(a\kappa + \lambda) - (1 + a\lambda)\left(z + 5z^2 + 3z^3\right)\right]}{48(1 + a\eta)(1 + a\lambda)}$$

$$\tag{2.48}$$

$$-\frac{b(z + 1)}{1 + a\lambda};$$

$$\Theta_2^{(0)}(z) = \frac{\eta M(a - z)\left[a^2(a\kappa + \lambda) + \kappa(1 + a\kappa)\left(a^2 z - 5az^2 + 3z^3\right)\right]}{48a(1 + a\eta)(1 + a\lambda)}$$

$$\tag{2.49}$$

$$-\frac{b(\lambda z + 1)}{1 + a\lambda};$$

$$B_1^{(0)} = -\frac{3a\eta M}{2(1 + a\eta)}; \tag{2.50}$$

$$B_2^{(0)} = -\frac{3M}{2a(1 + a\eta)}. \tag{2.51}$$

The velocity is negative near the interface and positive near the rigid walls. On the interface, the absolute value of the velocity is equal to

$$v_s = |U_1^{(0)}(0)| = a\eta M/4(1 + a\eta). \tag{2.52}$$

This parallel flow is an analogue of the return thermocapillary flow considered in the one-layer approach (Davis, 1987).

The temperature profiles (2.48) and (2.49) are generated by the combined action of the heat advection by the parallel flow (terms proportional to $M$) and the external heating from below (terms proportional to $b$). The functions $\Theta_1^{(0)}(z)$ and $\Theta_2^{(0)}(z)$ can be non-monotonic. For instance, the vertical component of the temperature gradient in the lower fluid has equal minima in the points $z = -1$ and $z = 0$:

$$\Theta_1^{(0)''}(-1) = \Theta_1^{(0)''}(0) = -\frac{a\eta M(1 - \kappa a^2)}{48(1 + a\eta)(1 + a\lambda)} - \frac{b}{1 + a\lambda} \tag{2.53}$$

and a maximum in the point $z = -1/3$:

$$\Theta_1^{(0)''}(-1/3) = \frac{a\eta M[a(a\kappa + \lambda) + (7/9)(1 + a\lambda)]}{48(1 + a\eta)(1 + a\lambda)} - \frac{b}{1 + a\lambda}. \tag{2.54}$$

Thus, if $1 - \kappa a^2 > 0$, the vertical component of the temperature gradient is negative near the boundaries $z = -1$ and $z = 0$, but it is positive in the middle of the layer, if $b/M$ is less than a certain value. The sign of the interfacial temperature

$$\Theta_s = \Theta_1^{(0)} = \frac{\eta M a^2 (a\kappa + \lambda)}{48(1 + a\eta)(1 + a\lambda)} - \frac{b}{1 + a\lambda} \qquad (2.55)$$

is also determined by the ratio $b/M$. If $1 - \kappa a^2 < 0$, the temperature profile is non-monotonic in the upper fluid.

## 2.2 Nonisothermal flows in thin liquid layers

Let us generalize the problem considered in the previous section. Consider a thin liquid layer placed on a horizontal rigid plane which is inhomogeneously heated. For thin liquid films the assumption $Ga \gg 1$ is not valid. Therefore we shall consider finite values of $Ga$. Hence, we shall neglect buoyancy ($R \ll 1$) and take into account deformation of the free surface.

In the general case, such a problem cannot be solved analytically. Let us assume in addition that the depth of the layer is much smaller than the characteristic size of the inhomogeneity of the temperature field kept on the rigid plane. Thus, the temperature boundary condition on the rigid boundary is:

$$z = 0 : T(x, y) = \Theta(\epsilon x, \epsilon y), \qquad (2.56)$$

where $\epsilon \ll 1$. We use non-dimensional variables choosing the mean depth of the layer and characteristic temperature difference in the *horizontal* direction as new units.

Let us define

$$X = \epsilon x, \, Y = \epsilon y. \qquad (2.57)$$

After the transformation of variables (2.57), we should replace the operators $\partial/\partial x$ and $\partial/\partial y$ by $\epsilon \partial/\partial X$ and $\epsilon \partial/\partial Y$ correspondingly in all the equations and boundary conditions describing the system. For small $\epsilon$, we can construct the solution in the form of a series in powers of $\epsilon$. The deformation of the boundary in a thin layer may be of the same order as its mean depth, thus $h = O(1)$. From the boundary condition (1.25) for tangential stresses on the free boundary we find that the horizontal components of the velocity $\mathbf{u} = (v_x, v_y)$ are of order $O(\epsilon)$. Hence, from the continuity equation (1.23) we conclude that the vertical velocity $w = v_z$ is $O(\epsilon^2)$. For sake of simplicity, let us consider the case where $Bi = 0$ on the free surface. In this case, the local difference between $T(X, Y, z)$ and $\Theta(X, Y)$ caused by the convective heat transfer and the horizontal diffusion, is $O(\epsilon^2)$. Taking into account the characteristic temporal scale of the described processes, we shall introduce the temporal variable $\tau = \epsilon^2 t$.

Thus, we put

$$\mathbf{u} = \epsilon \mathbf{U}, \; w = \epsilon^2 W, \; T = \Theta + O(\epsilon^2). \qquad (2.58)$$

In the leading order in powers of $\epsilon$, we obtain the following boundary problem determining the stationary flow in an inhomogeneously heated thin layer:

$$0 = -\nabla_2 p + \mathbf{U}_{zz}; \qquad (2.59)$$

$$0 = -p_z; \qquad (2.60)$$

$$\nabla_2 \cdot \mathbf{U} + W_z = 0; \qquad (2.61)$$

at $z = h$ :

$$p = Gah; \qquad (2.62)$$

$$\mathbf{U}_z + M \nabla_2 \Theta = 0; \qquad (2.63)$$

$$h_\tau + \mathbf{U} \cdot \nabla_2 h = W; \qquad (2.64)$$

at $z = 0$ :

$$\mathbf{U} = 0; \; W = 0. \qquad (2.65)$$

Here the subscript denotes differentiation with respect to the corresponding variable, and $\nabla_2 = (\partial/\partial X, \partial/\partial Y)$.

The solution in the leading order is:

$$p = Gah; \qquad (2.66)$$

$$\mathbf{U} = \frac{1}{2} Ga \nabla_2 h \, z^2 - \frac{1}{2} Ga \nabla_2 (h^2) z - M \nabla_2 \Theta \, z; \qquad (2.67)$$

$$W = -\frac{1}{6} Ga \nabla_2^2 h \, z^3 + \frac{1}{4} Ga \nabla_2^2 (h^2) z^2 + \frac{1}{2} M \nabla_2^2 \Theta \, z^2. \qquad (2.68)$$

Thus, the motion in the liquid layer is influenced by *local* horizontal temperature and pressure gradients. The evolution of the relief of the surface is governed by the conservation equation

$$h_\tau + \nabla_2 \mathbf{Q} = 0, \qquad (2.69)$$

where the flux

$$\mathbf{Q} = -\frac{1}{3}Ga\nabla_2 \cdot (h^3\nabla_2 h) - \frac{1}{2}M\nabla_2 \cdot (h^2\nabla_2\Theta). \tag{2.70}$$

The closed system of equation (2.69) - (2.70) may be used for an analysis of the evolution of the layer's relief. Here we consider only the stationary relief generated by a one-dimensional temperature distribution, $\Theta = \Theta(X)$, in a closed box, $0 \le X \le L$. In this case, the flux $Q = 0$, and we obtain the following relation (Zuev and Pshenichnikov, 1987):

$$\frac{Gah^2(X)}{3M} + \Theta(X) = C; \tag{2.71}$$

Thus one has

$$h(X) = \sqrt{\frac{3M}{Ga}}(C - \Theta(X))^{1/2}, \tag{2.72}$$

where the constant $C$ is obtained from the condition that the fluid volume is conserved by the deformation of the surface:

$$\int_0^L h(X)dX = L. \tag{2.73}$$

As an example, let us consider the case of a constant horizontal temperature gradient: $\Theta(X) = -X$. The form of the surface is described by the equation

$$h(X) = \sqrt{\frac{3M}{Ga}}(C + X)^{1/2}, \tag{2.74}$$

where $C$ is a solution of the equation

$$2\sqrt{\frac{M}{3Ga}}[(C + L)^{3/2} - C^{3/2}] = L. \tag{2.75}$$

Obviously, the equation (2.75) has a solution only if

$$M < M_* = \frac{3Ga}{4L}. \tag{2.76}$$

As $M = M_*$, $C =$ and

$$h(X) = \frac{3}{2}\sqrt{\frac{X}{L}}.$$

Thus, a *dry spot* in the *hot* region appears as $M = M_*$. If $M > M_*$, we can assume that the rigid surface is dry ($h = 0$) in a certain region $0 < X < d(M)$, while in the region $d(M) < X < L$ the solution (2.74) is valid. Using the condition (2.75), we find that

$$d(M) = L(1 - \frac{M}{M_*}), \qquad (2.77)$$

$$h(X) = \frac{3}{2}\sqrt{\frac{X - d(M)}{L - d(M)}}. \qquad (2.78)$$

It should be noted that the long-scale approximation used in this section is actually not valid near the boundary of the dry spot where the flow is essentially non-parallel, and the effect of the neglected contact angle is significant. Nevertheless, the obtained solutions are reasonable outside the boundary region and describe well the experiment (Zuev and Pshenichnikov, 1987).

## 2.3  Nonparallel flows

In preceding sections, we considered parallel flows ($v_z = 0$, section 2.1) or almost parallel flows ($|v_z| << |v_x|$, section 2.2). However, in some cases the thermocapillary flow may be significantly non-parallel. For instance, in a closed cavity the flow is non-parallel near the lateral rigid boundary. The flow is non-parallel also in the case where the free boundary is non-flat because of capillarity. In the present section we consider some examples of this kind. Let us emphasize that we do not discuss here any phenomena connected with instability effects that will be considered in subsequent chapters.

### 2.3.1  Two-dimensional flows

#### i. Stokes flows near corners

We shall start our analysis by the consideration of a two-dimensional non-parallel thermocapillary flow in the vicinity of the corner built by the lateral rigid boundary (say, $x = 0$) and the fluid surface $z = h(x)$. We may assume that the free surface is flat and nondeformable, and the angle between this surface and the rigid boundary is equal to the contact angle $\theta_c$. Because of the boundary condition, $\mathbf{v} = 0$, on the rigid boundary, the flow near the corner is slow. Thus, in a sufficiently small vicinity of the corner we may *linearize* the equations of motion and use the *Stokes approximation* for the calculation of the flow. In this approximation, the stationary thermocapillary flow is governed by the equation (Moffat, 1964)

$$-\nabla p + \nabla^2 \mathbf{v} = 0. \tag{2.79}$$

For a two-dimensional flow, it is convenient to eliminate $p$ by taking the curl of the expression in the left-hand side of (2.79), and to introduce the stream function, $\psi$, such that

$$v_x = \frac{\partial \psi}{\partial z}, \; v_z = -\frac{\partial \psi}{\partial x}.$$

The function $\psi$ satisfies the biharmonic equation

$$\nabla^4 \psi = 0. \tag{2.80}$$

It is convenient to introduce the polar coordinate system $(r, \theta)$. In the vicinity of the corner point the surface is flat, and the angle between this surface and the rigid boundary is equal to the contact angle $\theta_c$. The boundary conditions are:

$$\theta = 0 : \; \psi = \frac{\partial \psi}{\partial \theta} = 0; \tag{2.81}$$

$$\theta = \theta_c : \; \psi = 0, \; \frac{1}{r^2} \frac{\partial^2 \psi}{\partial \theta^2} + M \frac{\partial T}{\partial r} = 0. \tag{2.82}$$

Near the corner point, we can take

$$\frac{\partial T}{\partial r} = \left( \frac{\partial T}{\partial r} \right)_c = const,$$

where the subscript "c" means that the derivative is calculated in the corner point. The exact solution of the problem (2.80-2.82) is (Shevtsova *et al.*, 1996)

$$\psi = r^2 (A \cos 2\theta + B \sin 2\theta + C\theta + D) \tag{2.83}$$

where

$$A = \frac{1}{4} M \left( \frac{\partial T}{\partial r} \right)_c \frac{2\theta_c - \sin 2\theta_c}{\sin 2\theta_c - 2\theta_c \cos 2\theta_c}$$

$$B = \frac{1}{4} M \left( \frac{\partial T}{\partial r} \right)_c \frac{1 - \cos 2\theta_c}{\sin 2\theta_c - 2\theta_c \cos 2\theta_c},$$

$$C = -2B, \ D = -A.$$

Note that the vorticity

$$\omega \equiv (curl\,\mathbf{v})_y = \frac{\partial v_x}{\partial z} - \frac{\partial v_z}{\partial x} = \nabla^2 \psi$$

near the corner point is governed by the formula

$$\omega = 4(C\theta + D). \tag{2.84}$$

Thus, the vorticity is not continuous in the corner point.

### ii.  Flows in a rectangular cavity

Let us consider now the *nonlinear* two-dimensional flow in a cavity $-L/2 \leq x \leq L/2$, $0 \leq z \leq h(x)$. It is assumed that the vertical endwalls are rigid: $\mathbf{v}(\pm L/2) = 0$, and are maintained at different temperatures: $T(\pm L/2) = \pm L/2$. The boundary conditions for $h$ may be $h(\pm L/2) = 1$ (fixed positions) or $\partial h/\partial x(\pm L/2) = \cot\theta_c$ (prescribed contact angle $\theta_c$).

If $L >> 1$, one can expect that in the central part of the cavity an almost parallel flow ("core flow") similar to that described in Sec. 2.1 will arise. Further we shall assume that $Ga \gg 1$, thus the surface is only weakly distorted because of the inhomogeneity of the pressure. Such an almost parallel flow was indeed observed in experiments (Kirdyashkin, 1984) and found in numerical simulations (Strani *et al.*, 1983) with $L \geq 4$.

However, the flows near the vertical walls ("boundary layers") are significantly non-parallel. Generally, these flows should be calculated *numerically* and then *matched* with the parallel flow in the central region. Such solutions were constructed by Sen and Davis (1982) and Strani *et al.* (1983).

For smaller values of $L$, the flows generated by both thermo-capillarity and buoyancy were investigated numerically for different values of the Prandtl number and for different temperature distributions on the surface in numerous papers (see, e.g., Polezhaev *et al.*, 1981; Strani *et al.*, 1983; Wilke and Löser, 1983; Napolitano *et al.*, 1984; Zebib *et al.*, 1985; Cuvelier and Driesson, 1986; Bergman and Ramadhyani, 1986; Bergman and Keller, 1988; Carpenter and Homsy, 1989). An extensive investigation of flows with large $L$ for fluids characterized by low values of the Prandtl number was done

by Ben Hadid and Roux (1990, 1992). Experiments have been performed by Kirdyashkin (1984), Ochiai *et al.* (1984), Camel *et al.*
(1986), Villers and Platten (1987), Metzger and Schwabe (1988) and
others.

The most impressive feature of stationary flows generated by the
horizontal temperature gradient due to thermocapillarity is a strong
asymmetry between flows near a *hot corner* and a *cold corner* of
the surface. This effect appears if the Marangoni number is large
enough. Near a cold corner the flow is drastically stronger, and the
temperature field is characterized by a much smaller spatial scale
than in the case of a hot corner. This nonlinear effect is caused by the
fact that the surface flow, which is directed *towards* the cold corner,
*compresses* the temperature field on the surface, thus enhancing the
tangential stresses generating the surface flow and hence there is some
kind of *positive feedback*. This effect is absent near a hot corner.

An asymptotic analysis of high Marangoni flows near a hot corner
was performed by Cowley and Davis (1983) for large Prandtl number
fluids. An asymptotic analysis of the flow near a cold corner was done
by Zebib *et al.* (1985). Because of the existence of boundary layers,
the numerical calculation of thermocapillary flows near a cold wall
can be carried out only by means of sophisticated numerical methods
incorporating nonuniform grids (Canright, 1994).

### 2.3.2   Axisymmetric flows

#### i. Flows in an axisymmetric half-zone

Another configuration of interest for applications is a cylindrical layer
situated between two plates kept at different albeit constant temperatures. This is a simplified model of the floating zone that is used to
grow single crystals of semiconductors. In the present chapter we do
not discuss the transitions between flows with different spatial structures caused by instabilities, and restrict ourselves to consideration
of the primary steady axisymmetric thermocapillary flow generated
by the surface tension gradient.

As well as in the case of a planar system, the flow in the central ("core") region of the zone may be calculated in a parallel flow
approximation (Xu and Davis, 1983), if the zone is long enough
($L \gg 1$). However, a long cylindrical column is actually unstable
because of the capillary Rayleigh instability.

Numerous investigations were performed for finite aspect ratios, mainly by means of numerical methods. Generally, the shape of the surface was determined by the equation (1.35), where the term containing $\delta_\alpha$ was neglected, or the solution was constructed by means of series expansion in $\delta_\alpha$ (Rybicki and Florian, 1987; Kuhlmann and Rath, 1993). Most publications were devoted to the case $\theta_c = \pi/2$ where the radius of the cylinder does not depend on the longitudinal coordinate. Typical results show that in long enough half-zones ($L \geq 1$, if $Re = M/P$ is small) the axisymmetric flow consists of one toroidal vortex near the free surface. In shorter zones ($1/L \geq 1.7$, if $Re = M/P$ is small) an additional vortex ring appears near the axis of the cylinder. If the height of the cylinder is small in comparison with the radius, a sequence of vortex rings with alternating directions of rotation is observed. The smaller the radius of the ring, the weaker its intensity. The axisymmetric motions in half-zones with different values of $\theta_c$ were studied by Shevtsova et al. (1996).

### ii. Flows in a full zone

The flows in a "full" cylindrical zone ("liquid bridge") modelling a crystal growth system, are generated by inhomogeneous heating of the surface, rather than by the temperature difference between rigid plates. The surface temperature may be prescribed (say, a parabolic temperature profile, $T_\Gamma = 1 - (2z/L)^2$, is fixed), or governed by the boundary condition (1.34) with inhomogeneous distribution, $\bar{T}_g(z)$. Also, the radiative heat transfer on the interface may be taken into account.

For axisymmetric flows, in the case of a cylinder with constant radius and a parabolic temperature profile, Marangoni flows were studied by Chang and Wilcox (1976). For relatively small Marangoni numbers, two toroidal vortices were observed. For large values of Marangoni number, an intensive motion is concentrated near cold corners, and additional weak vortex rings, with alternating rotation directions, appear closer to the axis of the cylinder.

The investigation of flows in the case where the radius of cylinder, $R(z)$, depends on the longitudinal coordinate $z$, was done by Kozhoukharova and Slavchev (1986).

As Kuhlmann (1999) has recently provided an extensive description of the phenomena occuring in liquid bridges with and without the Marangoni effect, we shall not dwell on this matter here.

# Chapter 3

# Thermocapillary and solutocapillary migration of drops (and bubbles) and their spreading due to the Marangoni effect

In this chapter we shall deal with basic and striking features of drop or bubble (self-propulsion and autonomous motion and otherwise) migration and drop spreading on another liquid, driven by the Marangoni effect.

## 3.1 Hydrodynamic drag on a solid sphere, a drop, or a bubble

From Newton's experiments in 1710, theory and later observations, the magnitude of the drag force on solid spheres or drops in steady motion of viscous fluid was given as

$$F_D = 0.22\pi R^2 \rho U_\infty^2 \qquad (3.1)$$

where $U_\infty$ is the relative velocity of the particle and fluid, $R$ is the particle radius, and $\rho$ is the fluid density. This kinetic theory relation is for 'large' values of $U_\infty$, for which inertial effects are predominant.

Stokes (1851) suggested that at very low relative velocities the inertial effects are so small that they can be omitted from the hydrodynamic (Navier-Stokes) equations. Under these conditions, the viscous hydrodynamic force on a solid sphere is

$$F_D = 6\pi\eta R U_\infty. \tag{3.2}$$

with $\eta$ denoting, as in earlier chapters, the (dynamic) shear viscosity of the liquid.

Stokes introduced a velocity potential called the *stream function*, $\psi$, using the fact that the motion for a spherical particle or a drop is axisymmetric. Accordingly, using the vorticity $\boldsymbol{\omega} = \text{rot } \mathbf{v}$, disregarding the buoyancy term, the Navier-Stokes equations for the liquid flow around a sphere reduce to

$$\frac{\partial}{\partial t}\boldsymbol{\omega} + (\mathbf{v} \cdot \nabla)\boldsymbol{\omega} = \nu\nabla^2\boldsymbol{\omega} + (\boldsymbol{\omega} \cdot \nabla)\mathbf{v} \tag{3.3}$$

which is the vorticity balance equation where the kinematic viscosity, $\nu = \eta/\rho$, appears as the coefficient of vorticity diffusion, and $(\boldsymbol{\omega} \cdot \nabla)\mathbf{v}$ is responsible for the stretching of vortex lines with consequent production of vorticity.

For steady motions, in the linear approximation, hence for zero Reynolds number flow (Stokes flow or creeping flow ), Eqs. (3.3), and consequently, the Navier-Stokes equations, reduce to the Laplace equation

$$\nabla^2\boldsymbol{\omega} = 0. \tag{3.4}$$

Thus the flow is due solely to steady diffusion of vorticity. The sphere is a source of vorticity due to the no slip at its surface. Vorticity decreases in space as $1/r^2$, as originated from a dipolar source that produces for each component of $\boldsymbol{\omega}$ equal positive and negative quantities at the surface of the sphere. Then from the velocity field $\mathbf{v} = (v_r, v_\theta, 0)$ (Fig. 3.1) $\psi$ is, by definition, such that

$$v_r = \frac{1}{r^2\sin\theta}\frac{\partial\psi}{\partial\theta}, \; v_\theta = -\frac{1}{r\sin\theta}\frac{\partial\psi}{\partial r} \tag{3.5}$$

and, thus, using Eq. (3.4) yields the equation

$$E^4\psi = 0 \tag{3.6}$$

with

$$E^2 \equiv \frac{\partial^2}{\partial r^2} + \frac{\sin\theta}{r^2}\frac{\partial}{\partial\theta}\left(\frac{1}{\sin\theta}\frac{\partial}{\partial\theta}\right).$$

Note that the stream function (i) exists in all cases of two-dimensional incompressible flow and in three-dimensional flow if it is axially symmetric; (ii) its existence is not limited to steady motions; and (iii) its existence for axisymmetric motions depends entirely on the kinematic assumption of incompressibility and, hence, arises in both inviscid and viscous flows, for these two flows differ only in their *dynamical* properties.

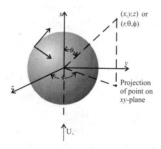

Figure 3.1: Geometry of the drop problem when fluid approaches from below with velocity $U_\infty$. Tangential and (inward) normal vectors indicate that at every point there are pressure and friction forces (stresses) acting on the sphere surface.

The solution of equation (3.6) with the appropriate no slip boundary condition on the sphere yields $\psi(r, \theta)$ from which using (3.5) the velocity field is obtained. Then one way to obtain the hydrodynamic drag force on the sphere is to estimate the power dissipated by viscous stresses, and the relationship (3.2) is obtained.

A shortcoming of the Stokes approach is that the disturbance due to the sphere extends to infinity as $|\mathbf{v}| \sim 1/r$ ($r \rightarrow \infty$). Hence the presence of a boundary at a finite distance or of another sphere can modify the flow when placed at a distance of diameters from the other. Furthermore, the energy of the disturbance flow induced by the sphere diverges as the volume of liquid around the sphere increases, which is physically unacceptable.

Rybczynskii (1911) and Hadamard (1911), independently, solved the Stokes problem for a liquid drop with flows outside and inside. Their extension of (3.2) is

$$F_D = 4\pi \frac{1 + 3\beta/2}{1 + \beta} \eta_1 R U_\infty \qquad (3.7)$$

with $\beta = \eta_2/\eta_1$, where the subscripts $i = 1, 2$ correspond to the liquids outside and inside the drop, respectively. Clearly, the limit $\eta_2$ going to infinity yields back Stokes' law (3.2), while $\eta_1 \gg \eta_2$ yields the corresponding law for a bubble, with the factor 6 being replaced by 4 in (3.2). Incidentally, the factor 4 also appears when we consider a solid sphere with slip boundary conditions, as in the case of a sphere with a polymeric or oily coating. Boussinesq (1913a-c) extended the theory to account for surface shear and dilational viscosities. Both Rybczynskii and Hadamard considered that the tangential velocity and the normal and shear stresses at both sides of the interface were equal. Thus Boussinesq's generalization included changes in the interfacial stresses resulting from surface rheology (for a recent discussion about the subject see Edwards *et al*, 1991; Miller *et al*, 1996). Noteworthy is that Bond (1927) studied both theoretically and experimentally the Stokes flow for bubbles and drops accounting for inner flow circulation. He provided correction to the Millikan's determination of the electron charge as the latter directly applied Stokes drag law (3.2) and not (3.7).

Another noteworthy study is that carried out by Weber (1854) who, presumably unaware of Stokes theory, provided flow fields around bubbles due to solutocapillarity, i.e., the surfactant or solutal Marangoni effect. He depicted return flows and hence inverse flow circulation relative to the far field flow similar to the one we shall describe later on when discussing *active* drops.

Oseen (1910, 1913) pointed out that at a great distance from the sphere the inertial terms become as important as the viscous terms, and suggested a possible improvement of the Stokes law (3.2) by taking inertial terms partly into consideration. Indeed, the ratio of inertial to viscous terms is negligible when we are near the sphere at zero Reynolds number, $Re = U_\infty R/\nu$. However, for distances of the order of the inverse of the Reynolds number the inertial forces become comparable to viscous forces and cannot be neglected. Thus, Stokes' solution defines a local flow around the sphere that, although, corresponds to all quantities (flow velocity, etc.) quantitatively small or even negligible, cannot be accepted on conceptual grounds. Oseen's drag force is

$$F_D = 6\pi\eta RU_\infty \left(1 + \frac{3}{8}Re\right) \qquad (3.8)$$

Today we know that neither Stokes' nor Oseen's laws are uniformly valid. The Oseen approximation, although incorporating inertial terms, is still a *linear* theory, as he did not consider in full the inertial part of the Navier-Stokes equations but simply the linear coupling between the known far field velocity with the unknown flow disturbance due to the sphere. As already said, the Stokes analysis is valid in a small enough neighborhood of the sphere and Oseen's analysis though valid for matching at the flow velocity values far from the sphere is not valid when approaching it. Several authors did attempt, not always successfully, an improvement upon the Oseen and Stokes analyses. It was not until 1957 that Stokes' and Oseen's results were properly put in context and generalized, thanks to the works of Proudman and Pearson (1957) who, following the seminal ideas of Poincare around 1886 and, more effectively, Kaplun and colleagues (Lagerstrom and Cole, 1955; Kaplun and Lagerstrom, 1957; Kaplun, 1957; Lagerstrom and Casten, 1972; Van Dyke, 1975; Benjamin, 1993), developed the matched-asymptotic-expansions method for the flow around the sphere, thus opening a domain in Applied Mathematics. The basic concept was to consider Stokes' solution as a local (inner) solution of the problem and Oseen's as a regular (outer) solution rather than considering Oseen's as an improvement upon Stokes. The inner solution was assumed to be valid in a spherical region of radius $1/Re$ around the sphere while the outer solution was valid from infinity down to the $1/Re$ neighborhood. In the overlapping zone both solutions were accepted as valid, hence the need for appropriately matching them. Proudman and Pearson solved the steady nonlinear extension of Stokes' equation (3.6), i.e.

$$\frac{Re}{r^2} \left\{ \frac{\partial(\psi, E^2\psi)}{\partial(r, \delta)} + 2E^2\psi L_r \psi \right\} = E^4\psi, \qquad (3.9)$$

where

$$\delta = \cos\theta, \qquad (3.10)$$

$$L_r = \frac{\delta}{1 - \delta^2} \frac{\partial}{\partial r} + \frac{1}{r} \frac{\partial}{\partial \delta}. \qquad (3.11)$$

For flows with non-vanishing albeit low Reynolds number ($Re \ll 1$), they found that the hydrodynamic drag on the sphere is

$$F_D = 6\pi\eta RU_\infty \left( 1 + \frac{3}{8}Re + \frac{9}{40}Re^2 \ln Re + O(Re^2) \right) \qquad (3.12)$$

which shows the non-analytic form of expansion. The scheme of Proudman and Pearson was also applied to the problem of heat and mass transfer to (from) a sphere. Acrivos and Taylor (1962) considered the convective terms in the heat equation hence with small albeit non-zero Peclet number ($Pe = U_\infty R/\kappa$) while using the Stokes flow. $\kappa$ is the thermal diffusivity. They obtained a Peclet number expansion of the average Nusselt number,

$$Nu = -\int_{-1}^{1} \left(\frac{\partial T}{\partial r}\right)_{r=1} d\delta,$$

with $T$ the normalized temperature:

$$Nu = 2 + \frac{1}{2}Pe + \frac{1}{4}Pe^2 \ln Pe + 0.03404 Pe^2 + \frac{1}{16}Pe^3 \ln Pe + O(Pe^3)$$
(3.13)

which also shows a non-analytic form. Needless to say, a similar expansion holds for mass transfer, using the appropriate diffusion related Peclet number. The first two terms are independent of the Reynolds number. Incidentally, results also exist in the extreme opposite limit for $Re \gg 1$ when flow boundary layer theory ought to be used (Chao, 1962; Moore, 1963; Gupalo *et al.*, 1972; Balasubramaniam and Subramanian, 2000).

As already emphasized, with surface tension gradient-driven (Marangoni) flows what actually matters is the heat equation, hence the Peclet number, and the Stokes approximation is quite a valid starting point. Occasionally, the Peclet number is called the Marangoni number and thus we shall use $Ma(Pe)$ in such cases where the velocity scale is given by the Marangoni effect. Please note that, as said in Chapter 1, we shall be using $Ma$ or $M$ to denote the Marangoni number. Subramanian (1981) and later Merritt and Subramanian (1988) solved the problem of the drift of a drop in an external temperature gradient using the matched-asymptotic-expansions method but not leading in this case to a logarithmic term.

Taylor and Acrivos, 1964 (see also Haj-Hariri *et al.*, 1990, 1997) also considered the deformation of the sphere (a bubble) and showed that for it to be noticeable the capillary number here defined using $V \equiv U_\infty$, the known far field flow velocity, $C = U_\infty \eta/\sigma$, or more appropriately, the Weber number, $We = ReC$, must be non-negligible, as expected. For a slightly deformable bubble in a flow field $U_\infty$ they obtained the radius in terms of the angle $\theta$:

$$R(\theta) = R\left[1 - \frac{5}{48}We(3\cos^2\theta - 1)\right]$$
(3.14)

for $Re \ll 1$ and $C \ll 1$. Clearly, at $Re = 0$ (i.e. in the creeping flow approximation) a drop or a bubble remains spherical irrespective of the low or high value of the (constant) surface tension. However, deformation may be relevant even when inertial effects are ignored if, as earlier noted, the surface tension $\sigma$ is sufficiently non-uniform, and hence there is the Marangoni effect.

## 3.2  Passive drops and the Marangoni effect

Young, Goldstein, and Block (YGB) (1959) were the first to realize the possibility of levitating a drop or a bubble by means of Marangoni stresses. They started with the known fact that a drop or bubble placed in a temperature gradient tends to move towards the hotter point. This is the motion of the drop relative to the flow induced along its surface by the lowering of surface tension at its leading pole (hotter than the rear pole). The Marangoni effect creates a net force on a drop (bubble) which can compensate for buoyancy, hence levitation for a sufficiently high temperature gradient, and hence strong enough Marangoni stress. Using the Stokes-Rybczynskii-Hadamard approximation they computed the terminal velocity of a drop or a bubble in the field of gravity ($Re = Pe = 0$ ), and experimentally checked the theoretical prediction within reasonable accuracy (within 20%) with an experiment using rising bubbles in a liquid layer heated from below (diameters $2R = 10^{-3} - 2 \cdot 10^{-2}$ cm; $dT/dz = 10$ - $90$ K cm$^{-1}$). Later on Bratukhin *et al.* (1982) did a similar experiment with rising bubbles in a laterally heated vertical liquid layer (see also Bratukhin, 1975). They experimented using neutrally buoyant liquid water at 4°C [$R = (2 - 4) \cdot 10^{-2}$ cm, $|\nabla T| = 10$K cm$^{-1}$, $v \sim 1$ cm s$^{-1}$]. For the YGB problem the balance between capillary, buoyancy, and drag forces is

$$F_\sigma + F_g + F_D = 0 \qquad (3.15)$$

where the (Marangoni) capillary force is

$$F_\sigma = -\frac{4\pi R^2}{(1+\beta)(2+\gamma)}\frac{d\sigma}{dT}\frac{dT}{dz}\bigg|_\infty \qquad (3.16)$$

with $\gamma = \lambda_2/\lambda_1$ denoting the ratio of thermal conductivities. The buoyancy force is

$$F_g = \frac{4}{3}\pi R^3 g(\rho_2 - \rho_1). \qquad (3.17)$$

The drag force $F_D$ is (3.7). Incidentally, the YGB flow is an exact solution of the Navier-Stokes equations for any Reynolds number provided $Ma(Pe) \ll 1$.

In their experiments, Young *et al.* (1959) used as a liquid bridge the tiny gap between the anvils of a machinist's micrometer of customary use in introductory physics labs. Typical temperature gradients to hold the bubble stationary, hence levitating (3.15) were e.g. 10 Kcm$^{-1}$ for a bubble of radius $R = 10^{-3}$ cm, and about $10^2$ Kcm$^{-1}$ for $R = 10^{-2}$ cm, with linear interpolation between these two values. With air bubbles, the liquids used included n-hexadecane and silicone oils of various viscosities. An improvement eliminating possible capillary convection at the open sides was carried out by Hardy (1979). He used a closed cavity with silicone oil and air bubbles ($2R = (5 - 25) \cdot 10^{-3}$ cm; $dT/dz = 40 - 140$ Kcm$^{-1}$). The role of bubble deformation in its migration along the temperature gradient was studied by Chen and Lee (1992) and by Nas and Tryggvason (1993). Hardy remarked that the disagreement between theory and experiment in his and Young *et al.* (1959) work was to be in part attributed to the role of contamination by surfactants at the surface of the bubble. Indeed, it was known that small spherical bubbles in ordinary water supply appear to act like solid spheres. The apparent rigidity of the bubbles was soon attributed to pollution by surfactants, dissolved in the water. Clearly surfactant diffusion from the water to the bubble, adsorption-desorption kinetics at the interface, and diffusion and convection of surfactants along the interface are factors affected by the internal flow circulation in the bubble and in turn affect it. Surfactants tend to be swept around the rear of the bubble, forming a spherical cap, until the surface-tension gradient set up by its accumulation balances the tangential viscous (surface) stress around the bubble. With enough surfactant concentration, if the bubble is small the cap may cover the entire surface and can be brought to rest relative to the bubble, which then behaves like a solid particle. The argument applies verbatim to a drop (Levich, 1962).

Further improvement came with an experiment by Merritt and Subramanian (1988); see also Shankar and Subramanian, 1988). Experimentalists then started using drops rather than bubbles. Barton and Subramanian (1989) used neutrally buoyant drops ($2R = 20 - 600\mu$m, $dT/dz = 2.4$ K mm$^{-1}$). Recent Earth-based and low effective-$g$ (TEXUS and D2 missions) work by Braun *et al.* (1993) on thermocapillary migration of drops provided a most accurate verification of the YGB prediction. They used 2-butoxyethanol-water mixture with liquid-liquid phase separation at 61.14°C on the lower branch of the closed miscibility gap; $2R = 11$mm, $dT/dZ = 36.9$ K m$^{-1}$, which exhibits an anomalous thermocapillary effect: $d\sigma/dT > 0$. Flows were observed with Peclet numbers based on the surface tension gradient (Marangoni) velocity in the range $10^{-5}$ - $10^{-6}$.

Thompson et al. (1980) also conducted experiments with nitrogen bubbles in a drop tower at the NASA Lewis Research Center with experiment duration of between four and five seconds with Reynolds numbers in the range $10^{-1}$ - $10^{-3}$, Marangoni (Peclet) numbers in the range $10$ - $10^3$ and capillary numbers about $10^{-3}$. More recently, Balasubramaniam et al. (1996) have done a thorough systematic study of thermocapillary migration of bubbles and drops for moderately large Marangoni (Peclet) numbers in Stokes flows. They used air bubbles ($R = 1-7$ mm) and Fluorinert drops ($R = 1-7$ mm) migrating in silicone oils. Bubbles and drops remained spherical within the uncertainty of the diameter measurements made. Their experiments were conducted aboard the (1994) IML-2 mission of the NASA Space Shuttle in orbit. In their report reference is duly given with details to similar work done by other authors.

Noteworthy are the experimental results by Neuhaus and Feuerbacher (1987). Their experiments with bubble motions aboard the Spacelab (D1 mission) seem to support the role of surface dilational viscosity long ago advocated and earlier mentioned by Boussinesq (1913a-c).

The above evidence shows how the slightest thermal gradient does induce, via the Marangoni effect, drop motions outside and inside. The case is indeed very much like that of the flow induced by the slightest pressure gradient in the bulk of a liquid. Fig. 3.2 illustrates drop or bubble motions for two possible dependences on temperature of the surface tension.

The terminal velocity for a drop driven by the thermal gradient in the gravity field computed by Young et al. (1959) is following Rybczynskii and Hadamard (Levich, 1962; Happel and Brenner, 1983)

$$v = \left(\frac{2}{3\eta_1}\right)\left(\frac{1}{2\eta_1 + 3\eta_2}\right)$$

$$\times \left\{(\rho_2 - \rho_1)gR^2(\eta_1 + \eta_2) - \frac{3\eta_1 R}{2 + \lambda_2/\lambda_1}\frac{d\sigma}{dT}\frac{dT}{dz}\bigg|_\infty\right\}. \qquad (3.18)$$

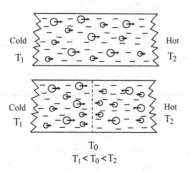

Figure 3.2: Thermocapillary migration of drops or bubbles for two different surface tension laws. In the upper picture $\sigma = \sigma_0 + (d\sigma/dT)_0(T - T_0)$, $(d\sigma/dT)_0 < 0$, while in the lower one $\sigma = \sigma_0 + \gamma(T - T_0)^2$, $\gamma > 0$, and as $(d\sigma/dT)_0 = 0$ the surface tension exhibits a minimum value for $T = T_0$.

For a bubble it suffices to set to zero all quantities with index 2. When the effective gravity or the density contrast are negligible only the second term survives. Fig. 3.3 shows the experimental results obtained by Braun *et al.* (1993).

Subsequently, experiments have been conducted with two interacting drops and bubbles in a thermal gradient under low or variable effective gravity. Noticeable experimental results have been obtained by Hähnel *et al.* (1989) using a Plateau tank and by Wei and Subramanian (1994). Balasubramaniam and Subramanian (1996) have studied bubble migration under zero effective-gravity conditions for large Marangoni (Peclet) numbers and either $Re \ll 1$ (Stokes flow) or $Re \gg 1$ (boundary layer flow asymptotics). To fix ideas note that the former limit is the case of a highly viscous flow of a large Prandtl number fluid, $Pr = \nu/\kappa$ while the latter corresponds to an inertia-dominated flow of a fluid of low to moderate Prandtl number. An example of the first is the movement of a bubble of radius $R = 5$ mm in a 50 cS silicone oil with a temperature gradient of 1 K cm$^{-1}$ ($Re \sim 0.5$; $Ma(Pe) \sim 250$; $Ca \sim 10^{-2}$ and deformation is negligible). An example of the small Prandtl number case is that of a bubble with $R = 2$ mm in a melt of semiconductor silicon with a temperature gradient of 2.5 K cm$^{-1}$ ($Re \sim 1700$; $Ma(Pe) \sim 50$; $Ca \sim 10^{-2} - 10^{-3}$ but $We$ is not small and deformation cannot be neglected). For the analogous mass-transfer process $Pr$ is to be replaced by the Schmidt number, $Sc = \nu/D$, with $D$ denoting mass surfactant or solute dif-

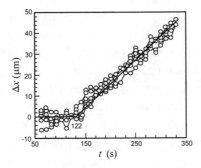

Figure 3.3: Plots of drop displacement ($\Delta x$ in $\mu$m) versus time (t in s). $\Delta x = 0$ is the position of a drop at the instant of time at which the temperature gradient of 0.369 Kcm$^{-1}$ was imposed and $t = 0$ is the beginning of the reduced gravity lapse. The slope of the solid line (0.22$\pm$0.03) $\mu$m/s is equal to the mean value (over 44 drops) of the velocity induced by the Marangoni effect (TEXUS 29 Sounding rocket experiment by Braun *et al.*, 1993).

fusivity. We shall, according to circumstances, use $Pe_D$ or $Sc$ interchangeably. With this convention we also want to emphasize the similarity, albeit not full, between heat and mass diffusion processes. For reviews on further work, including the effect of moderate values of the Reynolds and Peclet numbers, see Wozniak *et al.* (1988), Subramanian (1992) and Sadhal *et al.* (1997).

## 3.3 Active drops and instability; from drag to self-propulsion

### 3.3.1 Active versus passive drops and the Marangoni effect

Let us now consider *active* drops or bubbles, i.e. drops or bubbles with *internal* volume heat sources, with a surface where chemical reactions may occur, or where there is drop dissolution with heat release or heat absorption (e.g. about $10^3 - 10^4$ cal g$^{-1}$mol$^{-1}$ for the pair hexane - aniline; Davies and Rideal, 1961; Perez de Ortiz, 1992; Sawistowski, 1971; Ryazantsev, 1985; Mendes-Tatsis and Perez de Ortiz, 1996). Authors have referred to this as chemocapillarity but we shall stick to solutocapillarity. Take a drop at rest in a homogeneous other fluid and assume that there is (uniform) internal heat

generation or a surface chemical reaction. Let us figure out how the
state of rest can be made unstable. Consider, for instance, the latter
case with given uniform temperature and surfactant composition far
off the drop. We initially have radially symmetric distribution of
temperature and reactant. A composition fluctuation at the surface
of the drop brings the Marangoni effect which yields flows inside and
outside the drop. These flows may start with arbitrary direction in
the original radially symmetric fields and they can be sustained if
the Marangoni effect is strong enough relative to viscous drag and
heat diffusion, i.e. past an instability threshold. Indeed, as the drop
starts moving and spontaneously breaks the radial symmetry, the
flow along the outer surface brings to the rear pole surfactant hence
increasing there its concentration and depleting it at the leading
pole. However, if the concentration far off the drop is high enough
and if the reaction rate is strong enough, consumption of the surfactant is thus enhanced, increasing the strength of the flow disturbance
and helps overcoming viscous drag and the possible surface tension
gradient-driven (Marangoni) counterflow from the rear pole where
surfactant is also consumed, thus sustaining the flow fluctuations ini-
tiated by the composition fluctuation. The drop moves as a whole,
one possibility being its steady self-propelled motion. Note that this
argument differs little from the heuristic explanation for the onset of
steady (cellular) Benard convection to be discussed in a later chapter. If, on the other hand, the initial state is that of a uniform,
constant drop velocity or there is an externally imposed temperature
or composition gradient, as in the experiment of Young et al. (1959),
then the spherical symmetry of temperature and reactants is externally broken with consequent flow and forced drop motion to the
hot area. Instability of the forced flow is also possible, leading to a
different drop motion as a whole. Other instabilities like oscillatory
or pulsating motions may be enhanced by drop deformability and
time-dependence in disturbances.

### 3.3.2  Nonlinear equations and linear stability results

Following Proudman and Pearson (1957), Rednikov et al. (1994 a-e)
have considered various cases of an undeformable drop at rest or
moving with constant velocity in a temperature and composition
homogeneous, infinitely extended surrounding medium. They also
studied the case of an already existing imposed thermal gradient,

thus generalizing the YGB results. Both the inner and outer fluids were taken immiscible. The surface tension was assumed to vary linearly with temperature or surfactant concentration. In the outer fluid a solute or surfactant distributed with uniform concentration was allowed to react exo- or endothermally at the surface of the drop. Stefan flow, i.e. convective flow of the reacting components in a direction normal to the surface where the reaction is taking place, is generally a small effect for most chemical reactions, and it was neglected. Far off the drop concentration of solute was assumed to be constant. Accordingly, if the initially motionless state becomes unstable then it spontaneously breaks the radial symmetry, e.g., with steady motions of the drop as a whole, in an arbitrary direction.

For illustration, let us consider the case of surfactants with the simplest surface (first-order) reaction in the drop where it is consumed isothermally. Adding heat, hence endothermal or exothermal reactions, makes the problem more realistic, but then we ought to solve both Fick and Fourier equations with the Stokes equation (or higher-order corrections) coupling the surfactant concentration and temperature fields at the surface of the drop to the flow. Then too many parameters and asymptotic expansions make the problem unnecessarily complex beyond the scope of our purpose here, which is to merely illustrate how instability and, eventually, spontaneous flow motions appear. To further simplify, one can assume that the surface excess concentration, $\Gamma$, is in equilibrium with the bulk concentration, $C$, hence

$$C = \bar{\alpha}\Gamma, \quad \bar{\alpha} = \text{const} \tag{3.19}$$

with $\bar{\alpha}$ denoting an adsorption parameter. For the surface tension, let us once more take the linear equation of state

$$\sigma = \sigma_0 + \frac{d\sigma}{d\Gamma}(\Gamma - \Gamma_0), \quad \frac{d\sigma}{d\Gamma} = \text{const} \tag{3.20}$$

where $\sigma_0$ and $\Gamma_0$ are related reference values.

The mathematical problem for steady disturbances demands consideration of the continuity, Navier-Stokes and Fick diffusion equations (Sect. 1.2.2) that we recall here for the steady case:

$$(\mathbf{v}_i \cdot \nabla)\mathbf{v}_i = -\frac{1}{\rho}\nabla p_i + \nu_i \nabla^2 \mathbf{v}_i, \tag{3.21}$$

$$\mathbf{v}_1 \cdot \nabla C = D\nabla^2 C, \tag{3.22}$$

$$\nabla \cdot \mathbf{v}_i = 0, \tag{3.23}$$

with corresponding boundary conditions. For steady motions, with $i = 1$ denoting the outer fluid and $i = 2$ the inner one, in the appropriate axi-symmetric coordinate system (Fig. 3.1) we set for the velocity field

$$\mathbf{v}_1 \to \mathbf{U}_\infty \text{ at } r \to \infty; \; v_{r1} = v_{r2} = 0, \; v_{\theta 1} = v_{\theta 2} \text{ at } r = R, \tag{3.24}$$

and for the tangential stresses with the surfactant or solutal Marangoni effect

$$\eta_1 \left( \frac{\partial v_{\theta 1}}{\partial r} - \frac{v_{\theta 1}}{r} \right) - \eta_2 \left( \frac{\partial v_{\theta 2}}{\partial r} - \frac{v_{\theta 2}}{r} \right) + \frac{1}{R} \frac{d\sigma}{d\Gamma} \frac{d\Gamma}{d\theta} = 0. \tag{3.25}$$

For the bulk surfactant concentration one takes

$$C \to C_\infty \text{ at } r \to \infty \tag{3.26}$$

and (3.19) at $r = R$. The surface excess concentration may diffuse and convect according to the following surface balance (Sect. 1.2.2)

$$D_S \nabla_S^2 \Gamma - \nabla_S \cdot (\Gamma \mathbf{v}_S) + D \frac{\partial C}{\partial r} - k\Gamma = 0, \tag{3.27}$$

where $D_S$ and $D$ denote diffusivities on the surface and in the bulk, $\mathbf{v}_S$ is the velocity along the surface. The parameter $k$ is the reaction rate ($k < 0$, reaction produces surfactant; $k > 0$, reaction consumes surfactant; $k = 0$, surfactant does not react on the surface; and $k \to \infty$, the surface excess concentration tends to zero and no Marangoni effect exists).

The limitation to steady motions greatly simplifies the task of finding solutions to the problem but rules out unsteady flows which may include drop oscillatory motions. On the other hand, the stability analysis of steady solutions of the problem demands consideration of the full time-dependent evolutionary problem and was not studied by Rednikov $et$ $al.$ (1994a-e). Quantities and equations can be made dimensionless by measuring lengths with the drop radius, $R$, and introducing

$$\gamma = \Gamma/C_\infty R, \; c = C/C_\infty, \; \alpha = \bar{a}R,$$

$$\nu = \nu_2/\nu_1, \; \beta = \eta_2/\eta_1, \; d = D_S/D, \; Re = U_\infty R/\nu_1, \tag{3.28}$$

$$Pe_D = \frac{U_\infty R}{D}, \; \kappa = \frac{kR^2}{D}, \; Ma = \frac{d\sigma}{d\Gamma} \frac{C_\infty R}{\eta_1 U_\infty}.$$

Note that here $Ma$ is the Marangoni number as it appears in the tangential stress boundary condition (3.25) when made dimensionless (3.31) while the Peclet number, $Pe_D$, is defined using $U_\infty$. To avoid unnecessary complication in this Section we shall omit the subscript $D$. Then using the axisymmetry of the problem and hence (3.5), and scaling the stream function by $R^2 U_\infty$, the extension to active drops of Eq. (3.9) is

$$\frac{Re(\nu_1/\nu_i)}{r^2} \left\{ \frac{\partial(\psi_i, E^2 \psi_i)}{\partial(r, \delta)} + 2E^2 \psi_i L_r \psi_i \right\} = E^4 \psi_i \tag{3.29}$$

with

$$\psi_1 \to r^2 (1 - \delta^2)/2 \text{ at } r \to \infty$$

$$\frac{\psi_2}{r^2} < \infty \text{ at } r \to 0 \tag{3.30}$$

$$\psi_1 = \psi_2 = 0, \; \frac{\partial \psi_1}{\partial r} = \frac{\partial \psi_2}{\partial r} \text{ at } r = 1$$

together with

$$\left( 2\frac{\partial}{\partial r} - \frac{\partial^2}{\partial r^2} \right) (\psi_1 - \beta \psi_2) = (1 - \delta^2) Ma \frac{\partial \gamma}{\partial \delta} \tag{3.31}$$

(the definitions of $\delta$ and $L_r$ are given by Eqs. (3.10) and (3.11), respectively). Eq. (3.31) accounts for the Marangoni effect.

For mass diffusion

$$\frac{Pe}{r^2} \frac{\partial(\psi_1, c)}{\partial(r, \delta)} = \nabla^2 c \tag{3.32}$$

with

$$c \to 1 \text{ at } r \to \infty, \; c = \alpha\gamma \text{ at } r = 1. \tag{3.33}$$

and the corresponding dimensionless form of (3.27) also at $r = 1$,

$$d\frac{d}{d\delta} \left[ (1 - \delta^2) \frac{d\gamma}{d\delta} \right] - Pe \frac{\partial}{\partial \mu} \left( \gamma \frac{\partial \psi_1}{\partial r} \right) + \frac{\partial c}{\partial r} - \kappa\gamma = 0 \tag{3.34}$$

Then, following Proudman and Pearson (1957) and Acrivos and Taylor (1962), the problem was solved for $Pe \sim Re \ll 1$, and $MaPe \sim 1$.

In the *outer* region ($\Psi \equiv Pe^2\psi_1$, $\rho = Pe\,r$, $r \geq O(Pe^{-1})$)

$$\Psi(\rho, \delta) = \Psi_0 + Pe\Psi_1 + o(Pe), \tag{3.35}$$

$$c(\rho, \delta) = c^{(0)} + Pe\,c^{(1)} + Pe^2c^{(2)} + o(Pe^2), \tag{3.36}$$

while in the *inner* region ($1 \leq r \leq O(Pe^{-1})$)

$$\psi_1(r, \delta) = \psi_{10} + Pe\psi_{11} + o(Pe), \tag{3.37}$$

$$c(r, \delta) = c_0 + Pe\,c_1 + Pe^2 \ln Pe\,c_2 + Pe^2c_3 + o(Pe^2), \tag{3.38}$$

On the surface ($r = 1$)

$$\gamma(\delta) = \gamma_0 + Pe\,\gamma_1 + Pe^2 \ln Pe\gamma_2 + Pe^2\gamma_3 + o(Pe^2). \tag{3.39}$$

Inside the drop ($0 \leq r \leq 1$)

$$\psi_2(r, \delta) = \psi_{20} + Pe\,\psi_{21} + o(Pe^2). \tag{3.40}$$

Then the stream function in the outer region obeys the equation (3.29) as

$$\frac{1}{Sc\rho^2}\left\{\frac{\partial(\Psi, E_\rho^2\Psi)}{\partial(\rho, \delta)} + 2E_\rho^2\Psi L_\rho\Psi\right\} = E_\rho^4\Psi \tag{3.41}$$

where $Sc = \nu_1/D$ is here the earlier defined Schmidt number, and

$$\Psi \to \rho^2(1 - \delta^2)/2 \text{ at } \rho \to \infty, \tag{3.42}$$

and $E_\rho^2$ and $L_\rho$ as earlier defined using the variable $r$ and substituting it by $\rho$. Then to the lowest order the solution is

$$\Psi_0 = \rho^2(1 - \delta^2)/2.$$

For the inner region, the lowest order problem obeys Stokes equation (3.6) with $\psi_{i0}$ ($i = 1, 2$) together with

$$\psi_{10} = \psi_{20} = 0, \quad \frac{\partial\psi_{10}}{\partial r} = \frac{\partial\psi_{20}}{\partial r} \text{ at } r = 1. \tag{3.43}$$

Thus using the general solution of the two-dimensional Stokes equation (see e.g. Happel and Brenner, 1965) and the matching condition

$$\psi_{10} \to r^2(1 - \delta^2)/2 \text{ at } r \to \infty \tag{3.44}$$

one gets

$$\psi_{10} = \left(r^2 - \frac{1}{r}\right)G_2(\delta) + \sum_{n=2}^{\infty} A_n(r^{-n+3} - r^{-n+1})G_n(\delta), \qquad (3.45)$$

$$\psi_{20} = \frac{3(r^4 - r^2)G_2(\delta)}{2} + \sum_{n=2}^{\infty} A_n(r^{n+2} - r^n)G_n(\delta), \qquad (3.46)$$

with

$$G_2(\delta) = \frac{1 - \delta^2}{2}, \quad G_3(\delta) = \frac{\delta(1 - \delta^2)}{2} \cdots,$$

where $G_n(\delta)$ $(n = 2, 3, \ldots)$ are the Gegenbauer polynomials of the first kind of order $n$ and degree $-1/2$.

The constants $A_2$, $A_3$, $\ldots$ must be determined by imposing the boundary condition with the Marangoni effect (3.31). Before using this condition one has to consider the concentration field.

The problem for the concentration in the outer region is

$$\frac{1}{\rho^2}\frac{\partial(\Psi, c)}{\partial(\rho, \delta)} = \nabla_\rho^2 c, \qquad (3.47)$$

with

$$\rho \to \infty, \ c \to 1,$$

$\nabla_\rho^2$ is obtained from $\nabla^2$ by replacing $r$ by $\rho$. Then the solution is

$$c^{(0)} = 1. \qquad (3.48)$$

For the inner region, in the lowest order approximation

$$\nabla^2 c_0 = 0, \qquad (3.49)$$

with

$$r = 1, \ c_0 = \alpha\gamma_0,$$

$$d\frac{d}{d\delta}\left[(1 - \delta^2)\frac{d\gamma_0}{d\delta}\right] + \frac{\partial c_0}{\partial r} - \kappa\gamma_0 = 0$$

together with the matching condition to the outer solution

$$c_0 \to 1 \text{ at } r \to \infty. \qquad (3.50)$$

Then the solution is

$$c_0 = 1 - \frac{\kappa}{\alpha + \kappa}\frac{1}{r}, \quad \gamma_0 = \frac{1}{\alpha + \kappa}. \qquad (3.51)$$

This surfactant concentration distribution (3.51) is spherically symmetrical, and hence to obtain the unknown coefficient, $A_n$, in (3.45), (3.46) one must solve for the next-order approximation.

Lengthy algebra yields

$$A_2 = -\frac{3(4 - \kappa)m + 1 + 3\beta/2}{(8 - \kappa)m + 1 + \beta}, \qquad (3.52)$$

with

$$m = -Ma\,Pe\,\frac{1}{12(\alpha + \kappa)(2d + 2\alpha + \kappa)}, \qquad (3.53)$$

and

$$A_n = 0 \text{ if } m \neq m_n; \ A_n, \text{ arbitrary, if } m = m_n, \ n = 3, 4, \ldots, \quad (3.54)$$

with

$$m_n = -\frac{(2n - 1)(1 + \beta)[n(n - 1)d + n\alpha + \kappa]}{3[4n(n - 1) - \kappa](2d + 2\alpha + \kappa)}, \ n = 3, 4, \ldots, \quad (3.55)$$

Note that $m$ redefines the original Marangoni number (3.28) now coupled to the Peclet number, and contains $\kappa$ which here denotes the dimensionless reaction rate. Then if $|\kappa| \to \infty$, $A_2$ does tend to $-(1 + 3\beta/2)/(1 + \beta)$, as expected ($m = 0$ corresponds to the Rybczynskii-Hadamard case).

It appears that the meaningful lowest-order approximation corresponds to the zeroth order in Reynolds number for the velocity field and to the first order in Péclet number for the concentration field. Thus the linearity is due to the fact that the problem for the velocity field and the convective contribution to the concentration field (not the full concentration field) is linear (except for the region away from the drop in the scale $1/Pe$).

Explicit evaluation of all the terms yields the hydrodynamic drag force acting on the drop as in (3.7)

$$F_D = -4\pi\eta_1 RA_2 U_\infty. \qquad (3.56)$$

Note that if $A_2$ (3.52) is negative the force (3.56) is drag ($F_D$ and $U_\infty$ are of the same sign). However, if $A_2$ happens to be positive, the force (3.56) becomes thrust, i.e. $F_D$ and $U_\infty$ are of opposite signs, and consequently, the hydrodynamic force acts in the same direction the drop moves. It is important to notice that for such a remarkable situation to occur one does not need to have any anomalous thermocapillarity, $d\sigma/dT > 0$ $(m < 0)$. Indeed, owing to the chemical reaction, thrust can take place with $d\sigma/dT < 0$ $(m > 0)$. One only needs the reaction to be strong enough, with $\kappa > 4$ as can be established from (3.52). Then, $A_2$ changes sign as $m$ passes over $(1 + 3\beta/2)/3(\kappa - 4)$, i.e. when the Marangoni effect is also strong enough. Under the condition of thrust, application of an external force such as buoyancy would result in the drop moving *against* this force. Here notice that if $U_\infty$ is the velocity of the uniform stream far off the drop in the reference frame where the drop is at rest, the drop velocity in the laboratory frame of reference, where the fluid is at rest far off the drop, is $-U_\infty$.

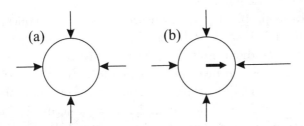

Figure 3.4: Active drops. Mass transfer to the drop due to diffusion and convection in the bulk and the Marangoni effect. (a) If the drop is motionless the mass flux is purely diffusional and spherically symmetric. (b) If the drop moves (in the direction of the thick arrow) the flow may bring to the leading pole the higher concentration of surfactant. Yet the role of the chemical reaction is crucial. Consuming surfactant at the surface, it makes the concentration far off the drop higher than that in its vicinity.

The physicochemical nature of a zero drag and, eventually, thrust can be explained in the following way. To reduce the drag down to zero or even to create thrust we need a mechanism for the concentration at the leading pole of the drop to be always larger than that at the trailing one. A high enough rate of the surface chemical reaction, with the subsequent flow, hence the Marangoni effect, does provide such a mechanism. One can also visualize this by plotting $A_2$ versus $\kappa$ for given values of the other parameters like, say, $\alpha = \beta = d = 1$. Then for high enough values of $|MaPe|$ thrust is ensured. In the case of surfactant consumption, the concentration at the surface is *lower* than the homogeneous value at infinity. Thus the flow brings to the leading pole the solution with higher concentration of surfactant (Fig. 3.4). Generally, there is competition between this and other mechanisms (for example, convective transport of the surfactant along the surface film to the trailing pole and subsequent counterflow from it).

At least as long as a chemical reaction is involved, it seems to be relevant to consider the case of anomalous surface tension dependence on the concentration $(d\sigma/d\Gamma > 0)$ as well. Really, some complex chemical process (may be together with adsorption-desorption processes) with a number of surfactants involved can lie behind the simplest one just considered. Indeed, the concentrations of the surface active agents are mutually interconnected and it can happen that the effective sign of $d\sigma/d\Gamma$ for one particular surfactant is positive although the surface tension falls with the overall concentration of all the surfactants.

The drop problem just discussed above bears conceptual similarity with the Benard problem where in a liquid layer at rest the temperature gradient first defines a steady diffusion process and, subsequently, past an instability threshold induces steady patterned convection due to the Marangoni effect. Note that high-order Gegenbauer polynomials provide a complex flow structure inside the drop which is reminiscent of Benard cells (Chapter 4; Benard, 1900, 1901; Koschmieder, 1993; Velarde and Normand, 1980; Van Dyke, 1982). The initial homogeneous and spherically symmetric case corresponds to the motionless heat diffusive steady state of the liquid layer heated from below, to be studied in Chapter 4.

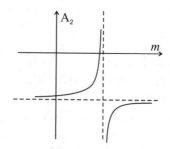

Figure 3.5: Active drops. Coefficient $A_2$ versus Marangoni parameter $m$.

The explicit value of $A_2$ is (3.52) and a graphic representation is provided in Fig. 3.5 as a function of $m$ (3.53) for $\kappa > 8$. For $\kappa < 8$ the asymptotes are located in the other semiplanes.

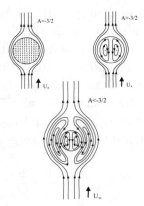

Figure 3.6: Active drops. Flow patterns (streamlines) (a) $A_2 >$ $-3/2$, (b) $A_2 = -3/2$, (c) $A_2 < -3/2$, Eqs. (3.45), (3.46) at $A_n = 0$ $(n = 3, 4, \ldots)$.

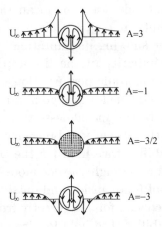

Figure 3.7: Active drops. Velocity profiles corresponding to typical flow patterns shown in Fig. 3.6.

The flow patterns and corresponding streamlines for different values of $A_2$ are sketched in Figs. 3.6 and 3.7 according to (3.45) and (3.46) with $A_n = 0$ $(n = 3, 4, \ldots)$. For $A_2 > -3/2$ it is qualitatively analogous to the case of a drop with no Marangoni effect. When $A_2 = -3/2$, there is no flow inside the drop and outside the flow looks like the flow around a rigid sphere. When $A_2 < -3/2$, the flow pattern becomes more complex, a circulation zone appears outside the drop while inside circulation changes the direction and the drag becomes higher (superdrag) than the value for a rigid sphere.

There are two critical values $m_1$ and $m_2$ of the parameter $m$ at which the coefficient $A_2$ vanishes and diverges to infinity, respectively. Equation (3.52) gives

$$m_1 = \frac{1 + 3\beta/2}{3(\kappa - 4)}, \tag{3.57}$$

$$m_2 = \frac{1 + \beta}{\kappa - 8} \tag{3.58}$$

hence

$$3(4 - \kappa)(m - m_1) + (8 - \kappa)(m - m_2)A_2 = 0. \tag{3.59}$$

From the viewpoint of the general relationship $F_D = F_D(U_\infty)$, not limited to $Re \ll 1$ and $Pe \ll 1$, the linear results obtained above actually yield the slope $dF_D/dU_\infty$ at $U_\infty = 0$ (where also $F_D = 0$). Thus, at $m = m_1$ it is $dF_D/dU_\infty = 0$, while at $m = m_2$ we obtain $dF_D/dU_\infty = \infty$. But this does not mean that the force $F_D$ itself is either zero or infinity at $U_\infty \neq 0$. It is so only within the linear approximation, hence showing its limitations.

In view of the similarity of the linear problem and the problem for the neutral monotonic perturbations, the critical values $m_n$ $(n = 1, 2, \ldots)$ of the redefined Marangoni number, $m$, correspond to the neutral stability of the motionless state of the drop and inner and outer fluids in the absence of buoyancy. Each $m_n$ represents the monotonic instability threshold for the corresponding mode in (3.45) and (3.46). For the higher-order modes $(n = 3, 4, \ldots)$, which describe the motion of fluids inside and outside the drop without its translation, this conclusion follows directly from (3.54). The critical value $m_1$ is the instability threshold for drop translations.

The interpretation of the critical values $m_2$ is less obvious. It is the instability threshold for a *fixed* drop, i.e., a drop which is forced to remain at rest as a whole for whatever flows develop around it, in contrast to a *free* drop, which is free for translations.

### 3.3.3    A few striking features of the nonlinear stability of spontaneous drop self-propulsion with the Marangoni effect

To overcome the difficulties left unsolved by the linear analysis without embarking on a true nonlinear stability study, Rednikov *et al.* (1994a) proceeded to a weakly nonlinear analysis for $m$ around $m_1$ and $m_2$. To illustrate the salient features found we just recall some of their findings.

The hydrodynamic drag force in the next order correction to (3.56) is

$$F_D = -4\pi\eta_1 R(A_2 + RB_2|U_\infty|/D)U_\infty, \qquad (3.60)$$

where $A_2$ is given by (3.52) and $B_2$ is a rather cumbersome expression, function of $m$ and other parameters of the problem. It appears that $A_2 \approx m - m_1$ and $B_2 \sim 1$.

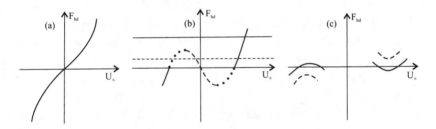

Figure 3.8: Active drops. Hydrodynamic force acting on the drop, Eqs. (3.63) and (3.64), versus the velocity, $U_\infty$, of the uniform stream far off the drop in the reference frame where the drop is at rest, in the case $m = m_1$. (a) Monotonic, and (b) multivalued dependence. If buoyancy exists (as indicated by the horizontal dotted line) then if it is too strong, a single velocity is selected, while if it is weak enough (as in a low effective-gravity laboratory) then up to three velocity values are possible for a single buoyancy level. Stable (solid line) and unstable (broken and dotted lines) fragments are also shown in accordance with the theory presented in the main text. (c) Details illustrating the possibility of sudden appearance of finite velocity values (solid line) as the Marangoni number, $m$, is varied (after Rednikov *et al.*, 1994a).

#### i. Behavior around $m = m_1$

Rednikov *et al.* (1994a) found that to the leading order around $m_1$, one has

$$B_2 \approx -\frac{3m}{2} \frac{2\kappa + q(4 - \kappa)}{(8 - \kappa)m + 1 + \beta} \qquad (3.61)$$

with

$$q = \kappa/(\alpha + \kappa). \qquad (3.62)$$

Accordingly, the force balance (3.60) can be written as

$$3(4-\kappa)(m-m_1)U_\infty + \frac{3mR}{2D}[2\kappa + q(4-\kappa)]|U_\infty|U_\infty - (8-\kappa)(m-m_2)A = 0. \qquad (3.63)$$

where $A$ is defined as a velocity scale of the hydrodynamic force

$$F_D = 4\pi\eta_1 RA \qquad (3.64)$$

The corresponding force is schematically plotted in Fig. 3.8. It is remarkable that for a value of $F_D$ one or more values of $U_\infty$ exist, and, in particular, that for zero value of $F_D$, autonomous motion ($U_\infty \neq 0$) is possible. This is once more a clear manifestation of how the Marangoni effect behaves like an "engine" transforming chemical energy into mechanical motion, with no need of an external force.

If buoyancy exists, $F_B = 4\pi(\rho_2 - \rho_1)R^3 g/3$, the velocity can be calculated from Eq. (3.63) using $F_D + F_B = 0$. In the absence of buoyancy, autonomous motion actually happens when $m$ passes through $m_1$ either supercritically ($\kappa > 4$) or subcritically (i.e. for $|m| < |m_1|$ if $0 < \kappa < 4$). Buoyancy merely shifts the bifurcation away from $m_1$. The velocity of the spontaneous autonomous motion is obtained from (3.63) by setting $A = 0$, i.e.

$$|U_{aut}| = \frac{2(\kappa - 4)(m - m_1)D}{[2\kappa + q(4 - \kappa)]mR}. \qquad (3.65)$$

The absolute value is due to the fact that in a homogeneous (and spherically isotropic) medium all directions in space are equally probable and the symmetry can only be broken by the spontaneous initial flow fluctuation.

## ii. Behavior around $m = m_2$

In the neighborhood of $m = m_2$, the analysis is similar to the previous case; for details we refer the reader to Rednikov *et al.* (1994a), with the difference that Eq. (3.60) yields the opposite multiplicity to the previous case, so that for a single value of $U_\infty$, one or more values of $F_D$ are possible. Then it follows the striking prediction that for $U_\infty = 0$ a non-vanishing value of $F_D$ may be experienced by the drop. Consequently, an active drop, with chemical reaction and the Marangoni effect, may very well levitate under a non vanishing value of buoyancy and such an active drop under different levels of buoyancy may have the same terminal velocity. Fig. 3.9 is a sketch of the $(F, U_\infty)$ relationship. Furthermore, one sees that there is the possible sudden appearance, as a subcritical instability, of a finite hydrodynamic force near $m = m_2$.

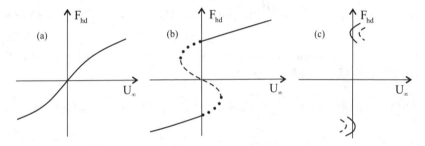

Figure 3.9: Active drops. Hydrodynamic force versus velocity, $U_\infty$, around $m = m_2$. (a) Monotonic, and (b) multivalued dependence. (c) Details illustrating the possibility of sudden appearance of finite values of the force (solid line) as the Marangoni number is varied (after Rednikov et al., 1994a).

In conclusion, it clearly appears that there is a qualitatively significant difference between the autonomous motion of an active drop described above and the earlier mentioned thermally induced, forced motion studied by Young et al. (1959) and others. When a temperature gradient is imposed, the passive drop migration is a response, using the Marangoni engine, to the external constraint. The drop motion predicted in the theory of Rednikov et al. (1994a) is that, even if no external thermal gradient exists, the Marangoni effect may be used by a drop to spontaneously initiate motion and remain moving, steady (or otherwise).

## iii. Stability analysis

Complete stability analyses for $m$ around $m_1$ and $m$ around $m_2$, respectively, have not been carried out. As mentioned earlier, they demand consideration of the time-dependent term left out when writing Eq. (3.29). Stability to particular cases of disturbances permits ruling out unstable states in the multiplicity available, as earlier noted.

The starting point is Eq. (3.63) which defines the force acting on the drop, and we look for the states of drop motion that admit a nontrivial neutral monotonic perturbation.

Take Eq. (3.63) in vector form, and assume small disturbances upon $\mathbf{U}_\infty$, hence $\mathbf{U} = \mathbf{U}_\infty + \mathbf{U}'$, with $|\mathbf{U}'| \ll |\mathbf{U}_\infty|$. As the quantity $\mathbf{A}$ is held fixed (e.g. an external force like buoyancy), then the disturbance obeys the linear equation

$$3(4-\kappa)(m-m_1)\mathbf{U}' + \frac{3mR}{2D}[2\kappa + q(4-\kappa)](|\mathbf{U}_\infty|\mathbf{U}' + |\mathbf{U}'|\mathbf{U}_\infty \cos\alpha) = 0,$$

$$(3.66)$$

where $\alpha$ is the angle between vectors $\mathbf{U}_\infty$ and $\mathbf{U}'$. Projecting (3.66) on $\mathbf{U}'$ and $\mathbf{U}_\infty$ yields, respectively,

$$\left\{3(4-\kappa)(m-m_1) + \frac{3mR}{2D}[2\kappa + q(4-\kappa)]|\mathbf{U}_\infty|(1+\cos^2\alpha)\right\}|\mathbf{U}'|^2 = 0,$$

$$(3.67)$$

$$\left\{3(4-\kappa)(m-m_1) + \frac{3mR}{2D}[2\kappa + q(4-\kappa)]|\mathbf{U}_\infty|\right\}|\mathbf{U}_\infty||\mathbf{U}'|\cos\alpha = 0.$$

$$(3.68)$$

From the compatibility condition we get that drop motions with the velocity $|\mathbf{U}_\infty| = |U_{aut}|/2$ admit a neutral monotonic perturbation with a direction parallel ($\cos\alpha = \pm 1$) to that of the base velocity vector $\mathbf{U}_\infty$, while the autonomous motion ($|\mathbf{U}_\infty| = |U_{aut}|$) admits a neutral monotonic perturbation in the orthogonal direction ($\cos\alpha = 0$). Note that the motion regime with $|\mathbf{U}_\infty| = |U_{aut}|/2$ corresponds to the extremum of the curve $F_D = F_D(U_\infty)$.

Let us further comment on the consequences of the results obtained. First, take the case $\kappa > 4$. Then one may expect that for subcritical $m$ ($m < m_1$) the single motion regime existing for any values of the force (Fig. 3.8) is stable. When $m > m_1$ there is multiplicity of velocity values. Then, as discussed earlier, drop motions with velocity of absolute value less than $|U_{aut}|$ are unstable with respect to orthogonal disturbances, while motions with velocity less than $|U_{aut}|/2$ are also unstable to parallel disturbances.

These results are illustrated in Fig. 3.8(b) where the regimes unstable to both parallel and orthogonal disturbances are indicated by a broken line, the regimes unstable only to orthogonal disturbances are marked by a dotted line, while the solid line shows the stable regimes.

In the case $0 < \kappa < 4$, as the non-monotonic dependence of the force already appears for a subcritical $m$ ($|m| < |m_1|$) one may expect that the conclusions about stability are the opposite. In addition, higher mode instabilities can arise in this case ($0 < \kappa < 4$) if the corresponding critical number in the scale of the parameter $m$ is smaller, in absolute value, than $m_1$.

Let us briefly discuss a striking and possible experimental observation of the theoretical predictions. The *autonomous* motion of the drop can be observed in the absence of buoyancy under appropriate physicochemical conditions. All directions in space are allowed for this motion but the autonomous motion is neutrally stable to perpendicular disturbances. Then one can expect in an experiment that the autonomously moving drop changes the direction of its motion as a consequence of disturbances like biasing fields. For instance as soon as buoyancy is present, there is a preferred direction for the motion with, however, the limitation that the lower is the buoyancy the longer is the time to achieve this stationary state. In addition, when buoyancy is so small that its influence on the motion is in the scale of any other disturbances acting on the system, the stationary regime may never be attained ($g$-jitter in space due to various disturbances may very well be such a case). Yet the important qualitative difference distinguishing the stable regime studied by Rednikov *et al.* (1994a) from the motion of a drop without Marangoni effect is that in the latter case the velocity tends to zero as buoyancy vanishes. In the former case the velocity approaches a finite value, the *autonomous* motion velocity. Consequently, under low enough buoyancy (or effective gravity) in the first case the drop can be expected to move much faster than in the second case, which is a clear-cut qualitative prediction amenable to experimental test.

A simple argument permits one to show that the regimes corresponding to extrema of the curve $F_D(U_\infty)$ and the regimes of autonomous motion always admit a neutral monotonic mode of perturbation, not only for $|m - m_1| \ll 1$, but also for the general case $m - m_1 \sim 1$, even though the solution in this case is not available. Indeed, in the presence of, say, buoyancy the drop velocity satisfies the equation

$$F_D(U_\infty) + F_B = 0. \tag{3.69}$$

Now let us impose the small stationary perturbation $U'$ to the base velocity $U_\infty$ ($|U'| \ll |U_\infty|$) in the same direction. Linearizing (3.69) yields

$$F_D(U_\infty) + \frac{dF_D}{dU_\infty}(U_\infty)U' + F_B = 0,$$

or in view of (3.69)

$$\frac{dF_D}{dU_\infty}(U_\infty) = 0.$$

Thus only when

$$\frac{dF_D}{dU_\infty} = 0,$$

i.e., when we have extrema (maximum or minimum) there can exist a nontrivial perturbation.

For the regime of autonomous motion

$$|\mathbf{U}_\infty| = |U_{aut}|, \qquad (3.70)$$

which means that $\mathbf{U}_\infty$ is arbitrarily oriented in space with magnitude $|U_{aut}|$. Let us impose a small stationary perturbation $\mathbf{U}'_\infty$ to the autonomous motion. Then linearizing (3.70) results in

$$|\mathbf{U}_\infty| + |\mathbf{U}'_\infty| \cos \alpha = |U_{aut}|, \qquad (3.71)$$

where again $\alpha$ is the angle between $\mathbf{U}_\infty$ and $\mathbf{U}'_\infty$. From (3.71) it follows that the only acceptable perturbations are the orthogonal ones ($\cos \alpha = 0$) which shows that around the autonomous motion there exists a neutral monotonic perturbation. Note that this result is to be expected since all directions in space are allowed for the autonomous motion, hence neutrally stable to any infinitesimal orthogonal disturbance.

Finally, let us recall that Rednikov *et al.* (1995a) have also considered the role of a time-varying effective-gravity field mimicking, in a drastically simplified approach, what it is expected to occur in space ($g$-jitter) and in some Earth-based experiments. The evolution leads to a time-dependent, weakly nonlinear vector-form equation for the velocity of the drop, as expected in view of the symmetry of the problem. This equation can be used not only to find the stationary regimes and analyze their stability, but also to consider some time varying motions. In the case they considered of a small amplitude buoyancy force changing sinusoidally with time the striking and clearcut result found is that an *active* drop capable of *autonomous* motion actually tends to move at all times in a direction orthogonal to the time-varying force, something that is also amenable to experimental test either in a low effective-gravity laboratory or in a Plateau tank.

### iv. Numerical estimates

Let us conclude with a couple of quantitative predictions for, plausible, realistic experiments. For this purpose let us make estimations for the case of uniform internal heat generation of intensity $q$ ($q > 0$ heat generation; $q < 0$ heat absorption). The latter case was studied by Rednikov et al. (1994c) and we shall take advantage of their theoretical predictions applied here to two particular cases.

Let us consider first the case of autonomous motion. For simplicity let us assume $\beta = \kappa = \lambda = 1$ with $\kappa = \kappa_2/\kappa_1$ and $\lambda = \lambda_2/\lambda_1$, denoting thermal diffusivity and heat conductivity ratios, respectively. Then using the appropriate definition of $m$ in this case and the corresponding estimate of $m_1$, the threshold for instability is at

$$\Delta T = qR^2/3\lambda_2, \tag{3.72}$$

and hence

$$-\left(\frac{d\sigma}{dT}\right)\frac{R\Delta T}{\nu_2\kappa_2} \sim -30. \tag{3.73}$$

For water-like liquids $d\sigma/dT \sim 10^{-1}$, $\nu_2 \sim 10^{-2}$ and $\kappa_2 \sim 10^{-3}$, in CGS units. For a drop of size $R = 10^{-1}$ cm, it follows that $\Delta T \sim 3 \cdot 10^{-2}$K, which is a remarkably low temperature difference. Note that the result for the choice $\kappa = 1$ demands in practical terms that a drop exhibits anomalous thermocapillarity ($d\sigma/dT > 0$) or, alternatively, one should have heat absorption rather than heat generation.

The other quantitative estimate targeted refers to the striking prediction of (super) drag higher than predicted by the Stokes-Rybczynskii-Hadamard theory (3.7). Again for purpose of numerical illustration take $\beta = 1$, $\kappa = 1/20$, and $\lambda = 1/5$ which are values fitting the drop-liquid case. Then, following Rednikov et al. (1994c), one gets

$$-\left(\frac{d\sigma}{dT}\right)\frac{R\Delta T}{\nu_2\kappa_2} \sim 2 \cdot 10^3 \tag{3.74}$$

that, for the choice of values given earlier, yields $\Delta T \sim 2$K which is a reasonable temperature difference, much higher, however, than the previous value found for autonomous motion.

Let us mention a result found for a non-isothermal surface chemical reaction, with heat release, $Q > 0$, obtained by Golovin *et al.* (1986) [see also Kurdyumov *et al.*, 1994 and Balasubramaniam and Subramanian, 1996] in the case $Re \ll 1$ (Stokes flow) and $Pe_T = U_\infty R/\kappa_1 \ll 1$ with, however, $Pe_D = U_\infty R/D_1 \gg 1$ (see Eqs. 3.28). Their study reduced the surfactant problem to the boundary layer asymptotics with purely analytical results in the first significant approximation. Using the corresponding equation to (3.56), Golovin *et al.* (1986) obtained an estimate of the autonomous motion, $U_{aut} = U_\infty$. The appropriate definition of Marangoni number in this case is

$$Ma = -\left(\frac{d\sigma}{dT}\right)\frac{QC_\infty R}{\eta_1\lambda_1 Pe_D}. \qquad (3.75)$$

Then the approximate result found is

$$|U_{aut}| = 5 \times 10^{-2}DR[QC_\infty/\eta_1\lambda_1(1+3\beta/2)(2+\delta)]^2\left(\frac{d\sigma}{dT}\right)^2. \quad (3.76)$$

Using as reasonable estimates $D \approx 10^{-5}$ cm$^2$ s$^{-1}$, $\eta \approx 10^{-2}$ g cm$^{-1}$s$^{-1}$, $\beta \approx 5$, $C_\infty \approx 10^{-1}$ g cm$^{-3}$, $Q \approx 10^3$ cal g$^{-1}$ and $d\sigma/dT \approx 2 \times 10^{-1}$ g cm$^{-1}$K$^{-1}$, it follows $|U_{aut}| \approx 5 \times 10^{-1}Rs^{-1}$, with $R$ in cm. For a drop of $R = 10^{-1}$ cm, $|U_{aut}| \approx 5 \times 10^{-2}$ cm s$^{-1}$ which, just for reference, is about the convective velocity in a Benard cell slightly above instability threshold (Normand *et al.*, 1977; Koschmieder, 1993). Gupalo *et al.* (1989) also studied a similar problem when $d\sigma/dT$ vanishes, and hence $\sigma$ becomes a nonlinear function of the temperature (Fig. 3.2).

### v. Related experimental result with a mechanical analog of the Marangoni effect

Fig. 3.10 shows a peculiar wake structure and corresponding recirculation zone observed at moderate Reynolds number when a bubble ascends a slope (Bush, 1997). The wake entirely encircles the bubble, and is characterized by an intense "edge jet" which transports fluid from the rear pole to the front of the bubble. Bush used a simple device consisting of a thin (2 mm) gap of water bound between two rectangular glass plates glued together at their edges. When the gap was inclined at very shallow angles $(1-2°)$ relative to the horizontal, he observed the anomalous wake structure.

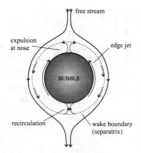

Figure 3.10: Mechanical analog of the Marangoni effect and superdrag. The peculiar wake structure corresponds to a moderate Reynolds number flow viewed in the frame of reference translating uniformly with the bubble. Compare to Figs. 3.6 and 3.7 with $A < -3/2$ (After Bush, 1997).

Bush provided a simple argument to explain his observations. At the leading edge of the bubble, the emplacement of surface material onto the channel walls leads to a region of surface divergence, where the concentration of surfactant material must be decreased. At the trailing edge, surface material is rolled off the glass plates, giving rise to an area of surface convergence and a local maximum in surfactant concentration on the trailing edge of the bubble. Assuming, as is generally the case, that surfactants act to decrease the local surface tension then the surfactant concentration gradient along the edge of the bubble leads to an azimuthal surface tension gradient, or surface traction, which drives the edge jet. Thus, in practical terms, we have a mechanical analog of the chemical reaction leading to Marangoni stresses with an experimental finding that accommodates well with the prediction from the more sophisticated theory by Rednikov *et al.* (1994a) of (super) drag well above the Stokes-Rybczynskii-Hadamard value (see Figs. 3.6 and 3.7, $A < -3/2$).

### vi. Heat of solution leading to Marangoni stresses

Another verification of the theoretical predictions made by Rednikov *et al* (1994a) is the following. Fig. 3.11 is a sequence of photos obtained by Agble and Mendes-Tatsis (2000) exhibiting flow reversal due to the Marangoni effect. A drop of ethylacetoacetate is suspended at the end of a syringe capillary and immersed in a cell con-

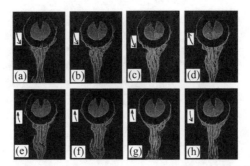

Figure 3.11: Schlieren photos where the arrows indicate the flow direction on the surface of a suspended drop of ethylacetoacetate (saturated with water) which is undergoing mass transfer into the outer water phase. Experiment is unstationary and shows flow reversal $(d, h)$ twice (Photos kindly provided by Dr. Alcina Mendes-Tatsis).

taining pure water. The ethylacetoacetate phase was saturated with water so that mass transfer occurs from the drop to the outer water phase (negative heat of solution about $-10^4$ cal g$^{-1}$ mol$^{-1}$). The mass transfer process was filmed with a Sony digital camera in a Schlieren optical set-up. The arrows indicate the direction of the interfacial convection on that side of the drop. The time elapsed between pictures (a) and (e) in the sequence is one second. In picture (d) the flow is reversed with corresponding eddy formation and recirculation zone. This eddy develops further in picture (e) and the flow continues to be in the upward direction around it, until picture (g). Picture (h) shows the flow returning to the initial direction. From picture (d) to picture (g) the eddy moved upwards. The above mentioned flow reversal has also a corresponding flow inside the drop very much like that shown in Figs. 3.6, 3rd picture, and 3.7, 4th picture.

## 3.4   Spreading of surfactant drops and films and the Marangoni effect

### 3.4.1   Static phenomena. Spreading of drops

Let us consider a drop of a fluid (L) placed on a flat horizontal solid surface (S) and contacting with an ambient fluid, say, vapor (V) (Fig.

3.12).

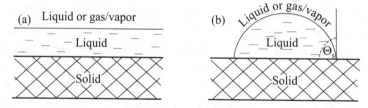

Figure 3.12: Three–phase equilibrium between solid, liquid, and immiscible fluid (liquid or gas [vapor]) (a) complete wetting, (b) drop forming a contact angle, $\Theta_s$.

The interface between each $(L, V, S)$ two media is characterized by a certain interfacial tension $\sigma_{LV}$, $\sigma_{SL}$, and $\sigma_{SV}$. The balance of stresses on the contact line, where all three media come into contact, determines the *static contact angle* $\theta_S$ (Young, 1805; Marangoni, 1865; Dupre de Rennes, 1869; Plateau, 1873; Harkins, 1952; Di Pietro *et al*, 1978; Landau and Lifshitz, 1987; de Gennes, 1985; Cox, 1986a,b; Cazabat, 1987; Leger and Joanny, 1992; Ross and Becher, 1992; Schrader and Loeb, 1992; Myers, 1999):

$$\sigma_{SL} + \sigma_{LV} \cos \theta_S = \sigma_{SV}. \tag{3.77}$$

The equation (3.77) has a solution $0 < \theta_S < \pi$ only if $\sigma_{SL} - \sigma_{LV} < \sigma_{SV} < \sigma_{SL} + \sigma_{LV}$; this case is usually called *partial wetting*. Otherwise, the static situation shown in Fig. 3.12 is impossible, and the system tends to eliminate the contact line. For instance, in the case $\sigma_{SV} > \sigma_{SL} + \sigma_{LV}$ (*complete wetting*) the unbalanced molecular forces produce a *precursor film* of the fluid 1, which tends to cover the solid surface. Thus

$$S = \sigma_{SV} - (\sigma_{SL} + \sigma_{LV}) \tag{3.78}$$

is the spreading coefficient. Let us recall (Becher, 1990) that the spreading coefficient is a measure of the ability of a liquid to spread on the surface of another liquid or solid. It is the difference between the work of adhesion between the two phases and the work of cohesion of the spreading liquid. In view of the argument above, with $S < 0$, there is a partial wetting, as the solid wants to remain dry while if $S > 0$ we have spontaneous complete wetting. Eq. (3.77) is named after Young (1805) and Harkins (1952). This phenomenon cannot be described using only macroscopic fluid dynamics, but it is essential for the construction of a macroscopic theory describing the spreading of the fluid drop in the case of a complete wetting.

Note that there is no reason to use the formula (3.77) for determining the *dynamic contact angle* which is observed when the *macroscopically visible* contact line is moving on the solid surface. Indeed, measurements show that the "macroscopic" dynamic contact angle depends on the direction of the contact line motion (Huh and Scriven (1971) suggested to distinguish between *advancing* and *receding* or *recessing* contact angles), as well as of the velocity of the contact line (Dussan V, 1979). The dependence of the contact angle on the velocity is especially pronounced in thin capillaries (Sobolev *et al.*, 2000), and it is strongly influenced by the radius of the capillary. It was shown that the processes connected with the formation of an adsorption film in front of the moving meniscus were crucial for this phenomenon.

The attempt to use the standard no-slip condition for the fluid velocity near the contact angle leads to a singularity (Dussan V and Davis, 1974), which is usually eliminated by an artificial slip condition on the liquid-solid interface near the contact line (Hocking, 1977). An up-to-date review of applications of this phenomenological approach, which is combined with the lubrication approximation (Hocking, 1990) and includes thermocapillarity (Ehrhard and Davis, 1991), evaporation (Anderson and Davis, 1995) and gravitational flow (Lopez *et al.*, 1996), can be found in the review by Oron *et al.* (1997).

The phenomenological approach described above agrees well with experiments, though the physical origin of the effective slip and the difference between the static and dynamic contact angles is not clear. Specifically, they can be caused by the *roughness* of the interface (Huh and Mason, 1977; Cox, 1983, 1986a,b, 1998; Cazabat and Cohen Stuart, 1986) and its *chemical heterogeneity* (Joanny and de Gennes, 1984; Pomeau and Vannimenus, 1985; Raphaël and de Gennes, 1989; Leger and Joanny, 1992; Alexeev and Vinogradova, 1996).

Several physical mechanisms have been suggested for explanation of the slippage in the case of a liquid flowing over a *lyophobic/hydrophilic* solid surface (for review, see Vinogradova, 1999), which include a real molecular slippage caused by the fact that the strength of attraction between the liquid molecules is greater than that between the liquid and the solid ones (Tolstoi, 1952a, 1952b; Blake, 1984), and the hypothesis of the existence of a gas gap between the solid and the liquid. The most convincing explanation links the apparent slippage effect with a *decrease* in liquid's viscosity in a transition region between the liquid and the lyophobic surface (Churaev *et al.*, 1984). It should be noted however that on the *lyophilic/hydrophobic* solid surface the liquid viscosity *increases* near the boundary (Churaev *et al.*, 1971; Kiseleva *et al.*, 1979). In that case the physical nature of those phenomena should be diverse.

It is worth recalling how macroscopic fluid dynamics fails to describe how a liquid can advance over a solid while displacing another fluid. When steady, incompressible, and Newtonian flow is enforced right down to the apparent wetting line - together with boundary conditions that impose no slip along a smooth, homogeneous, and impenetrable solid surface, continuity of traction and no mass transfer at the fluid/fluid interface, and a local contact angle, $\theta$, in the range $0 < \theta < \pi$ - the velocity field turns out to be multivalued at the wetting line. As a result, viscous stresses and pressure are predicted to increase as $1/r$, where $r$ is the distance from the line, and the force that would be required to advance the liquid becomes logarithmically divergent. Likewise, the total viscous dissipation near the wetting line becomes unbounded. This theoretically predicted singularity cannot be experimentally found, for if the singularity was real, it would give rise to drastic changes in temperature and material properties. These, in turn, would contradict the assumptions of constant density and viscosity made when deriving the Navier-Stokes equations. Recently, there have been some attempts to go beyond standard phenomenological theory based on the consideration of the *kinetics* of interface formation on the contact line (Shikhmurzaev, 1994, 1997; Seppecher, 1996; Pismen and Pomeau, 2000; Thiele *et al.*, 2001a-c).

According to estimations (Hocking, 1977) the liquid-solid slippage occurs when the thickness of liquid layers becomes very small (around $10^{-5}$ cm). However inside such thin liquid layers surface forces action can not be neglected (Derjaguin *et al.*, 1987). A manifestation of these forces is the presence of an additional (disjoining) pressure in a narrow region close to the three phase contact line. In this (transition) region both capillary and disjoining pressure act simultaneously, which results in an augmented Laplace (Derjaguin) equation

$$\sigma_{LV} K + \Pi(h) = P_e \tag{3.79}$$

where $K$ is the curvature of the liquid profile. $\Pi(h)$ is the disjoining pressure isotherm and $P_e$ is the difference between pressure in the ambient air and in the liquid. Equation (3.79) results in the following equation for the equilibrium contact angle (Derjaguin *et al.*, 1987; Starov, 1992)

$$\cos \theta_s = 1 + \frac{1}{\sigma_{LV}} \int_{h_e}^{\infty} \Pi(h) dh \tag{3.80}$$

where $h_e$ is the thickness of the equilibrium liquid film in front of the drop/meniscus. The latter equation shows that the integral in the right hand site should be negative in the case of partial wetting (Fig. 3.13). In the case of complete wetting the right hand side of Eq. (3.80) becomes larger than unity which corresponds to $S > 0$, when drops spread completely. In the case of partial wetting the disjoining pressure isotherms have s-shape (Fig. 3.13) and equation (3.80) can be approximately rewritten as (Starov, 1992)

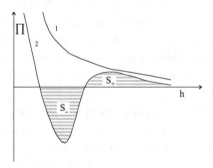

Figure 3.13: Disjoining pressure isotherm. 1 = complete wetting case. 2 = partial wetting case (s-shaped isotherm).

$$\cos \theta_s = 1 - \frac{S_- - S_+}{\sigma_{LV}} \tag{3.81}$$

where $S_-$ and $S_+$ are areas below and above the $h$ abcissa axis (Fig. 3.13). Roughness also changes the equilibrium film thickness (Starov and Churaev, 1976).

No doubt that the presence of roughness and chemical non-uniformity causes contact angle hysteresis. However, contact angle hysteresis is observed on molecularly smooth homogeneous substrates.

The $S$-shape of the disjoining pressure isotherm allows explaining the presence of hysteresis in the contact angle on smooth homogeneous surfaces. In the case under consideration apart from the unique equilibrium state (and the unique equilibrium contact angle) the motion of liquid always takes place. This motion can be "microscopic," within very thin liquid layers or "macroscopic," when thick layers become involved. Transition from "microscopic" to "macroscopic" motion determines the values of the receding and advancing contact angles (Starov, 1992).

The disjoining pressure action removes the singularity at the three phase contact line because in this case the liquid profile tends asymptotically to zero thickness while the disjoining pressure, in its turn, diverges to infinity. In the case of complete spreading, theory predicts a power law: $r(t) = At^{0.1}$, where the prefactor is obtained using the disjoining pressure (Starov, 1992).

Capillary spreading of liquid drops over pre-wetted surfaces has also been considered (Kalinin and Starov, 1986). The same power law for this case has been established and the prefactor $A$ has been calculated using the pre-wetted layer thickness.

Let us discuss briefly the problem of *stability* of the contact line. Even in absence of thermal and solutal inhomogeneities, the contact line is subject to *fingering instabilities*, both in the case of a gravitationally driven motion (Huppert, 1982; Dussan V, 1985; de Bruyn, 1992) and in the wetting case (Williams, 1977; Troian *et al.*, 1989a). The fingering instabilities are found also with heat and mass transfer (Troian *et al.*, 1989b; Cazabat *et al.*, 1990). A specific instability leading to pattern formation was found in experiments on the evaporation of thin wetting films (Elbaum and Lipson, 1994, 1995).

In conclusion, let us mention recent experiments on the spreading of a drop on a surface of a porous substrate under the action of the drop's own weight. It turns out that there are two different stages of this process. During the first stage, the propagation of the liquid on the surface and inside the porous medium takes place with equal velocities. In the second stage, the propagation of the liquid inside the porous medium becomes faster than on the first stage, while the radius of the drop on the surface grows slower than on the first stage. Then the drop starts to shrink and eventually disappears. Deviation from power laws for the liquid spreading are observed. The presently available theory is able to explain only part of these phenomena.

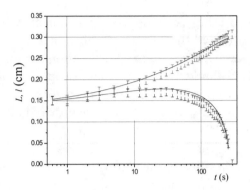

Figure 3.14: Drop spreading (open circles) and imbibition (open squares) processes on a porous substrate. Liquid: silicone oil (5 St). Membrane with mean pore size 3 $\mu$m. The solid lines are the theoretical predictions (after Starov *et al.*, 2001b).

The spreading of different silicone oils with viscosity 26.83, 55.4, 112.0, and 558.2 St over thin porous substrates have been studied by Starov *et al.* (2001a,b) using cellulose nitrate membrane filters purchased from Sartorius (type 113) with mean pore size 0.2 and 3 $\mu$m. The membrane thickness ranges from 0.0130 to 0.0138 cm and the porosity of membranes is in the range 0.65 – 0.87. During the spreading process the drop spreads and simultaneously imbibes the dry porous support. Both radii of saturated region inside the membrane and the spreading drop were monitored using an optical device with two observation directions. In addition, the height and shape were measured in order to study the evolution of the contact angle over time.

In Fig. 3.14 a typical result is shown for the case of a very viscous silicone oil and for a membrane with mean pore size 3 $\mu$m. It appears that the initially fast spreading drop reaches some maximal radius and then slowly collapses. The solid lines in Fig. 3.14 correspond to the theoretical predictions using parameter values obtained in separate independent experiments.

To ascertain if universal behavior exists in the spreading process a new dimensionless scale was introduced. The time scale selected was the total time of imbibition into the porous substrate; the radius of

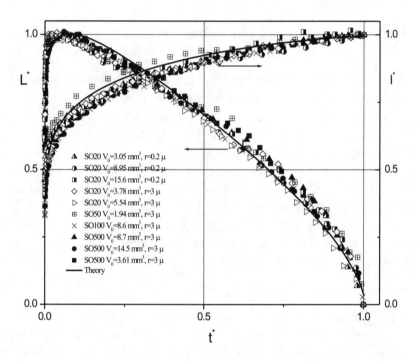

Figure 3.15: Drop spreading and imbibition processes on a porous substrate. Universal behavior in the time evolution of the dimensionless drop radius $L$ (left ordinate) as well as radius of imbibition front $l$ (right ordinate). The points correspond to different liquids used, two different membrane types, and several drop volumes. The solid lines indicate the theoretical prediction (after Starov *et al.*, 2001b).

imbibition is a fraction of the maximum radius of imbibition, and the spreading radius a fraction of the maximal radius of spreading drops. Experimental data are presented in Fig. 3.15. It is clearly seen, that although the viscosity of the liquids used differs in two orders of magnitude and the pore sizes for two types of the membranes used differs by a factor of 15, all experimental data exhibit a common behavior.

### 3.4.2    Liquid-liquid spreading of partially miscible liquids and the Marangoni effect

Let us now consider the case of a drop deposited on the surface of another liquid where it might be partially miscible or immiscible.

When a drop of a surfactant or some other liquid (denoted here, for convenience, o), is deposited on the surface of an *immiscible* liquid (here denoted w), gravitational forces cause spreading which is instantaneously opposed by inertia forces arising from the acceleration of the slick. As time proceeds, viscous drag with the substrate replaces inertia as the retarding force. This is also the initial situation when the effective gravity is not significant. Whether "o" continues to spread or not depends on the sign of the earlier defined (3.78) spreading coefficient, here expressed as

$$S = \sigma_w - (\sigma_o + \sigma_{o/w}),  \qquad (3.82)$$

where here $\sigma_w$ and $\sigma_o$ are the surface tensions of the phases w and o, and $\sigma_{o/w}$ is their interfacial tension. When a surfactant concentration gradient exists the spreading driving force derives from the Marangoni shear stress integrated over the length of the spreading film,

$$S = \int_0^{R(t)} \frac{\partial \sigma}{\partial x} dx.$$

For a spherical drop, axisymmetrically spreading on an immiscible liquid layer of thickness $H$, it has been established, both theoretically and experimentally, that when capillary forces dominate with viscous dissipation in the underlying liquid substrate, then the drop radius, $R$, grows according to a power law (Hoult, 1972; Joos and Pintens, 1976)

$$R = At^n.  \qquad (3.83)$$

The prefactor $A$ depends on material properties. For thick enough drops when gravity competes with inertia, neglecting dissipation,

$n = 1/2$. If, however, gravity competes with viscous drag, then $n = 1/4$. Finally, when capillary forces compete with dissipation, two possible cases have been studied. If dissipation mostly takes place in the viscous boundary layer of the underlying liquid (the case with a thick enough substrate; Dussaud and Troian, 1998)

$$n = 3/4 \text{ and } A \propto S^{0.5}/(\eta\rho)^{0.25}, \tag{3.84}$$

while if the underlying liquid is so thin that dissipation takes place over the whole bulk (as in the lubrication approximation; Fraaije and Cazabat, 1989; Starov et al., 1997), then for immiscible liquids

$$n = 1/2 \text{ and } A \propto (HS/\eta)^{0.5}. \tag{3.85}$$

Here, as earlier, $\rho$ and $\eta$ are, respectively, the density and the (dynamic) shear viscosity of the liquid in the underlying layer. This analysis ignores the drop finite size.

Bacri et al. (1996) have provided theory and experimental data for the spreading of drops on a wettable liquid layer when dissipation inside them predominates over dissipation in the liquid substrate. Their results are size-dependent and hence for drops of size smaller than the capillary length, spreading is driven by capillary forces balancing dissipation and follows the power law $n = 1/4$. For larger drops, the spreading is gravity-driven and then $n = 1/2$.

For *miscible* liquids there is dissolution during the spreading. When capillary forces for rather thick layers, neglecting, however, inertia and gravity, compete with viscous drag, $n = 1/2$. If, however, the layer is rather thin, $n = 3/4$ (Lopez et al., 1976; Brochart-Wyart et al., 1990; Redon et al., 1992). As dissolution causes bulk and surface concentration gradients, if the drop is of denser material than the liquid substrate then buoyancy-driven convection may accompany the mixing process, thus altering the spreading process. Moreover, if due to the surface concentration gradient and the Marangoni effect, interfacial motions develop at the interface between the two liquids, these can give rise to interfacial instability also altering the spreading process. The mentioned convective phenomena occurs when evaporative processes accompany spreading. Dussaud and Troian (1998) noted that the dissipation induced by buoyancy-driven convection lowers in their experiment the theoretically expected value of $n = 3/4$ down to $n = 1/2$.

Although the spreading problem with miscible liquids is of great technological interest, it has received little attention (Suciu et al.,

1967, 1969, 1970; Ruckenstein *et al.*, 1970; Santiago-Rosanne *et al.*, 1997, 2001).  Here to succinctly illustrate phenomena we shall restrict consideration to the most recent work by Santiago-Rosanne *et al.* (2001) on the spreading of a nitroethane or an ethyl acetate drop carefully deposited on a water surface. Nitroethane is partially soluble in the water, it is denser and less viscous, and it has a much lower surface tension than water. Ethyl acetate is also partially soluble in the water but has a lower surface tension, and being lighter than water it yields in solution a stably stratified layer. Just after deposition, the drop spreads as a central cap surrounded by a primary film. In their earlier publication (Santiago-Rosanne *et al.*, 1997) they reported that after about one second, the (surfactant) Marangoni effect excites surface waves on the primary film whose properties will be discussed later on in Chapter 5.

In the experiments performed by Santiago-Rosanne *et al.* (2001) with mutually saturated nitroethane and water (they can therefore be considered as practically immiscible) on a rather thin substrate layer ($H =10.2$ mm), the spreading coefficient is $S =0.45$ mN m$^{-1}$, and the measured spreading law is for $R(t)$ expressed in mm,

$$R(t) = 40t^{0.56}, \tag{3.86}$$

which agrees well with the theoretical prediction (3.85), $n = 0.5$.

When dissolution occurs, the measured exponents were also close to $n = 0.5$, as long as $H$ is between 13 and 3.5 mm. They calculated the spreading prefactor assuming that the spreading coefficient (3.82) is $S = \sigma_W - \sigma_0$. Agreement between theory and experiment was not too good, as the prefactor was generally higher in theory than experiment. Eq. (3.85) corresponds to immiscible liquids. Furthermore, dissolution leads to mass loss and this and some other physicochemical factors may be at stake.

Earlier, Dussaud and Troian (1998) also reported on work with volatile liquids on thick layers a reduction in the spreading power law from $n = 3/4$ to $n = 1/2$. Their explanation was that buoyancy-driven convection added dissipation beneath the leading edge. Santiago-Rosanne *et al.* (2001) also found convective rolls near the edge and also along the whole region below the primary film. A difference exists, however, between the experiment with nitroethane and that with ethyl acetate. In the latter case small convective cells were observed and this was attributed to the fact that ethyl acetate is lighter than nitroethane and hence due to the dissolution of the drop buoyancy creates stabilizing density gradients that confine possible surface tension gradient-driven convection, with the Marangoni effect, to shallow regions near the interface while for nitroethane density gradients are rather destabilizing and generate convective cells that penetrate deep into the liquid substrate. Nitroethane ideally mixes with water since $\rho$ and $\sigma$ depend linearly on $C$ in the whole range of their mutual miscibility. As their interfaces behave as free surfaces, since there is no surfactant accumulation, the thickness of the aqueous liquid substrate decreases with time. Accordingly, viscous dissipation drastically increases, due to the additional friction imposed by the lower sticky boundary. This justifies a lowering of $n$ from the predicted value 0.5 down to the measured value $n = 0.1$, obtained when the sublayer thickness is about the size of the observed convective cells. For thinner layers, the power law $n = 0.1$ is the same as that found for the spreading of a liquid drop on a solid surface when viscous dissipation occurs in the drop (Tanner, 1979).

Santiago-Rosanne *et al.* (2001) also noted that since aqueous solutions of ethyl acetate are stably stratified when the surface tension gradient yields convection, due to the Marangoni effect, cell size would be below the limit of mechanical stability of the liquid substrate. The (solutal) Marangoni and Rayleigh numbers here are $Ma = (\partial\sigma/\partial C)\Delta Ch/\eta D$, and $Ra = (\partial\rho/\partial C)\Delta Cgh^3/\eta D$ where, $D$ is the diffusion coefficient of the organic solute in water, respectively. Confusion between the quantities $C$ and $\Gamma$ is of no importance here. The quantity $h$ denotes the vertical penetration depth in the liquid substrate of the dissolved amount of drop material, and $\partial\sigma/\partial C$ is obtained from Eqs. (3.84) and (3.85), while $\partial\rho/\partial C$ is obtained from refractive index measurements and the application of the Lorentz-Lorenz formula. Then their data fit well with the critical values for both buoyancy-driven convection ($Ra \sim 10^3$) and surface tension gradient-Marangoni-driven convection ($Ma \sim 10^2$). When the spreading liquid is lighter than the liquid substrate, as with ethyl acetate on water, the Marangoni effect dominates and hence convective cells remain near the liquid-liquid interface. For the opposite case with nitroethane, both buoyancy and the Marangoni effect operate simultaneously, the former leading to a deep penetration of convective cells in the liquid substrate. When the depth of the liquid substrate is about the size of the convective cells, bottom friction comes into play and it drastically lowers the spreading parameter.

### 3.4.3   Spreading of a drop of practically insoluble surfactant due to Marangoni stresses

Let us now consider the case studied by Starov *et al.* (1997) about the spreading of a drop of a practically insoluble surfactant solution deposited on the surface of a liquid substrate. As already noted in Chapter 1, two time scales are associated with the surfactant transfer: $\tau_d$, accounting for desorption, hence transfer from the liquid surface to its bulk, and $\tau_a$ that accounts for adsorption with transfer from the bulk back to the surface. These two scales depend on energy and entropy (of configuration, due to the orientation of the tails of the surfactant molecules), and differences between corresponding states and their interactions (Ravera *et al.*, 1993, 1994). For instance, if we take water and a standard surfactant with hydrophilic head and hydrophobic tail, then $E_{hl}$, $E_{ta}$, and $E_{tl}$ are the corresponding energies (in $kT$ units) of head-water, tail-air, and tail-water interactions. The energy of a molecule adsorbed at the interface liquid-air is $E_{ad} = E_{ta} + E_{hl}$, while for the same molecule in the bulk $E_b = E_{tl} + E_{hl}$. Accordingly, $\tau_d \sim \exp(E_b - E_{ad}) = \exp(E_{tl} - E_{ta})$, and $\tau_a \sim \exp(E_{ad} - E_b) = \exp[-(E_{tl} - E_{ta})]$. Generally $\tau_d/\tau_a \sim \exp[2(E_{tl} - E_{ta})] \gg 1$, and hence desorption appears as a much slower process than adsorption. If the duration of the spreading process is shorter than $\tau_d$ then the surfactant is considered as insoluble during the experiment.

#### i. Experiment

Observation of the spreading process was done by Starov *et al.* (1997) on a thin layer of liquid using aqueous solutions of sodium dodecyl sulfate (SDS) at a concentration $C = 20$ g l$^{-1}$, above CMC (the critical micellar concentration for SDS is 4 g l$^{-1}$). The thin layer of liquid was prepared by coating the bottom of a borosilicate glass Petri dish of diameter 20 cm with 10 ml of distilled water. The resulting thickness was then $H = 0.32 \pm 0.01$ mm and did not dewet during the time of the experiment. With a syringe, a drop of the surfactant solution, volume 3 $\mu$l, was placed on the surface of the water layer. When touching the water surface, the surfactant spread on it and this motion was followed using a small amount of talc powder as a marker and a 25 Hz video camera to record it.

The spreading of the surfactant makes the water flow away from the initial location of the drop, thus creating a depression where only

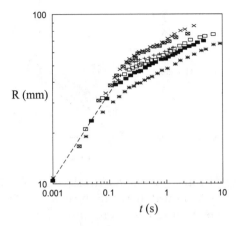

Figure 3.16: Drop spreading due to the Marangoni effect. Evolution of the hydraulic jump formed at the outer periphery of the drop (radius, $R = r_f$, in mm versus time, t, in seconds) exhibiting two spreading stages with different power laws: first, 0.6 (expt.) - 1/2 (theory) and, subsequently, 0.17 (expt.) - 1/4 (theory) (after Starov et al., 1997).

a thin film of liquid subsists for quite some time. The periphery of this depression, i.e., the liquid front, has a sharp increase in thickness. If the layer is horizontal and the drop is carefully placed, there is no preferred direction and the peripherical edge is circular, but in practice some modulation always appears in the experiment after a few seconds. Note that the surfactant occupies more surface area than the depressed zone since the talc powder is pushed ahead of it. The time dependence of the radius of the surfactant patch is given in Fig. 3.16 using a log-log plot. Two successive stages are clearly seen. First is a short time interval when the surfactant spreads following the power law $r_f(t) \sim t^{0.60 \pm 0.15}$. The exponent is not determined with a high precision because this first stage is too fast (about 0.1 s, which is only three times the time resolution, 0.04 s). At the end of this first stage the motion abruptly slows down and the moving front, a hydraulic jump, follows a different power law, $r_f(t) \sim t^{0.17 \pm 0.02}$.

In Fig. 3.17, the radius of the shallow region, i.e., the dry spot radius, is plotted for the same six experimental runs shown in Fig. 3.16. The power law dependence $r_d(t) \sim t^{0.25 \pm 0.05}$ was observed. Experimentally, the ratio $r_d(t)/r_f(t)$ is about 1/3, although it slowly

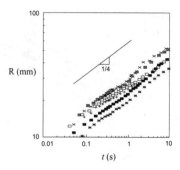

Figure 3.17: Drop spreading due to the Marangoni effect. Evolution of the hole formed at the drop center (radius, $R = r_d$, in mm versus time, t, in seconds) with power law 1/4 (after Starov et al., 1997).

changes with time during the second stage.

## ii. Lubrication theory with Marangoni effect

### ii.1 Generalities

A theory was also developed by Starov et al. (1997). Although agreement is not perfect, as it captures the essence of the spreading process with Marangoni stress, let us summarize it.

The relevant dimensional and dimensionless variables are $r^* = R_0 r$, $z^* = H_0 z$, $\Gamma^* = \Gamma_m \Gamma$, $\sigma^* = \sigma_m + S\sigma$, $u^* = Uu$, $w^* = Ww$, $t^* = Tt$, and $p^* = Pp$, where $R_0$ is the initial radius of the drop, $\Gamma_m$ is the CMC, $S = \sigma_0 - \sigma_m$ with $\sigma_0$ the liquid substrate–air surface tension in the absence of the drop, $\sigma_m = \sigma(\Gamma_m)$, $U = SH_0/\eta R_0$ (velocity scale due to the Marangoni effect), $W = SH_0^2/\eta R_0^2$, $T = \eta R_0^2/SH_0$ and $P = S/H_0$.

With $Re = UR_0/\nu$ denoting the Reynolds number here and $\epsilon = H_0/R_0$, Starov et al. (1997) limited their analysis to the case $\epsilon \ll 1$, and $\epsilon^2 Re \ll 1$ (lubrication approximation). Thus the Navier-Stokes equations reduce to

$$\frac{1}{r}\frac{\partial(ru)}{\partial r} + \frac{\partial w}{\partial z} = 0, \tag{3.87}$$

$$-\frac{\partial p}{\partial r} + \frac{\partial^2 u}{\partial z^2} = 0, \tag{3.88}$$

$$\frac{\partial p}{\partial z} + Bd = 0, \tag{3.89}$$

where $Bd = \rho g H^2/S$ denotes here the *dynamic* Bond number. These hydrodynamic equations are augmented with the already defined (subsection 1.2.2) surfactant balance equation along the surface that in this case reduces to

$$\frac{\partial \Gamma}{\partial t} = \frac{1}{Pe_D} \frac{1}{r} \frac{\partial}{\partial r}(r\Gamma) - \frac{1}{r}\frac{\partial}{\partial r}(r\Gamma u)|_{z=h}, \qquad (3.90)$$

where the Schmidt number is here $Pe_D = U R_0/D_s$. $D_s$ is the surface diffusion coefficient, as in earlier sections.

As it is usually assumed that surface tension is a linear function below CMC, then in dimensionless form we can define

$$\sigma(\Gamma) = \begin{cases} 1 - (1-\lambda)\Gamma, & \text{below CMC} \\ \lambda, & \text{above CMC} \end{cases} \qquad (3.91)$$

with $\lambda = \sigma_m/\sigma_0 < 1$.

At the bottom of the liquid layer with a solid support there is no slip,

$$u = w = 0 \text{ at } z = 0 \qquad (3.92)$$

whereas at the open surface one has the kinematic condition,

$$\frac{\partial h}{\partial t} = w - u\frac{\partial h}{\partial r} \text{ at } z = h(r,t) \qquad (3.93)$$

or, alternatively,

$$\frac{\partial h}{\partial t} + \frac{1}{r}\frac{\partial}{\partial r}\int_0^h ru(r,z,t)dz = 0 \qquad (3.94)$$

together with the normal stress balance (incorporating the Laplace overpressure) at $z = h(r,t)$:

$$-p + 2\epsilon^2\frac{\partial w}{\partial z} = \gamma\epsilon^2\frac{1}{r}\frac{\partial}{\partial r}\left(r\frac{\partial h}{\partial r}\right) \qquad (3.95)$$

that to the leading order yields

$$p = -\frac{\gamma}{r}\frac{\partial}{\partial r}\left(r\frac{\partial h}{\partial r}\right) \qquad (3.96)$$

with $\gamma$ denoting the dimensionless surface tension. The tangential stress balance with the Marangoni effect is

$$\frac{\partial u}{\partial z} = \frac{\partial \sigma}{\partial r} \qquad (3.97)$$

that when $\sigma$ is below the CMC yields

$$\frac{\partial u}{\partial z} = -\frac{\partial \Gamma}{\partial r} \tag{3.98}$$

at $z = h(r,t)$.

If the liquid layer is thin enough, a millimeter or less $(Bo \approx 0)$ and surface diffusion is a slow process relative to convection (in practical terms, $D_s = 0$), then one has

$$p(r,t) = -\frac{\gamma}{r}\frac{\partial}{\partial r}\left(r\frac{\partial h}{\partial r}\right) \tag{3.99}$$

and hence no $z$-dependence. Accordingly,

$$u(r,z,t) = \frac{\partial}{\partial r}\left[\frac{\gamma}{r}\frac{\partial}{\partial r}\left(r\frac{\partial h}{\partial r}\right)\right]\left(zh - \frac{z^2}{2}\right) - z\frac{\partial \Gamma}{\partial r} \tag{3.100}$$

Thus the hydrodynamic problem becomes

$$\frac{\partial h}{\partial t} = -\frac{1}{r}\frac{\partial}{\partial r}\left\{r\left[\frac{h^3}{3}\frac{\partial}{\partial r}\left(r\frac{\gamma}{r}\frac{\partial}{\partial r}\left(r\frac{\partial h}{\partial r}\right)\right) - \frac{h^2}{2}\frac{\partial \Gamma}{\partial r}\right]\right\} \tag{3.101}$$

$$\frac{\partial \Gamma}{\partial t} = -\frac{1}{r}\frac{\partial}{\partial r}\left\{r\Gamma\left[\frac{h^2}{2}\frac{\partial}{\partial r}\left(r\frac{\gamma}{r}\frac{\partial}{\partial r}\left(r\frac{\partial h}{\partial r}\right)\right) - h\frac{\partial \Gamma}{\partial r}\right]\right\}. \tag{3.102}$$

Theoretical analyses of similar spreading processes with due account of diffusion and convection have been provided by Grotberg and collaborators (Borgas and Grotberg, 1988; Gaver and Grotberg, 1990, 1992; Jensen and Grotberg, 1992, 1993; and Jensen, 1994, 1995). To guide the reader in comparing the above given theory with these other studies let us once more emphasize that the validity of the approximation used rests on the fact that for most purposes the diffusion over the liquid scales as $D_s\Gamma_s/L_s^2$ with $\Gamma_s$ a scale over the surface where the surfactant is spreading and $L_s$ the scale of the clean surface where the spreading process is considered. Convective flow is estimated according to the depth of the liquid, $h$, hence the characteristic flow velocity is $V_s \sim \sigma h/\eta L_s$. Then the corresponding Schmidt number is $Pe_D = V_s L_s/D_s \gg 1$ for most common liquids and surfactants acting on shallow layers and with , $\epsilon = h/L \ll 1$, diffusion may be neglected with the exception of a diffusion boundary layer not considered in the above given theory.

The above given system of equations is of the fourth order in the space (radial) derivatives. Let us see the boundary conditions to be satisfied. First, note that far away from the initial location of the drop the liquid layer is undisturbed, hence

$$h(r,t) = 1 \text{ at } r \to \infty, \tag{3.103}$$

$$\Gamma(r,t) = 0 \text{ at } r \to \infty \tag{3.104}$$

while at the center of the drop

$$\frac{\partial h}{\partial r} = 0 \text{ at } r = 0, \tag{3.105}$$

$$\frac{\partial \Gamma}{\partial r} = 0 \text{ at } r = 0. \tag{3.106}$$

The symmetry of the problem also demands that all *odd* derivatives of $h$ and $\Gamma$ also vanish.

To simplify the solution of the nonlinear problem (3.101) and (3.102) we first note that two distinct spreading stages are expected. First there is a fast stage, during which there is dissociation of micelles, and all micelles are dissolved (region above CMC); hence, the surface concentration, $\Gamma$, is practically constant: $\Gamma(t) = 1$ at $r = 0$. Then, the subsequent stage in the spreading process occurs with the total mass of surfactant practically conserved. Consequently,

$$\int_0^\infty r\Gamma dr = 1.$$

*ii.2 First spreading stage* $[\Gamma(t = 0) = 1]$

Ignoring capillary forces during the evolution of the first spreading stage. Then in the first stage the hydrodynamic problem is drastically reduced to

$$\frac{\partial h}{\partial t} = \frac{1}{2r}\frac{\partial}{\partial r}\left(rh^2\frac{\partial \Gamma}{\partial r}\right) \tag{3.107}$$

$$\frac{\partial \Gamma}{\partial t} = \frac{1}{r}\frac{\partial}{\partial r}\left(r\Gamma h\frac{\partial \Gamma}{\partial r}\right) \tag{3.108}$$

which is controlled by Marangoni stresses only. Clearly, as the order of the equations goes down to the second order in space derivatives the above given boundary conditions (for $r = 0$ and $r \to \infty$) cannot be satisfied by a regular solution, hence a shock or a hydraulic jump appears. Let $r_f(t)$ be a position of this hydraulic jump and for $r < r_f(t)$, $\Gamma \neq 0$ while for $r > r_f(t)$, $\Gamma = 0$. The conditions at the shock are obtained integrating the evolution equations between $r = r_f^-(t)$ and $r = r_f^+(t)$. We have

$$r_f(h_- - h_+) = \frac{1}{2}\left(h_+^2\frac{\partial\Gamma_+}{\partial r} - h_-^2\frac{\partial\Gamma_-}{\partial r}\right), \qquad (3.109)$$

$$r_f(\Gamma_- - \Gamma_+) = \Gamma_+h_+\frac{\partial\Gamma_+}{\partial r} - \Gamma_-h_-\frac{\partial\Gamma_-}{\partial r} \qquad (3.110)$$

where $X_\pm = \lim_{r\to r_f^\pm} X(r,t)$.

As the liquid layer is unperturbed downstream, $h_+ = 1$ and $\Gamma_+ = 0$. Accordingly, the shock conditions become

$$r_f(h_- - 1) = -\frac{1}{2}h_-^2\frac{\partial\Gamma_-}{\partial r}, \qquad (3.111)$$

$$\Gamma_-(r_f + h_-\frac{\partial\Gamma_-}{\partial r}) = 0. \qquad (3.112)$$

Self-similar solutions were found using $\xi = r/t^\alpha$, $h(r,t) = f(\xi)$, and $\Gamma(r,t) = g(\xi)$. It turns out that $\alpha = 1/2$. Thus, the hydrodynamic problem reduces to

$$f'(\xi) = -\frac{1}{\xi^2}\frac{d}{d\xi}(\xi f^2 g'), \qquad (3.113)$$

$$g'(\xi) = -\frac{2}{\xi^2}\frac{d}{d\xi}(\xi f g'). \qquad (3.114)$$

For the upstream flow (in the downstream region $f = 1$ and $g = 0$) one has $0 < \xi < \nu$. Then the jump is located at

$$r_f(t) = \nu t^{1/2}. \qquad (3.115)$$

The boundary conditions become

$$\nu[f(\nu) - 1] = -f^2(\nu)g'(\nu), \qquad (3.116)$$

$$g(\nu) = 0, \qquad (3.117)$$

$$\nu/2 = -f(\nu)g'(\nu), \qquad (3.118)$$

hence $f(\nu) = 2$, $g(\nu) = 0$ and $g'(\nu) = -\nu/4$.

To eliminate $\nu$, it suffices to again change variables,

$$\eta = \xi/\nu, \; f(\xi) = F(\eta), \; g(\xi) = G(\eta) \qquad (3.119)$$

together with $F(1) = 2$, $G(1) = 0$, $G'(1) = -1/4$. Note that $F$ and $G$ obey the same evolution equations that $f$ and $g$.

These results deserve some comments. As $G$ diverges near the center of the original drop ($r = 0$), the self-similar solution found is not valid in this region where $\Gamma(0, t) = 1$. On the other hand, the liquid layer becomes thinner and thinner for small $r$. Thus near the center it eventually becomes depleted and a dry spot must form. This is a purely macroscopic effect and no appeal to microscopic (Van der Waals) forces is made. It is also a consequence of mass conservation in the system, contrary to an argument developed by Gaver and Grotberg (1994) who attributed it to microscopic forces.

*ii.3 Second spreading stage*

The end of the first spreading stage just described comes when all surfactant micelles have disappeared and then one goes from the region above CMC to that below CMC where the surface tension is no longer constant as the surfactant molecules come to the surface. Yet to the leading order approximation one can approximate $\gamma$ by an effective value $S = \epsilon^2 \sigma_m/s$. Otherwise due to the eventual dissolution of the surfactant, $\sigma$ approaches $\sigma_0$ and variation of $\sigma$ can be neglected in the evolution equation although we must keep surface tension gradients in the boundary conditions. Accordingly, the hydrodynamic problem is the same as before with the substitution of $\gamma$ by $S$, taken constant, together with the condition of mass conservation

$$\int_0^{+\infty} r\Gamma(r, t)dr = 1. \qquad (3.120)$$

As done for the first spreading stage, self-similar solutions were also sought to locate the spreading front in the second stage. Setting $\xi = r/t^m$, self-consistency in the evolution yields $n = 1/4$. Accordingly, $\xi = r/t^{1/4}$, $h(r, t) = f(\xi)$ and $\Gamma(r, t) = g(\xi)t^{-0.5}$. Then for the hydrodynamic equations to be satisfied either $g(\xi) = 0$ or

$$g' = -\frac{\xi}{4f(\xi)} + \frac{S}{2}f(\xi)\left[\frac{1}{\xi}(\xi f'(\xi))'\right]'. \qquad (3.121)$$

The first condition shows that there is no surfactant, as expected. Then the location of the front is at

$$r_f(t) = \lambda t^{1/4} \tag{3.122}$$

that to the lowest order approximation corresponds to $\lambda = 2^{5/4}$. Consequently, matching of the solutions demands $r_f(t) = \lambda t^{1/4} = \nu t^{1/2}$ thus providing $\nu = \lambda t^{-1/4}$, to be satisfied at some time instant $t_1$, which is determined by the total mass of micelles. Conservation of matter demands that the radius of the inner dry spot also follows the same power law in its evolution.

Similar results have been found by Gaver and Grotberg (1992). Their case $Pe_D = 100$ is a reasonable approximation of the theoretical result by Starov *et al.* (1997) for $Pe_D \to \infty$ when diffusion becomes negligible. In the experiments reported by Starov *et al.* (1997) their Schmidt number is about $Pe_D = 10^5$. The concentration profile exhibits well the evolution to a jump profile. Even quantitatively the shock values, $h_- = 1$, $h_+ = 2.75$, found by Gaver III and Grotberg for $Pe_D = 100$ agree rather satisfactorily with the result found by Starov *et al.* (1997).

Grotberg and collaborators (Gaver and Grotberg, 1992; Jensen and Grotberg, 1992; Jensen, 1995) have studied in detail the spreading of a surfactant film. Jensen and Grotberg (1992) have shown that gravity significantly affects the jump, regularizing it over a boundary layer of thickness $\delta = 1/BdPe_D$ or either scale of the shock. As in the experiments reported by Starov *et al.* (1997), it is $\delta \sim 10^{-4}$, the assumption $Bo = 0$ seems well justified.

As seen on Fig. 3.17 the radius of the dry spot moves with the speed predicted by the theory of Starov *et al.* (1997) while the (outer) spreading front proceeds more slowly than theoretically predicted. As diffusion ($Pe_D = 10^5$) does not seem to play a significant role in the experiments, the disagreement may be due to the fact that for concentrated solutions the surface tension is concentration-dependent. The disagreement may also be due to the fact that in the experiment by Starov *et al.* (1997) the surfactant is not really insoluble. It is insoluble for all practical purposes provided the desorption time of the surfactant is not attained, as earlier noted. Furthermore, recall that gravity not only may be important at the higher edge of the propagating front but also gravity may be responsible for flow reversal that cannot be ruled out in the second spreading stage as mentioned by Gaver and Grotberg (1992).

# Chapter 4

# Stationary interfacial patterns in liquid layers

In this chapter we shall study under what conditions the mechanical equilibrium of a shallow liquid layer becomes unstable, and what kind of flows will appear as the result of an interfacial instability due to the Marangoni effect (Benard-Marangoni convection). We shall consider the paradigmatic problem experimentally studied by Benard about the evolution of a horizontal liquid layer uniformly heated from below that, however, yields an inhomogeneous, cellular, structure.

## 4.1 Stability of a thin horizontal layer heated from below (Benard-Marangoni convection)

### 4.1.1 Heuristic arguments

Let us start by considering the onset of surface tension gradient-driven instability leading to *Benard* cells (Benard, 1900, 1901, 1928, 1930) which appear in a thin horizontal layer heated from below (Fig. 4.1). Take a liquid layer situated on a rigid plane $z = -a$ with a free surface $z = 0$. For the sake of simplicity, we shall disregard here the surface deformations caused by the convection, as well as buoyancy. The conditions for the validity of these assumptions will be discussed later. We shall assume that the temperature of the rigid plane is fixed, while the temperature of the surface is subject to the

91

boundary condition with the Marangoni stress. In the absence of convection, the temperature field is

$$T_0(z) = T_b - A(z + a), \qquad (4.1)$$

where the temperature gradient $A$ can be calculated from the boundary condition (see also Sect. 4.1.3)

$$\lambda A = K(T_b - Aa - T_g). \qquad (4.2)$$

The static temperature drop across the layer is $\Theta = Aa$. The subscripts "b" and "g" denote bottom and gas, respectively.

Figure 4.1: Benard hexagonal convective cell. Exposure time is longer than the time it takes a particle to go from the center to the periphery. Silicone oil layer depth about a mm and velocity about $10^{-2}$ cm s$^{-1}$.

The temperature distribution (4.1) can be *unstable*. Let us imagine a positive temperature fluctuation in a certain region on the surface. The thermocapillary (Marangoni) forces are directed radially from the "hot spot" to the periphery. Because of the incompressibility of the liquid, the corresponding radial flow near the surface will produce a rising motion of the liquid towards the surface below the hot spot, which strengthens the initial temperature disturbance, if the vertical temperature gradient $A$ is large enough. The mechanism of instability (Dauzere, 1908; Block, 1956; Pearson, 1958) described above is the surface tension gradient driven (Marangoni) instability mechanism already used when studying drop motions.

The instability threshold is determined by the balance of thermocapillary forces and dissipation and, indeed, incompressibility. The time scale of heat diffusion is $\tau_\kappa = a^2/\kappa$, and the time scale of viscous dissipation of momentum is $\tau_\nu = a^2/\nu$. The Marangoni stress induced by the thermal gradient, $A$, creates a flow velocity disturbance, $(V, O, W)$, through the boundary condition here taken in its simplest form

$$\frac{d\sigma}{dT}\frac{dT}{dx} = -\eta\frac{dV}{dz}, \quad \text{hence} \quad \frac{d\sigma}{dT}\frac{d^2T}{dx^2} = \eta\frac{d^2W}{dz^2}, \qquad (4.3)$$

where T denotes a temperature disturbance along the interface. Assuming that vertical and horizontal scales of the temperature and velocity fields are similar, one finds $\alpha T \sim \eta W$ with $\alpha = -d\sigma/dT$. Then this temperature disturbance is affected by the flow and heat diffusion, and hence

$$\frac{dT}{dt} = -\frac{T}{\tau_\kappa} + AW. \qquad (4.4)$$

Consequently, steady disturbances may exist only if one has

$$\begin{vmatrix} \frac{1}{\tau_\kappa} & -A \\ \alpha & \eta \end{vmatrix} = 0 \qquad (4.5)$$

which, naturally, brings the Marangoni number defined here as

$$M = \frac{\alpha A a^2}{\eta\kappa}. \qquad (4.6)$$

*Instability* is possible whenever $M = O(1)$. A more realistic figure is obtained for a concrete problem when solving the corresponding hydrodynamic equations with boundary conditions as done further below. Recall that $M = V_s a/\kappa$, where $V_s = \alpha A a/\eta$ is the velocity defined by scale of motion due to the Marangoni effect at the open surface or interface. Thus, as earlier noted, $M$ can be a *thermal Reynolds number* or *Peclet number*. If we define the time scale related to the Marangoni effect as $\tau_\alpha = (\rho a^2/\alpha A)^{1/2}$, we find

$$M = \frac{\tau_\kappa \tau_\nu}{\tau_\alpha^2}. \qquad (4.7)$$

As with earlier heat problem the fluid itself is characterized by the *Prandtl number*

$$P = \frac{\nu}{\kappa}, \qquad (4.8)$$

which recall it is the ratio of the typical time scales of the heat diffusion, $\tau_\kappa$ and viscous dissipation of momentum, $\tau_\nu$. Thus $M = Re\,P$, with $Re = V_s d/\nu$. Note also that, in dimensionless form, the heat transfer on the surface open to air can be described as in earlier cases by the *Biot number*

$$Bi = \frac{Ka}{\lambda}. \qquad (4.9)$$

If we take into account the deformation of the free surface, we should introduce two more time scales associated with gravity (hydrostatic pressure) and surface tension that tend to suppress surface deformation, $\tau_g = (a/g)^{1/2}$ ($g$ is gravity acceleration) and $\tau_\sigma = (\rho g a^3/\sigma)^{1/2}$. These time scales are connected with *capillary-gravity waves* that may propagate on the surface (Chapter 5). The characteristic frequency of gravity waves, $\omega^2 \sim \tau_g^{-2}$, measured by means of the viscous time unit determines, naturally, the *Galileo number* here defined as

$$Ga = \frac{ga^3}{\nu^2} = \frac{\tau_\nu^2}{\tau_g^2}. \tag{4.10}$$

The competition between hydrostatic pressure and capillary (Laplace) pressure determines the *capillary length*, $a_c = (\sigma/\rho g)^{1/2}$, which is the length corresponding to the value unity of the *(static) Bond number* that we define here as

$$Bo = \left(\frac{a}{a_c}\right)^2 = \frac{\rho g a^2}{\sigma} = \frac{\tau_\sigma^2}{\tau_g^2}. \tag{4.11}$$

In some cases, instead of parameters (4.10) and (4.11), it is convenient to use the earlier introduced (modified) Galileo number $G = GaP = ga^3/\nu\kappa$, and then the *capillary* number is

$$C = Bo/G = \eta\kappa/\sigma a = \eta V/\sigma, \tag{4.12}$$

if we use $V = \kappa/d$ as the characteristic flow velocity. Clearly,

$$C = \frac{\tau_\sigma^2}{\tau_\nu \tau_\kappa} \tag{4.13}$$

is the ratio of three time scales involved in the problem, associated to deformability of the interface, viscosity and heat diffusion, respectively. Let us recall that the deformation of the boundary due to the Marangoni flow is negligible if gravity (hydrostatic pressure) or/and capillary forces preventing the deformation are large enough: $G \gg 1$, $C^{-1} \gg 1$. For completeness, let us recall the earlier used dynamic Bond number, here expressed as

$$Bd = \frac{G}{M} = \frac{\rho g a}{\alpha A} = \frac{\tau_\alpha^2}{\tau_g^2} \tag{4.14}$$

which is a measure of the ratio between the Marangoni and gravity forces.

Let us close this subsection by considering, for completeness, mass transfer rather than heat but otherwise about the same mathematical problem as the Benard-Marangoni instability described above. Indeed, as noted in the Preface, Marangoni, Thomson and others, earlier than Benard, albeit with no systematic study, observed interfacial cellular flows in mass transfer problems. Typical cases are the vaporization of ether from an unstirred aqueous solution or acetone evaporating from the liquid to air. Other cases of mass transport, absorption and desorption, are discussed in Chapter 5. It is possible that, due to some minor disturbance, e.g. the evaporation at a certain spot at the interface is reduced. Locally, the concentration of acetone increases somewhat. This effects a reduction of the surface tension. Due to the created surface tension gradient, and the corresponding Marangoni stress, liquid starts to flow away from the disturbed spot and is replaced by liquid from the bulk, which is richer in acetone. So, the disturbance is amplified, and as in the heat transfer problem just discussed above, here also convective cells form in the liquid. In this analysis, one also has to take the air phase into account. Indeed, in the atmosphere above the water, convection tries to stop the flow, as it brings "fresher" air with a low acetone concentration to the disturbed spot. Which transport process wins, or, in other words, whether the system is unstable or not, depends on, among other things, the appropriate ratios of kinematic viscosities and diffusivities in both phases as shown first by Sternling and Scriven (1959). Needless to say the configuration discussed above of a liquid layer heated from below and open to ambient (passive) air satisfies, indeed, the Sternling and Scriven (1959) criterion for instability.

### 4.1.2  Linear stability analysis

To express the above given heuristic arguments in quantitative terms, equations (1.28) - (1.30) have to be used. Written in their non-dimensional form disregarding buoyancy, which is valid approximation for a shallow (otherwise said thin) liquid layer, they are

$$\frac{1}{P}[\frac{\partial \mathbf{v}}{\partial t} + (\mathbf{v} \cdot \nabla)\mathbf{v}] = -\nabla p + \nabla^2 \mathbf{v}, \qquad (4.15)$$

$$\frac{\partial T}{\partial t} + (\mathbf{v} \cdot \nabla)T = \nabla^2 T, \qquad (4.16)$$

$$\nabla \cdot \mathbf{v} = 0. \tag{4.17}$$

Taking the (mechanical) equilibrium surface temperature $T_s = T_b - Aa$ as a reference temperature, and $\Theta = Aa$ as the temperature scale, we get the following boundary conditions at the bottom rigid boundary $z = -1$:

$$\mathbf{v} = 0, \ T = 1. \tag{4.18}$$

On the open surface to air we shall disregard its deformability which amounts to considering the limit $G \to \infty$. Then at $z = h = 0$ we get the following system of boundary conditions:

$$\tau'_{lz} - M \frac{\partial T}{\partial x_l} \tau_l = 0, \ l = 1, 2; \ x_1 \equiv x, \ x_2 \equiv y; \tag{4.19}$$

$$v_z = 0; \tag{4.20}$$

$$\frac{\partial T}{\partial z} = -1 - BiT; \tag{4.21}$$

The condition for normal stresses comes only in the case of finite $G$. The boundary value problem (4.15) - (4.21) has a solution

$$\mathbf{v}_0 = 0, \ p_0 = 0, \ T_0 = -z \tag{4.22}$$

which corresponds to the mechanical equilibrium state with (molecular) heat diffusion in the steady state. In order to investigate the *linear stability* of this solution, we add *small* disturbances $(\tilde{\mathbf{v}}, \tilde{p}, \tilde{T})$ to the solution (4.22):

$$\mathbf{v} = \tilde{\mathbf{v}}, \ p = \tilde{p}, \ T = T_0(z) + \tilde{T}, \tag{4.23}$$

substitute (4.23) into equations and boundary conditions (4.15) - (4.21) and *linearize* the equations with respect to tilded variables.

Since the original equations and boundary conditions are invariant with respect to translations in time $t$ and horizonal spatial coordinates $x$ and $y$, the disturbance fields can be written as a superposition of "normal modes" with exponential dependence on $t$, $x$ and $y$:

$$(\tilde{\mathbf{v}}, \tilde{p}, \tilde{T}) = (\hat{\mathbf{v}}(\mathbf{z}), \hat{p}(z), \hat{T}(z)) \exp(i\mathbf{k} \cdot \mathbf{x}_\perp + \lambda t), \tag{4.24}$$

$$\mathbf{x}_\perp = (x, y).$$

Because only solutions bounded in the whole space are physically meaningful, the wavevector $\mathbf{k} = (k_x, k_y)$ is real, while the growth rate $\lambda$ is generally complex. No confusion is expected with the use here of $\lambda$. Using the linearized horizontal Navier-Stokes equation (4.15) and the incompressibility condition, we obtain

$$\hat{p} = \frac{1}{k^2}\left(\frac{d^2}{dz^2} - k^2 - \frac{\lambda}{P}\right)\frac{d\hat{v}_z}{dz}, \quad \mathbf{k}\cdot\hat{\mathbf{v}}_\perp = i\frac{d\hat{v}_z}{dz}. \quad (4.25)$$

where $\hat{\mathbf{v}}_\perp = (\hat{v}_x, \hat{v}_y)$. These relations can be used to eliminate $\hat{p}$ and $\hat{\mathbf{v}}_\perp$. We then have to solve the following eigenvalue problem:

$$\left(\frac{d^2}{dz^2} - k^2\right)^2 \hat{v}_z - \frac{\lambda}{P}\left(\frac{d^2}{dz^2} - k^2\right)\hat{v}_z = 0; \quad (4.26)$$

$$\left(\frac{d^2}{dz^2} - k^2\right)\hat{T} + \hat{v}_z - \lambda\hat{T} = 0; \quad (4.27)$$

$$z = -1: \hat{v}_z = \frac{d\hat{v}_z}{dz} = \hat{T} = 0; \quad (4.28)$$

$$z = 0:$$

$$\frac{d^2\hat{v}_z}{dz^2} + k^2 M\hat{T} = 0; \quad (4.29)$$

$$\hat{v}_z = 0; \quad (4.30)$$

$$\frac{d\hat{T}}{dz} + Bi\hat{T} = 0. \quad (4.31)$$

The problem (4.26) - (4.31) determines the growth rate $\lambda$ as a function of the wavenumber $k$ amd the parameters $M$, $P$ and $Bi$, already defined.

If for given values of $P$ and $Bi$, and for the chosen value of $M$, $Re\lambda(k) < 0$ ($\leq 0$) for any eigenvalues, the mechanical equilibrium is *linearly stable* (*neutrally stable*) (the completeness of the set of eigenfunctions is implicitly assumed). If $Re\lambda(k) > 0$ for a certain eigenvalue corresponding to some $k$, the mechanical equilibrium is *linearly unstable*. Note that, as already seen in other problem, if a system is linearly unstable it is unstable indeed, while a system may very well be linearly stable hence stable to infinitesimal disturbances and yet unstable to finite amplitude perturbations.

Thus, our first goal is to calculate the *stability boundary* $Re\lambda(k, M) = 0$, which determines parametrically a curve $M = M_0(k)$ which is called the *neutral* (or *marginal*) *stability curve*. If $Im\lambda(k, M) = 0$

along this curve, it is a monotonic (also said stationary) instability, otherwise the instability is *oscillatory* (overstability).

For the problem (4.26) - (4.31), most probably, there exists only the monotonic instability (though that was never proved rigorously). It was investigated by Pearson (Pearson, 1958) who found the following neutral curve:

$$M_0(k) = \frac{8k(k\cosh k + Bi\sinh k)(\sinh k\cosh k - k)}{\sinh^3 k - k^3\cosh k} \qquad (4.32)$$

(Fig. 4.2, left plot). The Marangoni instability disappears in the limit $Bi \to \infty$, because open surface boundary becomes isothermical and hence there is no Marangoni effect.

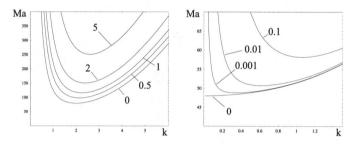

Figure 4.2: Benard-Marangoni convection. The neutral stability curves for a liquid layer lying on a rigid thermally conducting (left) and thermally insulating (right) plate, for various Biot numbers at the upper open surface to air.

The shape of the neutral curve resembles qualitatively that of the well-known buoyancy-driven (Rayleigh-Benard) instability (Chandrasekhar, 1961; Normand *et al*, 1977), but the short-wave asymptotics is different ($M_0 \propto 8k^2$ as $k \to \infty$, while the marginal Rayleigh number $Ra_0 = O(k^4)$ as $k \to \infty$ in the Rayleigh-Bénard problem). The main results however look quite similar. If the Marangoni number $M$ is less than some critical value $M_c = M_0(k_c)$ corresponding to the minimum of the neutral curve, any small disturbances decay on the background of the mechanical equilibrium. But if the layer is heated in such a way that $M > M_c$, the disturbances with wavevectors inside a certain ring $k_- < |\mathbf{k}| < k_+$ around the circle $|\mathbf{k}| = k_c$ (where $k_c$ is the critical wavenumber, corresponding to the minimum of the curve $M_0(k)$) start to grow with the corresponding growth rate $\lambda(k)$ which is determined by solving the problem (4.26) - (4.31).

Several questions appear which cannot be answered in the framework of the linear approach. First, what kind of motion will be eventually established in the system? One can imagine (i) some spatially periodic ordering with few basic non-zero Fourier harmonics $\{\mathbf{k}_j\}$, $j = 1, \ldots, N$ surviving as the result of the nonlinear competition of disturbances with different wavevectors, (ii) a quasiperiodic ordering with incommensurate Fourier harmonics and (iii) even a "spatial turbulence" (Newell and Pomeau, 1993; Bragard and Velarde, 1997). If as already shown by Benard (1900, 1901) and confirmed by Koschmieder and others (Koschmieder, 1967, 1991, 1993; Koschmieder and Biggerstaff, 1986; Koschmieder and Switzer, 1992; Cerisier *et al.*, 1987a,b) a periodic, cellular structure appears, what is its wavenumber selected from the interval $(k_-, k_+)$?

We shall discuss these questions in other sections of this chapter. In the rest of the present section, we consider some modifications and generalizations of the above formulated linear problem.

An exact solution of the problem can be easily found (Pearson, 1958) also in the case of a really poorly heat conducting $(Bi = 0)$ rigid boundary, where the condition $d\hat{T}/dz = 0$ is used on the boundary $z = -1$ instead of the isothermal condition $\hat{T} = 0$. The former is also misnamed an insulating boundary although there is heat transfer with heat flux held constant hence not allowed to fluctuate. The corresponding formula determining the neutral curve is

$$M_0(k) = \frac{32k(Bi \cosh(k) + k \sinh(k))(\sinh(k)\cosh(k) - k)}{(4k^2 - 1)\cosh(k) + \cosh(3k) - 4k(2 + k^2)\sinh(k)} \quad (4.33)$$

(Fig. 4.2, right plot). Here, let us note that in the limit $Bi \to 0$ (where both boundaries are heat-insulated) the critical wavenumber $k_c$ tends to 0, i.e. there is a *long-wave* instability. In this limit, there is a *neutral* (not growing and not decaying) mode at $k = 0$ ($\lambda(0) = 0$) which corresponds to a uniform change of the mean temperature in the layer. Weakly non-uniform deviations of the mean temperature generate a thermocapillary motion, and the coupling between the temperature gradient and the velocity field generates the instability. We shall consider this limit in more detail in the next sections, as it allows us to describe qualitatively a variety of phenomena observed even in the more realistic case $k_c = O(1)$.

Near the instability threshold, for long-waves ($k \sim \epsilon \ll 1$) and low enough heat transfer, with Biot number $Bi \sim \epsilon^4$, the neutral stability curve can be approximated as

$$M_0(k) \simeq 48\left(1 + \frac{Bi}{k^2} + \frac{k^2}{15}\right) \quad (4.34)$$

up to corrections of order $\epsilon^4$. Hence, the critical wavenumber and Marangoni number are given by

$$k_c \simeq (15Bi)^{1/4}, \quad M_c \simeq 48 \left( 1 + 2\sqrt{\frac{Bi}{15}} \right) \tag{4.35}$$

In fact, for such type of *stationary long-wave instability*, the linear growth rate may generically be written

$$\lambda(k) = -\alpha + \mu k^2 - k^4 + O(k^6) \tag{4.36}$$

where in our case $\alpha \sim Bi$ and $\mu \sim (M - 48)$, accounts for supercriticality or distance to the instability threshold.

In the following sections, we shall concentrate on such long-wave regimes, as many of their qualitative features are in good agreement either with experimental results, or even with theoretical approaches developed for $k_c = O(1)$.

An important question is the influence of buoyancy earlier left out in the stability analysis. If buoyancy is taken into account, the term $k^2 R \hat{T}$, $R = g\beta\Theta d^3/\nu\kappa$, should be added to the right-hand side of the equation (4.26). The corresponding problem was solved by Nield (1964, 1966), in the case of a rigid and perfectly heat-conducting bottom boundary. When the latter boundary is taken with heat flux held fixed and the upper free surface also has a vanishing Biot number, the critical wavenumber also vanishes, and the instability occurs on the line

$$\frac{M}{48} + \frac{R}{320} = 1 \tag{4.37}$$

Note that this result is qualitatively similar to the result of Nield (1964), who obtained the approximate relationship $M/79.6 + R/669 \simeq 1$, after minimization of the neutral stability curve with respect to $k$ (the result is that $k_c \simeq 2$ is approximately constant along the neutral line). In both cases, the instability region is located above the line. It should be noted that under the experimental conditions where the heating is increased while other characteristics of the system including the thickness of the layer are held fixed, the Rayleigh and the Marangoni numbers are not independent but proportional:

$$\frac{M}{R} = \left( \frac{d_c}{d} \right)^2, \tag{4.38}$$

with

$$d_c = \sqrt{\frac{\alpha}{\rho g \beta}}. \qquad (4.39)$$

In really "thin" layers ($d \ll d_c$) the instability is caused mainly by the (Marangoni) thermocapillary effect, while in "thick" layers ($d \gg d_c$) it is mainly buoyancy-driven. In the intermediate case $d \sim d_c$ both effects should be taken into account.

We shall postpone to Sect. 4.6.2 the discussion of the validity of another important limitation of the above given stability analysis, namely that of non-deformability of the open surface to air. As said above the deformation is negligible if the Galileo number is large enough. The latter condition is usually satisfied, except for very thin layers of highly viscous liquids.

Let us close this section by noting that global (finite amplitude) energy stability analyses of this surface tension gradient-driven (Benard-Marangoni) instability and its extensions accounting for e.g. double-diffusion phenomena provide thresholds significantly below the linear prediction (Davis, 1969; Davis and Homsy, 1980; Castillo and Velarde, 1982). This item is not a minor one and we shall return to argue why an hexagonal platform is expected subcritically rather than supercritically (Palm, 1960; Newell and Pomeau, 1993). Experiments show, indeed, the subcriticality of hexagons with corresponding hysteretic phenomena (Schatz et al., 1995; Schatz and Neitzel, 2001).

### 4.1.3  Experiments

Experiments allowing one to observe Benard cells are easy to carry out provided one merely wishes to have a qualitative result with convective patterns exhibiting plenty of defects (Velarde and Normand, 1993). Take a cup of hot tea and pour on it a bit of milk (not the other way around as done by most people when having tea). You would be able to see Benard cells even showing somewhat violent convective motions. For a quantitative study, the situation demands great care. A typical set-up built by a very careful experimentalist, illustrating how an experiment is designed to provide reliable data and, furthermore, help theorists in their analysis is the following (Koschmieder and Switzer, 1992; Koschmieder, 1993). An experimental set-up for mass transfer will be described in Chapter 5.

The liquid rests on a cromium coated silicon crystal 12.7 mm thick and 17.8 cm in diameter (Fig. 4.3). The surface of the coated crystal was optically plane to 20 wavelengths and served as the mirror for shadowgraph visualization. The silicon crystal has a thermal conductivity of $\lambda = 0.37$ cal cm$^{-1}$s$^{-1}$K$^{-1}$, that is 1/3 of the thermal conductivity of copper ($\lambda = 0.94$ cal cm$^{-1}$s$^{-1}$K$^{-1}$). Liquids used included silicone oil whose conductivity is of the order of $4 \cdot 10^{-4}$ cal cm$^{-1}$s$^{-1}$K$^{-1}$. The silicon crystal was heated from below by a copper block, 5 cm thick and 17.8 cm in diameter. The copper block was heated in turn by a regulated electrical current passing through a resistance wire. The arrangement with the silicon crystal and the copper block provides a practically uniform temperature at the surface of the crystal and hence at the liquid bottom.

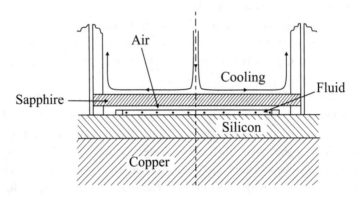

Figure 4.3: Schematic drawing of a set-up to study Benard-Marangoni convection (after Koschmieder and Switzer, 1992).

The liquid layer on top of the mirror was bounded by lucite rings of 10.48 cm inner diameter and different thicknesses, ranging from 3 to 1.2 mm machined with an accuracy of order of $10^{-2}$ mm. The nominal thermal conductivity of lucite differs from the thermal conductivity of the silicone oil by only about 10%. On top of the fluid layer was a thin air gap of 0.4 mm depth. The thickness of the air layer was held as low as possible in order to suppress motions of the air on top of the fluid. The air was contained above by a sapphire lid of 5 mm thickness and 13.34 cm diameter whose top and bottom surfaces were polished and optically plane. Thus we have a two-layer system. The purpose of the sapphire is to provide a transparent lid of good thermal conductivity, which is $\lambda = 0.088$ cal cm$^{-1}$s$^{-1}$K$^{-1}$ according to the manufacturer (Union Carbide). The thermal conductivity of the sapphire is about 200 times better than the thermal conductivity of the silicone oil. The sapphire was cooled from above by water of constant temperature ($\pm 10^{-2}\,°$C) circulated at a rate of about 60 cm$^3$s$^{-1}$. By the uniform temperature of the lid and by the small depth of the air gap the temperature of the air on top of the fluid was made practically uniform. The sapphire was held in place by a lucite ring surrounding the ring containing the fluid.

The experimental set-up complies in a very good approximation with the assumptions above made in theory like uniform temperatures on a plane perfectly-conducting bottom as well as uniform temperatures on top of the cover of the air layer, which leaves the liquid-air interface with non-uniform temperature distribution and hence the Marangoni effect. The other assumption made is that the fluid layer is of infinite horizontal extent. This assumption cannot be fulfilled in an experiment but can be well approximated by the small aspect ratio of the fluid layers, whose minimum value was about $10^{-2}$. The aspect ratio is defined as the ratio of the fluid depth divided by the diameter of the fluid layer. There are authors who define the inverse as aspect-ratio. However, in all experiments the consequences of the presence of the lateral boundary were always noticeable (Velarde and Normand, 1980; Koschmieder, 1993). It seems difficult to match both thermally and mechanically the lateral walls to the liquid, even if the container is filled brimfull or with liquid surrounding the walls on both sides (Ondarçuhu et al., 1993a,b; Schatz et al., 1995).

Note that in experiments where the surface is open to the unbounded ambient air (Benard, 1900, 1901; Cerisier et al., 1987a,b) heat transfer is difficult to characterize because buoyancy-driven (natural, Rayleigh) convection in the air exists. Thus although air is mostly, mechanically, passive relative to the liquid if flow motions become intense they are going to couple and, eventually, alter significantly, the thermocapillary (Marangoni) motions in the liquid. In this case a two–layer model system must be considered to account for experiment.

The actual temperature difference has to be calculated from the temperature difference between the copper bottom plate and the temperature of the cooling water on top of the sapphire because no temperature sensor can be placed in the narrow air gap or on the fluid surface itself without introducing a significant disturbance. Assuming that the surface of the liquid experiences small deformation the (mean) temperature difference across the fluid follows from the formula

$$\Delta T_{fl} = \Delta T / \left(1 + \frac{\bar{\lambda}_{fl}\Delta Z_a}{\bar{\lambda}_a\Delta Z_{fl}}\right),$$

where $\Delta T$ is the temperature difference between the top of the sapphire and the copper block, $\bar{\lambda}_{fl}$ the mean thermal conductivity of the liquid, $\bar{\lambda}_a$ the mean thermal conductivity of the air and $\Delta Z_{fl}$ and $\Delta Z_a$ are depth of the fluid layer in the (1-2 mm) range and the depth of the air (about 0.4 mm). After the onset of convection the vertical temperature distribution in the fluid is no longer linear; that should, however, have no influence on the value of $\bar{\lambda}$. Of great importance for the determination of $\Delta T_{fl}$ is the contribution of the air layer, because the thermal conductivity of air ($\lambda = 6.1 \cdot 10^{-5}$ cal cm$^{-1}$s$^{-1}$K$^{-1}$ at 20° C) is so poor. More than half of the temperature decrease between the sapphire and the copper plate occurs in the air layer, although it is thin. The contribution of the silicon crystal and the sapphire to the temperature decrease is negligible. The temperature difference $\Delta T_{fl}$ determined in this way is not known with an accuracy better than ±5%, mainly because the depth of the fluid layer is not known better than 1%, and the depth of the air layer is not known better than 2%. For a similar set-up see VanHook et al. (1997) who used liquid depths in the range 0.07-0.2 mm with gas (air, helium) layers between 0.2 and 1 mm in containers of (circular) horizontal extent about 4 cm in diameter. They also discussed the influence of surface deformation on the mean temperature difference. This is important for they studied both long and short wave excitations.

A straightforward alternative to obtain the quantity $\Delta T_{fl}$ is to consider $\Delta T_{bottom} - \langle T_s \rangle$ where $T_s$ is the nonuniform temperature at the open surface and the bracket represents the spatial average horizontally across the surface. $T_s$ can be obtained from noncontact optical measurements (Schatz et al., 1999).

The liquid needs a long time to establish equilibrium in the horizontal plane of extent $L$. This time is characterized by the horizontal relaxation time, which may be about a day. Koschmieder (1993) has conducted experiments with heating increasing in two steps during a day so that the temperature increase across the fluid was about 0.5°C per day. That meant that some of his experiments ran continuously for a couple of weeks e.g. in order to reach a maximal temperature difference of about 30 °C between the two solid supports of the enclosure containing the fluid and the air gap. He observed the onset of convection at $\Delta T$ about $7°C$ and hence a Marangoni number about 62 with a Rayleigh number of 40. The former is a bit smaller than theory predicts ($M \sim 80$) while the latter is clearly below the critical Rayleigh number ($R \sim 10^3$) for buoyancy-driven instability. This critical temperature difference corresponds to the traditional onset of Benard cells, i.e. to the short wavelength mode. Schatz *et al.* (1995) found $M_c = 83.6$ with a precision of $\pm 0.5$ but with an accuracy $\pm 11$ due to uncertainty in the thermal properties of the silicone oil used.

## 4.2 Nonlinear evolution equation for a horizontal layer heated from below

As remarked above, when a liquid layer is confined between two boundaries where the heat flux crossing them is held fixed, a uniform shift of the temperature is possible, which will neither be damped nor amplified. This corresponds to $\alpha = 0$ in the dispersion relation (4.36), and in the limit $k \to 0$, we have $\lambda = 0$, i.e. there is indeed a mode with zero eigenvalue (Goldstone mode). When slow horizontal modulations of the temperature field are allowed, flows may be generated by the surface tension gradient and/or buoyancy and instability can occur when $\mu$ is increased past some threshold value. In this case, Knobloch (1990) has shown that, starting with Eqs. (4.15)–(4.21), the nonlinear evolution of the temperature field, $\phi$, in surface tension gradient-driven (Benard-Marangoni) can be described by

$$\frac{\partial \phi}{\partial t} = -\alpha\phi - \mu\Delta\phi - \Delta^2\phi + \kappa\nabla \cdot \left[(\nabla\phi)^2\nabla\phi\right] + \beta\nabla \cdot [\Delta\phi\nabla\phi] + \delta\Delta\left[(\nabla\phi)^2\right] - \gamma\nabla \cdot [\phi\nabla\phi]$$

$$(4.40)$$

to be called the Knobloch equation, where one indeed recognizes the linear form (4.36), but in addition, nonlinear terms appear, leading to saturation of the linear growth as well as competition between the supercritically amplified modes. No confusion is expected with use done here of $\kappa$, $\gamma$, etc. Note that although initially we need several fields, velocity, temperature, etc. to describe flow, Eq. (4.40) deals with only one, which is considered as a slaving mode or order-parameter for the others. This is a consequence of the scaling and asymptotics used by Knobloch (1990). The terminology slaving mode comes from Synergetics (Haken, 1983a,b) and order-parameter from the Landau theory of phase transitions (Landau and Lifshitz, 1980).

The linear form of Eq. (4.40) is identical to that in the Swift-Hohenberg (1977) equation suggested for buoyancy-driven (Rayleigh-Benard) convection.

$$\partial_t \phi = -\alpha\phi - \mu\Delta\phi - \Delta^2\phi - \phi^3 \qquad (4.41)$$

Note that one starts with different approximate model equations like (4.40) or (4.41) and augmenting their compexity based on heuristic and symmetry arguments, one can finally end up by having two evolution equations which are about the same mathematically. Consequently, appropriate generalizations of Eqs. (4.40) or (4.41) can be chosen in order to approximate the actual nonlinearities in systems where the critical wavenumber is not small, and even in cases where the first instability is oscillatory, or when there are several order-parameters.

In the following, we shall restrict attention to Eq. (4.40), which can be rigorously derived from the equations (4.15-4.21) when both boundaries are sufficiently insulating, i.e. the last of Eqs (4.18) has to be replaced by $\partial T/\partial z = -1$, while $Bi$ is taken small in Eq. (4.21). Specifically, a multiscale expansion can be performed by assuming $Bi \sim \epsilon^4$, $\partial/\partial x \sim \epsilon$, $\partial/\partial t \sim \epsilon^4$, and small supercriticality $M - 48 \sim \epsilon^2$. While all these scalings can be justified by the linear stability analysis (e.g. all terms are of the same order in the linear dispersion relation (4.36)), the order at which expansions of the perturbation fields should start is determined by a balance between linear and nonlinear terms. Actually, the temperature field needs to be written as $T = -z + \phi + O(\epsilon^2)$, while lowest-order pressure, horizontal, and vertical velocity fields respectively scale as $p = O(1)$, $\mathbf{v}_\perp \sim \epsilon$ and $v_z \sim \epsilon^2$. The evolution equation for $\phi$ is then Eq. (4.40), at order $\epsilon^4$ in the perturbation scheme (Knobloch, 1990; see also Gertsberg and Sivashinsky, 1981; Shtilman and Sivashinsky, 1991).

In fact, equation (4.40) holds for one-layer or multi-layer systems with undeformable interfaces, including buoyancy and/or surface tension gradient and non-Boussinesq effects. Only the values of the various coefficients $\alpha$, $\mu$, $\kappa$, $\beta$, $\delta$ and $\gamma$ depend on the specific nature of the problem considered. A restriction, however, is that the Prandtl number(s) ought to be high enough, otherwise mean flow effects have to be taken into account (Sect. 4.6.1). Our approach in the following will be to consider Eq. (4.40) or some of its variants (subsection 4.6) as models of more general problems to allow a detailed study of cellular convection in two (rather than three) spatial dimensions, tailoring many of the qualitative features of realistic experimental set-ups.

When $\beta \neq 0$, $\delta \neq 0$ and/or $\gamma \neq 0$, the reflection symmetry $\phi \rightarrow -\phi$, inherited from the reflection symmetry in the layer midplane, is broken. This is the case when top and bottom boundaries are not identical ($\beta \neq 0$, $\delta \neq 0$), or when non-Boussinesq effects are taken into account ($\gamma \neq 0$). Note that $\delta = \beta$ when the linear problem is self-adjoint.

For instance, for purely surface-tension-gradient-driven (Benard-Marangoni) convection in a layer with undeformable interface and very high Prandtl number liquids, one has after some rescaling of length, time, and temperature

$$\alpha = 1 - \varepsilon^2, \ \mu = 2, \ \kappa = 1, \ \beta = -\frac{\sqrt{7}}{8}, \ \delta = -3\frac{\sqrt{7}}{4}, \ \gamma = 0 \quad (4.42)$$

where $\varepsilon$ is a smallness parameter, accounting for supercriticality.

Pontes, Christov and Velarde (1996, 1999) have provided a wealth of numerical solutions of Knobloch's equation (4.40) in square domains with boundary conditions $\phi = \partial\phi/\partial n = 0$ on the sidewalls, examining the role of terms proportional to $\beta$ and $\delta$ (Figs 4.4 and 4.5).

When $\delta \neq \beta$ the linear problem is not self-adjoint and this is, indeed, the case for the original equations (4.15)–(4.21) for Benard-Marangoni convection. The first pattern at threshold is always hexagonal as in most experiments (Koschmieder, 1993). Starting from random ("white-noise") small initial conditions, the pattern evolves towards a structure formed by domains of nearly perfect hexagons (Fig. 4.4) separated by defect lines (similar to grain boundaries). Point defects also frequently occur, most often like pairs formed by a pentagon and an heptagon, such as also seen in experiments (Koschmieder, 1993). Note that for Benard–Marangoni convection with very high Prandtl number liquids where $\beta > \delta$, the fluid motion is upwards at the center of hexagonal cells (up-hexagons), where the temperature is higher. In contrast, for fluids with a low Prandtl number (liquid metals), one has $\delta > \beta$, and in this case the typical patterns at threshold shows down-hexagons, i.e. the center of hexagons is colder and the fluid moves downwards there (note however that at low Prandtl numbers, the evolution of the pattern is strongly influenced by large-scale flows, as will be discussed later).

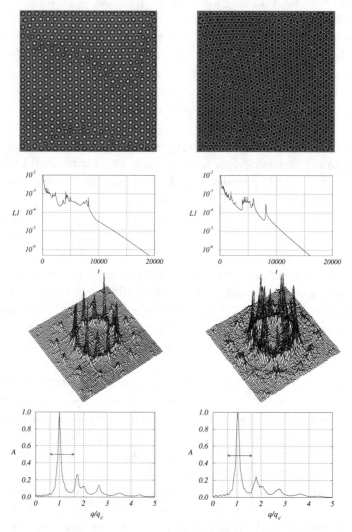

Figure 4.4:   Benard-Marangoni convection.   Hexagons in a square geometry, obtained from the numerical computations of Knobloch's equation (4.40) with $\alpha = 0.8$, $\mu = 2.7$, $\beta = -\sqrt{7}/8$, $\gamma = 0$ and $\kappa = 1$. Left panels: up-hexagons at $t = 19050$ for $\delta = -3\sqrt{7}/4$. Right panels:   down-hexagons at $t = 15990$ for $\delta = +3\sqrt{7}/4$. From top to bottom: the pattern; the time evolution of $L_1$-norm; the Fourier transform; the wavenumber content of the pattern (after Pontes *et al.*, 1999).

In Fig. 4.4, appears the evolution of a norm defined by $L_1 \sim \sum_{i,j} |\phi_{i,j}^{n+1} - \phi_{i,j}^n| / \sum_{i,j} |\phi_{i,j}^{n+1}|$, which measures the rate of change of the "distance" between two successive states $n$ and $n+1$ of the system (the sums run over all grid points), and therefore tends to zero on approach of the final steady state. Also represented in Fig. 4.4 are the Fourier transform and wavenumber content of the represented patterns, whose features will be commented on in later sections.

Figure 4.5: Benard-Marangoni convection. Square patterns obtained from direct numerical simulations of Knobloch's equation (4.40). Left panel: pattern at $t = 41100$, for $\alpha = 0.9$, $\mu = 2.5$, $\beta = \delta = -\sqrt{7}/4$, $\gamma = 0$ and $\kappa = 1$. Right panel : pattern at $t = 200000$, for $\alpha = 0.9$, $\mu = 2$, $\beta = \delta = -3\sqrt{7}$, $\gamma = 0$ and $\kappa = -1$.

Now, for $\beta = \delta$, one typically observes patterns formed by squares, rather than hexagons (Fig. 4.5), at the instability threshold.

Actually, even for $\beta \neq \delta$, where the first bifurcation leads to hexagons, a secondary transition to squares may occur, under some conditions to be discussed in the next section, when the constraint is increased past a threshold. This is also in qualitative agreement with experiments (see Fig. 4.6) performed by Eckert et al. (1998), and by Schatz et al. (1999; see also Schatz and Neitzel, 2001).

In Fig. 4.6, one can also observe that islands made of squares can coexist with hexagons, which is partly due to the constraint imposed by the circular geometry (Velarde and Normand, 1980). This coexistence was also obtained by Pontes et al. (1999). Eckert et al. (1998) have also found the hexagon-square transition from direct integration of the original equations (4.15-4.21). They remarked the role played by the mean flow (the Prandtl number is infinite) acting as a "lubricant" hence allowing fast rearrangement of the pattern leading to more regular square patterns.

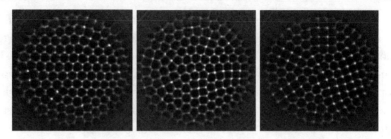

Figure 4.6:   Benard-Marangoni convection.   Hexagon-squares transition observed upon increasing the supercriticality measured by $\epsilon = (M - M_c)/M_c$ in a circular container with diameter 42 times larger than the liquid depth. Left : $\epsilon = 2.9$, Center : $\epsilon = 4.0$, Right : $\epsilon = 5.7$ (Courtesy of K. Eckert).

Even though Knobloch's equation (4.40) is valid for a number of situations, it has to be generalized when the Prandtl number is not large and mean flow effects are significant (subsection 4.6.1), or when surface deformation occurs (subsection 4.6.2). Before considering these complicating effects, we turn to an analysis of pattern selection using a methodology of widespread applicability in nonlinear science. We refer to amplitude equations, beyond the simplest case, obtained using an appropriate perturbative, multiscale analysis of the original equations with e.g. most parameters having finite values.

## 4.3   Selection of convective patterns near the instability threshold

The method to be described in this section applies near the threshold of any monotonic instability, even when $k_c = O(1)$. In this case however, the calculation may prove too cumbersome, because of the need of solving various systems of ODEs for the vertical dependence of the convective fields. Alternatively, recall that we said that rather than starting with the original equations (4.15-4.21), one can stick to Knobloch's equation (4.40), where the vertical dependence has already been eliminated due to the separation of scales occurring when $k_c \to 0$, as done in lubrication theory, and appropriately generalizing it. This ad hoc approach would be too demanding in mathematical complexity as Eq. (4.40), which is an amplitude equation, is far too simplified relative to Eqs. (4.15)–(4.21). However, as the amplitude equations ought to be about the same in both these cases, save the numerical values of coefficients, we will first describe the method at a general level, and then for simplicity particularize the analysis by using coefficients calculated from Eq. (4.40).

In the framework of the linear system (4.26) - (4.31), when $M$ is slightly above the critical value, $M_c$, the disturbances with any values of $k$ inside the interval $(k_-, k_+)$ exponentially grow independently of one another, because of the superposition principle. However, if we consider the original nonlinear problem (4.15) - (4.21), *interaction* occurs between disturbances with different wavevectors. First of all, the quadratic nonlinear terms in equations (4.15) and (4.16) corresponding to the convective transfer of momentum and heat, generate *combination* wavevectors. For instance, if the initial disturbance at $t = 0$ contains some Fourier components with wavevectors $\{k_j\}$, $j = 1, \ldots, N$, any combination $\sum_{j=1}^{N} n_j k_j$ will be present in the Fourier spectrum of the motion as $t > 0$. Also, the dependence of the fields in the vertical coordinate may be changed compared with the predictions of the linear theory. Generally, the solution of the original nonlinear problem, $f = (\mathbf{v}, p, T)$, can be written in the form of a series expansion

$$f = \sum_{\mathbf{k}} \sum_{n=0}^{\infty} a_n(\mathbf{k}, t) f_n(z; \mathbf{k}) e^{i\mathbf{k} \cdot \mathbf{r}},$$

where $\{f_n = (\mathbf{v}_n, p_n, T_n), n = 0, \ldots, \infty\}$ is a set of eigenfunctions of the problem (4.25) - (4.31) ($n = 0$ corresponds to the mode which generates the instability). The next step is to discriminate between slowly evolving *active* (hence slaving) modes with $n = 0$ and wavevectors belonging to the linear stability ring or located close to it, which are supported by the instability mechanism or decay slowly, and *passive* modes which include the modes with $n \neq 0$ and those with $n = 0$ but with wavevectors far away from the linear instability ring. The passive modes get their energy from the active modes and hence are "*slaved*" to them because their amplitudes are determined by the rapidly established balance between the nonlinear transfer of energy from the active modes and the linear decay of energy. The slaved modes can be eliminated under the assumption that this balance is constantly held (Haken, 1983, a,b). Then, the problem can be well approximated using the Fourier amplitudes of active modes only.

The derivation of the nonlinear amplitude equations describing the phenomena near the instability threshold is rather cumbersome, and we shall not perform it here (Busse, 1978; Cross and Hohenberg, 1993, and references therein). Let us write down the final form of equations which does not depend on the particular origin of the problem

$$\frac{da(\mathbf{k})}{dt} = \lambda(\mathbf{k})a(\mathbf{k}) + \frac{1}{2}\sum_{\mathbf{k}_1+\mathbf{k}_2=\mathbf{k}} \alpha(\mathbf{k}_1, \mathbf{k}_2)a(\mathbf{k}_1)a(\mathbf{k}_2)$$
$$- \frac{1}{6}\sum_{\mathbf{k}_1+\mathbf{k}_2+\mathbf{k}_3=\mathbf{k}} T(\mathbf{k}_1, \mathbf{k}_2, \mathbf{k}_3)a(\mathbf{k}_1)a(\mathbf{k}_2)a(\mathbf{k}_3), \tag{4.43}$$

where $a(\mathbf{k}) = a^*(-\mathbf{k})$ is a Fourier amplitude of the active (unstable or slowly decaying) mode (the subscript 0 is omitted), $\lambda(\mathbf{k})$ is the linear growth rate, $\alpha(\mathbf{k}_1, \mathbf{k}_2)$ and $T(\mathbf{k}_1, \mathbf{k}_2, \mathbf{k}_3)$ are coefficients of the corresponding quadratic and cubic nonlinear interaction between active modes.

Eq. (4.43) describing the interaction of a large number of modes looks very complicated. However, as seen in the previous section, the flow appearing as the result of the Benard-Marangoni instability is rather close to a nearly perfect *hexagonal pattern* close enough to the instability threshold. In Fourier space, only a finite number of peaks remain after a certain time. For instance, in the left part of Fig. 4.4, one can distinguish 12 large peaks corresponding to active modes; in addition, we also have lower secondary peaks corresponding to passive modes. In this case, the pattern is mostly composed by two "grains" of quite regular hexagons rotated by 30° with respect to each other. For a single regular hexagonal pattern, one would have in the Fourier space a superposition of 6 active modes with wavevectors $\pm\mathbf{k}_j$, $j = 0, 1, 2$ forming a resonant triad

$$\mathbf{k}_0 + \mathbf{k}_1 + \mathbf{k}_2 = 0, \quad |\mathbf{k}_0| = |\mathbf{k}_1| = |\mathbf{k}_2| = k \approx k_c \tag{4.44}$$

The combinations $\sum_{j=0}^{2} n_j \mathbf{k}_j$ with moduli different from $k$ are passive, and hence do not appear in Eq. (4.43). In this case, the high-dimensional system (4.43) is reduced to a three-dimensional system of equations (called after Landau; e.g. Scanlon and Segel, 1967; Kuznetsov and Spector, 1980; Cloot and Lebon, 1984; Bragard and Lebon, 1993):

$$\frac{da_j}{dt} = \lambda a_j + \alpha a^*_{[j-1]}a^*_{[j+1]} - a_j(T_0|a_j|^2 + T_1|a_{[j-1]}|^2 + T_1|a_{[j+1]}|^2), \tag{4.45}$$

where $[j \pm 1] = (j \pm 1)\mathrm{mod}\ 3$;

$$a_j = a(\mathbf{k}_j), \ a^*_j = a(-\mathbf{k}_j); \ \alpha = \alpha(\mathbf{k}_j, \mathbf{k}_l), j \neq l;$$

$$T_0 = T(0) = \frac{1}{2}T(\mathbf{k}_j, \mathbf{k}_j, -\mathbf{k}_j); \; T_1 = T(\pi/3) = T(\mathbf{k}_j, \mathbf{k}_l, -\mathbf{k}_l), j \neq l.$$

It should be noted that the system (4.45) in the simplest albeit significant model-problem generalizing Eq. (4.40). A more general case will be discussed later on. Eq. (4.45) can be written in the form

$$\frac{da_j}{dt} = -\frac{\partial F}{\partial a_j^*}, \; j = 0, 1, 2, \tag{4.46}$$

where the Lyapunov function $F(a_0, \ldots, a_{N-1}, a_0^*, \ldots, a_{N-1}^*)$, $N = 3$ is defined as

$$F = -\lambda \sum_j |a_j|^2 - \frac{1}{3}\alpha \sum_{jlm}(a_j a_l a_m + a_j^* a_l^* a_m^*) + \frac{1}{2} \sum_{jl} T_{jl} |a_j|^2 |a_l|^2;$$
$$\tag{4.47}$$

the cubic term contains only the *resonant triad* of modes with wavevectors $\mathbf{k}_j$, $\mathbf{k}_l$, $\mathbf{k}_m$ satisfying the condition $\mathbf{k}_j + \mathbf{k}_l + \mathbf{k}_m = 0$, and

$$T_{jl} = T(\mathbf{k}_j, \mathbf{k}_l, -\mathbf{k}_l) = T(\theta_{jl}) \; \text{ for } \; j \neq l$$

($\theta_{jl}$ is the angle between $\mathbf{k}_j$, $\mathbf{k}_l$),

$$T_{jl} = \frac{1}{2}T(\mathbf{k}_j, \mathbf{k}_j, -\mathbf{k}_j) = T(0) \; \text{ for } \; j = l.$$

Note that
$$\frac{dF}{dt} = -\sum_j \left|\frac{da_j}{dt}\right| \leq 0,$$

and it is constant only for stationary patterns (i.e. when all $a_j$ are constant). If the Lyapunov function $F$ is bounded from below, we can conclude that only *stationary patterns* can exist in the limit $t \to \infty$. Formally, the minima of $F$ correspond to the locally *stable* patterns, while maxima and saddle points correspond to *unstable* stationary patterns.

The most important stationary solution of the system (4.45) is characterized by real and equal values of amplitudes

$$a_0 = a_1 = a_2 \equiv a_h, \tag{4.48}$$

where $a_h$ satisfies the equation

$$\lambda + \alpha a_h - (T_0 + 2T_1)a_h^2 = 0, \tag{4.49}$$

and hence

$$a_h = \frac{\alpha \pm \sqrt{\alpha^2 + 4\lambda(T_0 + 2T_1)}}{2(T_0 + 2T_1)}. \tag{4.50}$$

It corresponds to the *hexagonal pattern* centered in the point $x = 0$, $y = 0$. This hexagonal pattern is periodic in space with period $L_1 = 4\pi/k$ in the direction of each vector $\mathbf{k}_j$ and with period $L_2 = 4\pi/\sqrt{3}k$ in the perpendicular direction; the side length of the hexagons is $L = 2\pi/k$. Incidentally, Palm (1960) provided a similar argument when using a non-Boussinesq temperature dependence of the viscosity tried to explain the hexagonal patterns. Newell and Pomeau (1993) have placed this discussion in a general setting.

The coefficient $T_0 + 2T_1$ is usually positive in our case of Benard-Marangoni convection, while the coefficient $\alpha$ can have either sign. Here also, it is $\alpha < 0$ for $P < P_c$ and $\alpha > 0$ for $P > P_c$, where the critical value $P_c$ depends on the other parameters of the problem (Dauby et al., 1993; Golovin et al., 1995; Thess and Bestehorn, 1995).

One can see that hexagons arise in the *subcritical* region (Fig. 4.7) as

$$\lambda = \lambda_1 \equiv -\frac{\alpha^2}{4(T_0 + 2T_1)} < 0 \tag{4.51}$$

with the *finite amplitude*

$$a_h^* = \frac{\alpha}{2(T_0 + 2T_1)}. \tag{4.52}$$

Experiments by Schatz et al. (1995) (see also Schatz and Neitzel, 2001) support this theoretical prediction.

Let us define

$$\gamma \equiv \frac{\lambda T_0}{\alpha^2}, \; B_1 \equiv \frac{T_1}{T_0}. \tag{4.53}$$

Then the lower boundary of the existence of hexagons is

$$\gamma = \gamma_1 \equiv -\frac{1}{4(1 + 2B_1)}. \tag{4.54}$$

There are two branches of stationary solutions to the equation (4.49), but only one of them is *stable* (Fig. 4.7). This result can be easily shown by following the evolution in time of equation (4.45) with $a_0(t) = a_1(t) = a_2(t) \equiv a_h(t)$. For the stable branch, $a_h$ and $\alpha$ have the same sign. In the case of the positive value of $\alpha$ $(P > P_c)$ the stable hexagonal pattern has an upstream flow in the center of the cell

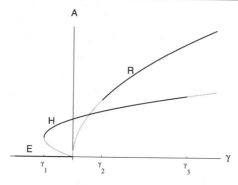

Figure 4.7: Benard-Marangoni convection. Bifurcation diagram: amplitudes of the hexagonal pattern (H) and of rolls (R) versus the control parameter $\gamma$. E = (mechanical) equilibrium state. Black (gray) curves correspond to stable (unstable) states. The diagram is drawn for $\alpha > 0$, $T_1 > T_0 > 0$.

and a downstream flow on the periphery (up/"$l$-hexagons"), while the negative value of $\alpha$ ($P < P_c$) leads to a motion in the opposite direction (down/"$g$-hexagons"). Here $l$ and $g$ are abbreviation of the words "liquid" and "gas".

Qualitatively, this difference can be explained as follows (Thess and Bestehorn, 1995). In the case of a high Prandtl number liquid the main nonlinear effect is *convection of heat* (cf. Eq. (4.15), (4.16)). The advection of the isolines of temperature along the interface leads to the growth of warm regions above the uprising flows at the expense of cold regions above the descending flows. This effect generates large "warm" cells on the interface separated by narrow "cold" layers, i.e. "$l$-hexagons". In the case of low Prandtl number liquids, the *convection of vorticity* is the dominant nonlinear effect. The thermocapillary advection of vorticity leads to its concentration in the cold regions, which produces a strong downward motion in those regions, i.e. "$g$-hexagons".

*Why* does just the hexagonal pattern arise at threshold in the Benard-Marangoni instability and not, say, a roll pattern or other? The answer is that only a hexagonal pattern is *stable* near the threshold if the quadratic interaction coefficient $\alpha$ does not vanish (Busse, 1967). First, let us consider the competition between hexagonal and roll patterns. The *roll* solution of the system (4.39) is

$$a_1 = a_2 = 0, \quad a_0 = \sqrt{\lambda/T_0}. \tag{4.55}$$

We shall assume that $T_0 > 0$, so that the rolls exist in the *supercritical region*, $\lambda > 0$, and assume also that $\alpha > 0$ (alternatively, $\alpha < 0$). Let us investigate the evolution of small disturbances $a_1(t)$, $a_2(t)$ imposed on the solution (4.55). Linearizing Eq. (4.45) with respect to such disturbances it appears that for the most dangerous (highest growth) mode $a_1 = a_2 = \tilde{a}$ is real, and its evolution is governed by the equation

$$\frac{d\tilde{a}}{dt} = (\lambda + |\alpha|a_0 - T_1 a_0^2)\tilde{a}. \tag{4.56}$$

Rolls are unstable if

$$\lambda + |\alpha|a_0 - T_1 a_0^2 = |\alpha|\sqrt{\frac{\lambda}{T_0}} + \lambda\left(1 - \frac{T_1}{T_0}\right) > 0.$$

Particularly, rolls are unstable in the vicinity of the instability threshold (as $\lambda$ is small), because for low enough values of $\lambda$, $\sqrt{\lambda} \gg \lambda$. If $B_1 = T_1/T_0 < 1$, the rolls never become stable; otherwise they are stable if

$$\lambda > \lambda_2 = \frac{\alpha^2 T_0}{(T_1 - T_0)^2}; \tag{4.57}$$

using (4.53), we can write that the stability region of rolls is

$$\gamma > \gamma_2 = \frac{1}{(B_1 - 1)^2}. \tag{4.58}$$

Thus, the rolls are never stable near the instability threshold, unless $\alpha = 0$. It should be noted that the region of stabilization of rolls determined by the inequality (4.58) may be outside the region of validity of Eq. (4.46) or Eq. (4.43), as both were obtained under the assumption of weak enough flow.

A similar analysis of the stability of the hexagonal patterns (4.48), (4.49) in the framework of the model (4.45), shows that the hexagons are stable in the interval

$$\gamma_1 < \gamma < \gamma_3, \tag{4.59}$$

where $\gamma_1$ is defined by (4.54), and

$$\gamma_3 = \frac{B_1 + 2}{(B_1 - 1)^2}. \tag{4.60}$$

if $B_1 > 1$. Otherwise, the hexagons are stable for any $\gamma$. Below the point $\gamma = \gamma_1$ the hexagonal solution does not exist. Beyond the point $\gamma = \gamma_3$ the hexagons are *internally* unstable with respect to disturbances violating the condition (4.48), which eventually destroy the hexagonal patterns and lead to establishing of rolls.

The analysis of the competition between hexagons and *square* patterns can be performed in a similar way, but then the system (4.43) is reduced to a *six-dimensional* system of equations describing the interaction of two resonant triads (i.e. two systems of hexagons) rotated by 30° with respect to each other.

Let us describe the results of the stability analysis (Kuznetsov and Spector, 1980; Malomed and Tribelskii, 1987; Malomed *et al.*, 1989; Golovin *et al.*, 1995; Bestehorn and Friedrich, 1998, 1999). In addition to the coefficient $B_1 = T(\pi/3)/T(0)$, the ratios $B_2 \equiv T(\pi/6)/T(0)$ and $B_3 \equiv T(\pi/2)/T(0)$ become significant. First of all, squares are stable versus rolls for any $\gamma$, if $B_3 < 1$, and are unstable otherwise. In Benard-Marangoni convection, typically $B_3 < 1$ for relatively large $P$ (Golovin *et al.*, 1997), while $B_3 > 1$ for moderate values of $P$. In the former case, the hexagons compete with square patterns while in the latter one can expect a competition between hexagons and rolls. Later on, we shall assume that $B_3 < 1$, so that rolls are always unstable, while squares can be stable.

Let us consider now the stability of hexagons. The criterion of the *internal* instability of hexagons (4.59) was obtained above. Now we shall consider the behavior of the disturbance with wavevector orthogonal to one of the wavevectors of hexagons. For such a mode, we can find that if $Q_3 = 1 - 2B_2 + 2B_1 - B_3$ is negative, this mode is stable, so that the hexagons never evolve *directly* into the square patterns. The transition can take place, for instance, through two stages: first, hexagons become internally unstable, and an unstable roll pattern (corresponding to a saddle point in the phase space of the dynamical system (4.46)) transiently appears and then decays into a square pattern. Other scenarios related to the spatial inhomogeneities caused by local defects will be described later. If $Q_3 > 0$, hexagons become unstable with respect to squares directly, as

$$\gamma > \gamma_4 = \frac{2B_2 + B_1}{Q_3^2}.$$

For any parameter values, the quantity $\gamma_4$ should be compared with $\gamma_3$ in order to predict the corresponding scenario of the transition.

Finally, let us discuss the instability of the square pattern with respect to hexagons. The significant parameter is $Q_2 = 1 - B_2 -$

$B_1 + B_3$. If $Q_2 > 0$, the square pattern is unstable with respect to hexagons for any $\gamma$. If $Q_2 < 0$, it becomes stable when

$$\gamma > \gamma_5 = \frac{1 + B_3}{Q_2^2}.$$

Hysteresis between hexagons and squares has also been predicted (Golovin et al., 1995, 1997).

Let us note that the system (4.46) with $N = 6$ can describe also a quasiperiodic dodecagonal pattern which is a superposition of two hexagonal structures (Busse, 1978). Though such a system can be locally stable in some cases (Malomed et al., 1989; Golovin et al., 1995; see subsection 4.6.2), it turns out that it does not provide the deepest minimum of the Lyapunov function. Probably, that is why it never was observed in experiments on the Benard-Marangoni convection.

Now, let us particularize the general results given above to analyze pattern selection using Knobloch's equation (4.40), valid for large-scale patterns. By some rescaling, Eq. (4.40) becomes

$$\begin{aligned}\frac{du}{dt} =& \mu u - (1 + \nabla^2)^2 u + \nabla \cdot [(\nabla u)^2 \nabla u] + \beta \nabla \cdot [\nabla^2 u \nabla u] \\ &+ \delta \nabla^2 [(\nabla u)^2] - \gamma \nabla \cdot [u \nabla u]\end{aligned} \quad (4.61)$$

In this case, the coefficients of amplitude equations derived for $k_c = 1$ (the horizontal coordinate has been rescaled) are given by

$$\lambda = \mu, \quad \alpha = \beta - \delta + \gamma, T_0 = 3 + \tfrac{2}{9}(\beta + 2\delta + \gamma)(2\beta + 4\delta - \gamma),$$
$$T_1 = 3 + \tfrac{3}{4}(3\delta - \gamma)(\beta + \delta + \gamma)$$

$$(4.62)$$

The additional coefficients necessary to study the competition between hexagons and squares are

$$T(\pi/2) = 2 - 4(\beta + \gamma)(\beta + \gamma - 2\delta), \quad T(\pi/6) = 5 + (\beta + 6\delta - 2\gamma)(\beta + \delta + \gamma)/2$$
$$(4.63)$$

Then, for values of $\beta$, $\gamma$, and $\delta$ given by (4.42), it appears from the discussion given above, that $B_3 < 1$ (squares are stable relative to rolls), and hexagons are preferred near the instability threshold, as $\alpha > 0$ (upflow at the center). The depth of the subcritical region of hexagons (hysteresis) is given by $\lambda_1 = -0.018$. Hexagons remain stable to squares for every $\lambda$, while they become unstable to rolls, but only at unrealistically large values of $\lambda$. Still, squares may be locally stable for $\lambda > 0.235$, and therefore may be observed, rather than hexagons, depending on the initial conditions (Beste-horn, 1996). Skeldon and Silber (1998) have found that near onset some more exotic spatially periodic planforms can also be preferred over the usual rolls, squares and hexagons. Exotic patterns, (limit cycle) bistability between hexagons and rolls and never settling to a steady state, a form of space-time chaoticity near the threshold of Benard cells, have been predicted by Bragard and Velarde (1997, 1998).

## 4.4  Modulations and instabilities of hexagonal patterns

Experiments show that reality is more complicated than what we have described using the system of ODEs (4.46). The real hexagonal pattern is not just spatially periodic. First of all, the hexagonal patterns are usually distorted near the lateral boundaries. Also, patterns often include defects, like the paradigmatic heptagon-pentagon pair (Velarde and Normand, 1980).

To account for such phenomena, let us generalize the system of equations (4.46) to accomodate for e.g. spatial modulations of hexagonal patterns and defects.

Starting from the equations (4.43), we shall assume (unlike done when proceeding to (4.45)) that the Fourier spectrum of the patterns is not reduced to the triad (4.44) but is concentrated in three small regions of radius $O(\epsilon)$ *around* the resonant wavevectors $\mathbf{k}_j$ (this widening of Fourier peaks is indeed visible in Fig. 4.4). Accordingly, the amplitude functions

$$A_j(\mathbf{x}_\perp, t) = \sum_{|\mathbf{k}-\mathbf{k}_j|=O(\epsilon)} a(\mathbf{k}, t) e^{i(\mathbf{k}-\mathbf{k}_j)\cdot\mathbf{x}_\perp}, \, j = 0, 1, 2 \qquad (4.64)$$

will be slow functions of both $\mathbf{x}_\perp$ and $t$. In order to derive their corresponding evolution equations we shall use the Taylor expansion of coefficients in (4.43) considered as functions of their arguments.

Our approach is similar to that used by Newell and Whitehead (1969) and Segel (1969), for buoyancy-driven (Rayleigh-Benard) convection. However, here there is the resonant interaction described by the quadratic terms in (4.43) which we are going to consider together with the cubic term (Busse, 1967).

The coefficient $\alpha$ is a function of the scalar invariants $k_1^2$, $k_2^2$ and $\mathbf{k}_1 \cdot \mathbf{k}_2$. Let $\mathbf{k}_j = k_c \mathbf{n}_j + \mathbf{q}_j$, $j = 1, 2$, where $k_c$ corresponds to the minimum of the neutral curve, $\mathbf{n}_1$, $\mathbf{n}_2$ are unit vectors with the angle $2\pi/3$ between them, and $|\mathbf{q}_j|$ are $O(\epsilon)$. Thus we have

$$\alpha(k_1^2, k_2^2, \mathbf{k}_1 \cdot \mathbf{k}_2) = \alpha_0 + C(\mathbf{n}_1 \cdot \mathbf{q}_1 + \mathbf{n}_2 \cdot \mathbf{q}_2) + D(\mathbf{n}_1 \cdot \mathbf{q}_2 + \mathbf{n}_2 \cdot \mathbf{q}_1) + O(\epsilon^2),$$
(4.65)

where $\alpha_0$, $C$, $D$ are constants. Using also the expansion $\lambda(k_j^2) = \lambda_0 - S q_j^2 + \ldots$, assuming $A_j = O(\epsilon)$ and keeping together the terms of the second and of the third order in $O(\epsilon)$, we obtain the following system of evolution equations for the amplitudes, $A_j(\mathbf{x}_\perp, t)$, hence the amplitude equations

$$\frac{\partial A_j}{\partial t} = \lambda_0 A_j + S(\mathbf{n}_j \cdot \nabla)^2 A_j + \alpha_0 A_{[j-1]}^* A_{[j+1]}^*$$

$$+ iC(A_{[j-1]}^* \mathbf{n}_{[j+1]} \cdot \nabla) A_{[j+1]}^* + A_{[j+1]}^* \mathbf{n}_{[j-1]} \cdot \nabla) A_{[j-1]}^*)$$

$$+ iD(A_{[j-1]}^* \mathbf{n}_{[j-1]} \cdot \nabla) A_{[j+1]}^* + A_{[j+1]}^* \mathbf{n}_{[j+1]} \cdot \nabla) A_{[j-1]}^*)$$

$$- A_j(T_0 |A_j|^2 + T_1 |A_{[j-1]}|^2 + T_1 |A_{[j+1]}|^2) \qquad (4.66)$$

which generalizes the system (4.45) and spans well beyond the phenomena embraced by Knobloch's equation (4.40). The system of equations (4.66) was derived for Benard-Marangoni convection by Bragard and Velarde (1998), Golovin $et\ al.$ (1997) and Colinet $et\ al.$ (1997). Similar equations also appear in other contexts (Brand, 1989; Gunaratne $et\ al.$, 1994; Kuznetsov $et\ al.$, 1995; Kuske and Milewski, 1999).

Using the scaling transformation

$$A_j \to \frac{\alpha_0}{T_0} A_j', \quad \frac{\partial}{\partial t} \to \frac{\alpha_0^2}{T_0} \frac{\partial}{\partial t'}, \quad \nabla \to \frac{\alpha_0}{\sqrt{S T_0}} \nabla'; \qquad (4.67)$$

dropping the primes, and expecting no confusion in the reader with the present use of $\gamma$, etc, we get

$$\frac{\partial A_j}{\partial t} = \gamma A_j + (\mathbf{n}_j \cdot \nabla)^2 A_j + A_{[j-1]}^* A_{[j+1]}^*$$

$$+ i\beta_1(A_{[j-1]}^* \mathbf{n}_{[j+1]} \cdot \nabla A_{[j+1]}^* + A_{[j+1]}^* \mathbf{n}_{[j-1]} \cdot \nabla A_{[j-1]}^*)$$

$$+ i\beta_2(A_{[j-1]}^* \mathbf{n}_{[j-1]} \cdot \nabla A_{[j+1]}^* + A_{[j+1]}^* \mathbf{n}_{[j+1]} \cdot \nabla A_{[j-1]}^*)$$

$$- A_j(|A_j|^2 + B_1 |A_{[j-1]}|^2 + B_1 |A_{[j+1]}|^2), \qquad (4.68)$$

where

$$\gamma = \frac{\lambda_0 T_0}{\alpha_0^2}, \; \beta_1 = \frac{C}{\sqrt{ST_0}}, \; \beta_2 = \frac{D}{\sqrt{ST_0}}. \tag{4.69}$$

The system (4.68) can also be written in the following form:

$$\frac{\partial A_j}{\partial t} = \gamma A_j + (\mathbf{n}_j \cdot \nabla)^2 A_j + A^*_{[j-1]} A^*_{[j+1]}$$

$$+ i K_0 (A^*_{[j-1]} \mathbf{n}_{[j+1]} \cdot \nabla A^*_{[j+1]} + A^*_{[j+1]} \mathbf{n}_{[j-1]} \cdot \nabla A^*_{[j-1]})$$

$$+ i K (A^*_{[j-1]} \boldsymbol{\tau}_{[j+1]} \cdot \nabla A^*_{[j+1]} - A^*_{[j+1]} \boldsymbol{\tau}_{[j-1]} \cdot \nabla A^*_{[j-1]})$$

$$- A_j (|A_j|^2 + B_1 |A_{[j-1]}|^2 + B_1 |A_{[j+1]}|^2), \tag{4.70}$$

where $\boldsymbol{\tau}_j$ are unit vectors mutually orthogonal to $\mathbf{n}_j$ (i.e. $\mathbf{n}_0 = (1,0)$, $\mathbf{n}_1 = (-1/2, \sqrt{3}/2)$, $\mathbf{n}_2 = (-1/2, -\sqrt{3}/2)$, $\boldsymbol{\tau}_0 = (0,1)$, $\boldsymbol{\tau}_1 = (-\sqrt{3}/2, -1/2)$, $\boldsymbol{\tau}_2 = (\sqrt{3}/2, -1/2)$). The values of the coefficients $B_1$, $\beta_1$ and $\beta_2$ (or $K$ and $K_0$) are not needed here and can be found in the literature (Bragard and Velarde, 1998; Golovin et al., 1997; Colinet et al., 1997).

Let us emphasize that the system (4.70), like the original Eqs. (4.15)– (4.21), is not variational, i.e. it cannot be presented in the form

$$\frac{\partial A_j}{\partial t} = -\frac{\delta F}{\delta A_j^*} \tag{4.71}$$

with a Lyapunov functional $F(A_j, A_j^*)$. The terms which bring this non-variational feature are those with the coefficients $\beta_1$ and $\beta_2$ (or $K$ and $K_0$). That is why we can expect that the dynamics of the system (4.70) is "richer" than that studied above with Eq. (4.40) or Eq. (4.46) and can include oscillatory instabilities, and chaotic flows.

Let us use the amplitude equations (4.70) to study modulational instabilities of hexagonal patterns, wavenumber selection due to the lateral boundaries and the evolution of defects in hexagonal patterns. We shall assume $B_1 > 1$.

## 4.4.1 Instabilities of hexagonal patterns

In the previous subsection we found that the hexagonal pattern is stable in the interval (4.59). However, the analysis included only the *internal* instability of the hexagonal patterns using the system (4.45), i.e. only with respect to disturbances which do not violate the perfect periodicity of the pattern (but may violate the equality

$a_0 = a_1 = a_2$). Now we are going to consider the *modulational instability* of patterns in the framework of the system (4.70), thus finding the domain of stability of periodic patterns as with Busse's balloon in Rayleigh-Benard convection.

Let us note that the stability of hexagonal patterns has received a great deal of attention (Busse, 1967; Kuznetsov and Spector, 1980; Pismen, 1980; Caroli *et al.*, 1984; Knobloch, 1990; Sushchik and Tsimring, 1994, Malomed *et al*, 1994; Bestehorn and Friedrich, 1998, 1999). Generally, those studies use a *variational* model that ignores the dependence of the three-wave interaction coefficient $\alpha(\mathbf{k}_1, \mathbf{k}_2)$ on the wave vectors of interacting waves (i.e. for $K_0 = K = 0$). The variational model describes reasonably well even the Rayleigh-Benard convection of a weakly non-Boussinesq fluid in the case of *identical* boundary conditions on the top and bottom horizontal boundaries. However, this approximation fails to apply to less symmetric cases, e.g., for Rayleigh-Bénard convection in the case of non-equivalent top and bottom boundary conditions (Kuznetsov *et al.*, 1995), for quasi-two-dimensional reaction-diffusion systems (Gunaratne *et al.*, 1994; Kuske and Milewski, 1999), and, indeed, for Benard-Marangoni convection (Bragard and Velarde, 1998; Golovin *et al.*, 1997; Colinet *et al.*, 1997).

Let us consider the family of solutions of (4.70)

$$A_j = a_h(q)e^{iq\mathbf{n}_j \cdot \mathbf{x}_\perp}, \; j = 0, 1, 2 \qquad (4.72)$$

corresponding to hexagonal patterns with wave numbers $k$, $k - k_c \propto \epsilon q$. The amplitude of the hexagonal pattern is determined by Eq. (4.50), where, in view of the above given scaling (4.67), we should introduce

$$\alpha = 1 + 2K_0 q, \; \lambda = \gamma - q^2, \; T_0 = 1, \; T_1 = B_1$$

Hence,

$$a_h(q) = \frac{1 + 2K_0 q \pm \sqrt{(1 + 2K_0 q)^2 + 4(\gamma - q^2)(1 + 2B_1)}}{2(1 + 2B_1)}. \qquad (4.73)$$

The coefficient $K_0$ determines the dependence of the quadratic interaction coefficient on $q$ and changes sign in the point $q = -1/2K_0$. Thus the branch with the sign $+$ in (4.73) can be stable as $1 + 2K_0 q > 0$, while the opposite branch can be stable as $1 + 2K_0 q < 0$. The lower boundary of the existence of hexagons is

$$\gamma_1(q) = q^2 - \frac{(1 + 2K_0 q)^2}{4(1 + 2B_1)} \tag{4.74}$$

(recall Eq. (4.54)), while the boundary of the internal instability of hexagons (4.60) can be written as

$$\gamma_3(q) = q^2 + \frac{(1 + 2K_0 q)^2 (B_1 + 2)}{(B - 1)^2}. \tag{4.75}$$

In order to investigate the stability of hexagons, it is convenient to use polar coordinates

$$A_j = \rho_j e^{i\phi_j}, \tag{4.76}$$

and hence the system (4.70) becomes:

$$\frac{\partial \rho_j}{\partial t} = \gamma \rho_j + (\mathbf{n}_j \cdot \nabla)^2 \rho_j - \rho_j (\mathbf{n}_j \cdot \nabla \phi_j)^2$$

$$+ \rho_{[j-1]}\rho_{[j+1]} \cos \Phi - \rho_j(\rho_j^2 + B_1 \rho_{[j-1]}^2 + B_1 \rho_{[j+1]}^2)$$

$$+ K_0(\rho_{[j-1]}\mathbf{n}_{[j+1]} \cdot \nabla \rho_{[j+1]} + \rho_{[j+1]}\mathbf{n}_{[j-1]} \cdot \nabla \rho_{[j-1]}) \sin \Phi$$

$$+ K_0 \rho_{[j-1]}\rho_{[j+1]}(\mathbf{n}_{[j-1]} \cdot \nabla \phi_{[j-1]} + \mathbf{n}_{[j+1]} \cdot \nabla \phi_{[j+1]}) \cos \Phi$$

$$+ K(\rho_{[j-1]}\boldsymbol{\tau}_{[j+1]} \cdot \nabla \rho_{[j+1]} - \rho_{[j+1]}\boldsymbol{\tau}_{[j-1]} \cdot \nabla \rho_{[j-1]}) \sin \Phi$$

$$+ K \rho_{[j-1]}\rho_{[j+1]}(\boldsymbol{\tau}_{[j+1]} \cdot \nabla \phi_{[j+1]} - \boldsymbol{\tau}_{[j-1]} \cdot \nabla \phi_{[j-1]}) \cos \Phi,$$

$$\rho_j \frac{\partial \phi_j}{\partial t} = 2(\mathbf{n}_j \cdot \nabla \rho_j)(\mathbf{n}_j \cdot \nabla \phi_j) + \rho_j(\mathbf{n}_j \cdot \nabla)^2 \phi_j - \rho_{[j-1]}\rho_{[j+1]} \sin \Phi \tag{4.77}$$

$$+ K_0(\rho_{[j-1]}\mathbf{n}_{[j+1]} \cdot \nabla \rho_{[j+1]} + \rho_{[j+1]}\mathbf{n}_{[j-1]} \cdot \nabla \rho_{[j-1]}) \cos \Phi$$

$$- K_0 \rho_{[j-1]}\rho_{[j+1]}(\mathbf{n}_{[j-1]} \cdot \nabla \phi_{[j-1]} + \mathbf{n}_{[j+1]} \cdot \nabla \phi_{[j+1]}) \sin \Phi$$

$$+ K(\rho_{[j-1]}\boldsymbol{\tau}_{[j+1]} \cdot \nabla \rho_{[j+1]} - \rho_{[j+1]}\boldsymbol{\tau}_{[j-1]} \cdot \nabla \rho_{[j-1]}) \cos \Phi$$

$$- K \rho_{[j-1]}\rho_{[j+1]}(\boldsymbol{\tau}_{[j+1]} \cdot \nabla \phi_{[j+1]} - \boldsymbol{\tau}_{[j-1]} \cdot \nabla \phi_{[j-1]}) \sin \Phi,$$

where the amplitudes and phases $\rho_j$, $\phi_j$ are real, and $\Phi = \phi_0 + \phi_1 + \phi_2$. To study stability we linearize the system (4.77) near the solution

$$\rho_0 = \rho_1 = \rho_2 = a_h(q), \quad \phi_j = q\mathbf{n}_j \cdot \mathbf{x}_\perp, \quad \Phi = 0. \tag{4.78}$$

and write perturbations $(\tilde{\rho}_j, \tilde{\phi}_j)$ in the form

$$(\tilde{\rho}_j, \tilde{\phi}_j) = (V_j, W_j)e^{\tilde{\lambda}t + i\tilde{\mathbf{q}} \cdot \mathbf{x}_\perp}, \quad j = 0, 1, 2. \tag{4.79}$$

The solvability condition (Fredholm alternative) of the linear system of equations for $V_j$, $W_j$ determines six eigenvalues $\tilde{\lambda}$ for any $\mathbf{k}$.

In the case $\tilde{\mathbf{q}} = 0$ which corresponds to the "internal" instability of patterns, there is a double-zero eigenvalue associated with two translational degrees of freedom (phase modes). Additional zeroes of the dispersion relation appear on the boundary of the internal instability (4.75).

Let us consider now the instability in the long-wave limit. Thus we take the following expansions

$$V_j(\tilde{q}, \beta) = V_j^{(0)}(\beta) + V_j^{(1)}(\beta)\tilde{q} + V_j^{(2)}(\beta)\tilde{q}^2 + \ldots,$$

$$W_j(\tilde{q}, \beta) = W_j^{(0)}(\beta) + W_j^{(1)}(\beta)\tilde{q} + W_j^{(2)}(\beta)\tilde{q}^2 + \ldots, \quad j = 0, 1, 2; \quad (4.80)$$

$$\lambda(\tilde{q}, \beta) = \lambda^{(0)}(\beta) + \lambda^{(1)}(\beta)\tilde{q} + \lambda^{(2)}(\beta)\tilde{q}^2 + \ldots, \qquad (4.81)$$

where $\beta$ is the angle between $\tilde{\mathbf{q}}$ and $\mathbf{n}_0$.

It can be shown that $\lambda^{(0)} = \lambda^{(1)} = 0$ everywhere except the boundaries (4.74), (4.75), and the coefficient $\lambda^{(2)}$, whose sign determines the long-wave phase stability of patterns, does not depend on the angle $\beta$. The equation $\lambda^{(2)}(q) = 0$ describes the boundaries of the Eckhaus instability of the hexagonal patterns and its explicit form can be found in the literature (Bragard and Velarde, 1998). The next non-zero coefficient $\lambda^{(4)}$ depends on $\beta$ and determines the direction of the wave vector of the most dangerous (highest growth) disturbance ($\beta = \pi/6$ in the lower part of the instability boundary, and $\beta = 0$ in the upper part of the instability boundary). Typical stability domains for equilateral hexagons can be found in Figs 4.8 and 4.9. For cases involving non-equilateral patterns, the reader is referred to the report by Nuz et al. (2000)

On the boundaries (4.74), (4.75) the expansion (4.81) contains non-vanishing $\lambda^{(1)}$. The boundaries of the Eckhaus instability (Eckhaus, 1965) touch the boundaries (4.74), (4.75) in some points. Between these points $\lambda^{(1)}$ is imaginary, so that there long-wave disturbances are oscillatory, i.e., $\lambda^{(1)} = \pm i\omega^{(1)}$,

$$\omega_1^2 = -2q^2 + \frac{K_0(\sqrt{3}K - K_0)}{4}\left(\frac{1 + 2K_0 q}{1 + 2B_1}\right)^2 +$$

$$\frac{q\sqrt{3}(\sqrt{3}K_0 - K)}{2}\left(\frac{1 + 2K_0 q}{1 + 2B_1}\right)$$

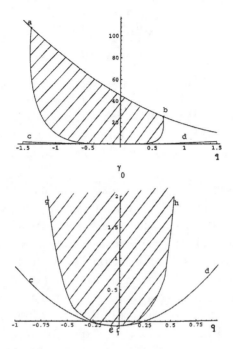

Figure 4.8: Benard-Marangoni convection. Stability domain of equilateral hexagons (hatched region): $K = 0.39$, $K_0 = -0.19$, $B = 1.26$. The lower plot comes from enlarging the upper one in the region of the origin.

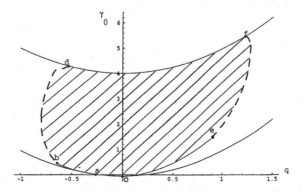

Figure 4.9: Benard-Marangoni convection. Stability domain of equilateral hexagons (hatched region) for $K = 3$, $K_0 = 0$, $B = 2$. Dashed lines correspond to the short-wave oscillatory instability.

on the boundary (4.74), and

$$\omega_1^2 = -q^2 + \left(\frac{\sqrt{3}K + K_0}{2}\right)^2 \left(\frac{1 + 2K_0 q}{B_1 - 1}\right)^2$$

on the boundary (4.75). In the interval where $\lambda^{(1)}$ is imaginary, $\lambda^{(2)}$ is real and negative (hence, disturbances decay oscillatorily); outside this interval, $\lambda^{(1)}$ is real, which corresponds to instability. Note that short-wave oscillatory instabilities, not described by expansions (4.80-4.81) are also possible, an example of which is depicted in Fig. 4.9.

When crossing one of the Eckhaus stability lines of the stability domain, the evolution of the pattern is generally expected to allow its return into the stable zone, via destruction or creation of new cells, as it is the case for roll patterns (Cross and Hohenberg, 1993). However, the nonlinear evolution of an hexagonal pattern initially quenched in the region where it is Eckhaus-unstable has not yet been studied in details, like using a nonlinear phase equation for the phase instability described above. Such phase instabilities may in some cases be subcritical, and lead to the formation of dislocations, i.e. points where the amplitude of some of the modes forming the hexagonal pattern vanish as we shall discuss further below. The motion of these defects then provides some mean to adjust the wavelength of the patterns. Wierschem *et al.* (1997) have shown that for Benard-Marangoni convection the wavelength of the pattern may adjust through creation or extinction of hexagonal cells, as illustrated in the two sets of pictures shown in Fig. 4.10.

Bestehorn (1993) has provided a detailed analysis of the stability domain of hexagonal patterns in Benard-Marangoni convection using the original equations (4.15-4.21) augmented with buoyancy. His results are in good agreement with the experiment (Benard, 1900, 1901; Dauzere, 1912; Koschmieder, 1991, 1993; Koschmieder and Switzer, 1992) as far as the tendency towards an increase of the selected wavenumber of the pattern with increasing values of $M$ is concerned.

## 4.4.2   Influence of lateral boundaries, fronts, and defects

Let us now succinctly describe a few other particular results obtained from the coupled system of non-variational amplitude equa-

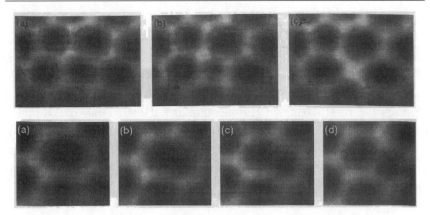

Figure 4.10: Infrared images of the surface-temperature distribution in Benard-Marangoni convection (silicone oil 50 cSt, layer depth 4.85 mm). The surface temperature is indicated by grey levels, darkening with relative higher temperature. The top sequence (from left to right) shows snapshots at $t = 151min$, $t = 210min$ and $t = 213min$, illustrating the extinction of a convective cell. The bottom sequence illustrates cell splitting process, at times $t = 38min$, $t = 106min$, $t = 110min$, and $t = 115min$ (after Wierschem et al., 1997).

tions (4.70) in view of possible comparison with the experiment.

## i. Influence of lateral boundaries

We have mentioned above that even in a set-up with lateral size much larger than the critical wavelength corresponding to the onset of the instability the distant side-walls affect the pattern nonlinear evolution by choosing the actual wavelength to be observed. For instance, it is known that for rolls parallel to the lateral walls, for which the amplitude function vanishes on the wall, the only solution that can be matched to the boundary condition has $q = 0$ (Daniels, 1977), i.e. a roll pattern with $k = k_c$ is selected. The influence of the higher order corrections in boundary conditions for the wavenumber selection of roll patterns was studied by Cross et al. (1980). On the other hand, experiments with hexagonal and square patterns in Benard-Marangoni convection show that the selected wavenumber is not $k_c$, (Cerisier et al., 1987a,b; Koschmieder, 1991; Koschmieder and Switzer, 1992; Eckert et al., 1998).

Let us show how the non variational terms in Eqs. (4.70) provide a mechanism albeit not unique for wavenumber selection.

For an hexagonal planform in a semi-space $x > 0$, $-\infty < y < \infty$, bounded by a rigid wall, $x = 0$, consider that the hexagonal planform is oriented with respect to the wall in such a way that one of the three planform wavevectors is orthogonal to the wall. The latter causes the modulation of the hexagonal pattern. Due to symmetry, this modulation can occur only in $x$-direction, and we can consider the amplitudes $A_j$, $j = 0, 1, 2$ in Eqs (4.70) as $A_0 = F(x,t)$, $A_1 = A_2 = G(x,t)$, where $F(x,t)$ is the complex amplitude of the system of rolls with the wavevector normal to the walls, $\mathbf{n}_0$, and $G(x,t)$ is the complex amplitude of the two other systems of rolls with the wavevectors $\mathbf{n}_1$ and $\mathbf{n}_2$. Then, from the system (4.70) one obtains:

$$\frac{\partial F}{\partial t} = \gamma F + \frac{\partial^2 F}{\partial x^2} + (G^*)^2 + i(K\sqrt{3} - K_0)\frac{\partial G^*}{\partial x}G^*$$
$$- |F|^2 F - 2B|G|^2 F, \tag{4.82}$$

$$\frac{\partial G}{\partial t} = \gamma G + \frac{1}{4}\frac{\partial^2 G}{\partial x^2} + F^*G^* - i(\frac{1}{2}K_0 + \frac{\sqrt{3}}{2}K)\frac{\partial G^*}{\partial x}F^*$$
$$+ iK_0\frac{\partial F^*}{\partial x}G^* - (1 + B)|G|^2 G - B|F|^2 G \tag{4.83}$$

The new system (4.82), (4.83) describes the modulation of a hexagonal pattern caused by the wall due to the non-variational terms (nonvanishing coefficients $K_0$ and $K$). The natural boundary conditions for $F$ and $G$ are $F = G = 0$ for $x = 0$, and they are assumed to be bounded at infinity.

Before describing numerical results, let us sketch how analytical results about wavelength selection can be obtained for $K \ll 1$ and $K_0 \ll 1$. Take $F(x,t) = P(x,t)\exp^{iqx}$, $G(x,t) = Q(x,t)\exp^{-iqx/2}$, introduce a small parameter $\epsilon \ll 1$, such that $K = O(\epsilon)$, $K_0 = O(\epsilon)$, $q = O(\epsilon)$, expand $P = P_0 + \epsilon P_1 + o(\epsilon)$, $Q = Q_0 + \epsilon Q_1 + o(\epsilon)$, and substitute the expansions in the system (4.82), (4.83). In the leading order one obtains the following equations for stationary modulations of the hexagonal pattern amplitudes :

$$\gamma P_0 + \frac{\partial^2 P_0}{\partial x^2} + Q_0^2 - P_0^3 - 2BQ_0^2 P_0 = 0, \tag{4.84}$$

$$\gamma Q_0 + \frac{1}{4}\frac{\partial^2 Q_0}{\partial x^2} + P_0 Q_0 - (1 + B)Q_0^3 - BP_0^2 Q_0 = 0. \tag{4.85}$$

The solutions $P_0(x)$ and $Q_0(x)$ of the system (4.84), (4.85) correspond to the modulated amplitudes of the hexagonal pattern when

$K = K_1 = 0$, and can be found numerically. For $x \to \infty$

$$P_0 = Q_0 = A_h \equiv \frac{1 + \sqrt{1 + 4\gamma/(1 + 2B)}}{2(1 + 2B)}, \qquad (4.86)$$

which corresponds to the amplitude of perfect hexagons with the wavenumber equal to that at the onset of the pattern forming instability.

In the order $O(\epsilon)$ one obtains two linear inhomogeneous equations for $P_1$ and $Q_1$, which are not needed here. The solvability condition of this system yields an equation for the selected wavenumber (or more precisely, the deviation of the wavenumber from its critical value), $q$:

$$q = \frac{8}{9} K_0 A_h - \frac{4(K_0 + K\sqrt{3})}{3 A_h^2} \int_0^\infty P_0 Q_0 \frac{\partial Q_0}{\partial x} \, dx. \qquad (4.87)$$

One can see that for $K_0 \ll 1$, $K \ll 1$ the selected wavenumber $q$ is also small and depends linearly on the (nonvanishing) values of the coefficients $K_0$ and $K$.

In order to find the selected wavenumber $q$ for arbitrary values of $K$ and $K_0$, the system (4.82), (4.83) was solved numerically by means of a finite difference method with central differences, using a semi-implicit Crank-Nicholson/Adams-Bachforth scheme, in an interval $x \in [0, L]$, with the boundary conditions $F(0) = G(0) = F(L) = G(L) = 0$ and with $L$ much larger than the length of the boundary layer near the ends of the interval. Fig. 4.11 shows the solution of the system (4.82), (4.83) for $K = K_0 = 0$.

It appears that $F(x) = |F(x)|e^{i\delta}$, $G(x) = |G(x)|e^{-i\delta/2}$, and far from the walls $|F| = |G| = A_h$, so that the selected wavenumber is zero, hence a constant phase shift $\delta$ depending on initial conditions. Thus, in the variational case the expected pattern is a perfect hexagonal pattern with the wavenumber corresponding to the onset of instability.

Fig. 4.12 shows the solution of the system (4.82), (4.83) for $K = 0.429$, $K_0 = -0.139$, $B = 1.179$, $\gamma = 5.0$. The chosen numerical values of the parameters have been computed using the theory developed by Golovin et al. (1997), to match an experiment with a silicone oil – air system (Eckert et al., 1998). Then the spatial distributions of the complex amplitudes $F$ and $G$ are harmonic functions (except in the thin boundary layers near the walls), namely,

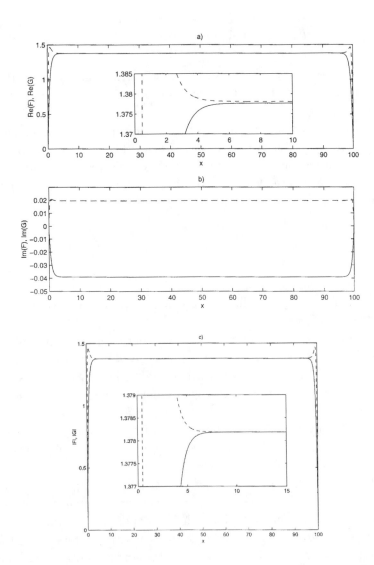

Figure 4.11: Benard-Marangoni convection. Spatial distributions of the complex amplitudes $F$ and $G$ of the hexagonal pattern between two distant side walls in the variational case ($K = K_0 = 0$) for $\gamma = 5$, $B = 1.179$; (a) real parts of $F$ and $G$; (b) imaginary parts of $F$ and $G$; (c) absolute values $|F|$ and $|G|$.

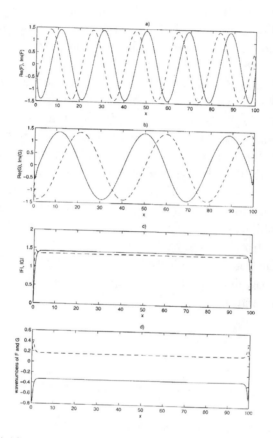

Figure 4.12: Benard-Marangoni convection. Spatial distributions of the complex amplitudes $F$ and $G$ of the hexagonal pattern between two distant side walls in the non variational case ($K = 0.429$, $K_0 = -0.139$, $\gamma = 5$ and $B = 1.179$); (a) real (solid line) and imaginary (dashed line) parts of $F$; (b) real (solid line) and imaginary (dashed line) parts of $G$; (c) absolute values $|F|$ (solid line) and $|G|$ (dashed line); (d) wave numbers $q_F$ (solid line) and $q_G$ (dashed line).

$F = F_0 \exp(iq_F x)$ and $G = G_0 \exp(iq_G x)$. Outside the thin boundary layers, $q_G = -q_F/2$. This corresponds to the selection of the hexagonal pattern with wavenumber $k_c + q$, hence shifted by $q \equiv q_F$ from the wavenumber $k_c$. For the particular case shown in Fig. 4.12, the selected wavenumber $q = -0.3276$. One can see that far from the walls the amplitudes $|F|$ and $|G|$ are constant, but *different*. Thus, the non variational terms lead to the formation of a *skewed* hexagonal pattern, even far from the side walls. Note that the amplitudes $|F|$ and $|G|$ in this case are determined by the wavenumber $q$ and the supercriticality $\gamma$ and can be found by substituting $F = F_0 \exp(iqx)$, $G = G_0 \exp(-iqx/2)$ into the system (4.82), (4.83). The solutions of the algebraic equations obtained for $F_0$ and $G_0$ agree with results of direct numerical simulations.

The solution of the system (4.82), (4.83) was also obtained for $K = -0.0173$ and $K_0 = 0.02$. Real and imaginary parts of $F$ and $G$ are also found to be harmonic functions outside the thin boundary layers. In this case the selected wavenumber $q = 0.269$, is in good agreement with Eq. (4.87) which gives $q = 0.265$.

The behavior of the selected wavenumber $q$ with the increase of the supercriticality $\gamma$ is shown in Fig. 4.13 for the above mentioned values of the parameters.

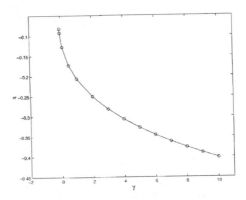

Figure 4.13: Benard-Marangoni convection. Dependence on the supercriticality $\gamma$, of the selected wavenumber $q$ of a hexagonal pattern between two distant side walls in the non variational case ($K = 0.429$, $K_0 = -0.139$ and $B = 1.179$).

One can see that $q$ is negative and grows in absolute value with the increase of $\gamma$, which means that the wavelength is increasing. This is in a qualitative agreement with the experimental results of

Eckert *et al.* (1998), high enough in the supercritical region. However, near enough threshold for small supercriticalities, an increase of the wavenumber was observed by Eckert *et al.* (1998), in agreement with previous experiments (Cerisier *et al.*, 1987; Koschmieder, 1991; Koschmieder and Switzer, 1992) and in both cases with findings already reported by Benard (1900, 1901) and Dauzere (1912). Accordingly, non variational terms in Eq. (4.70) are, probably, not the only factor responsible for wavenumber selection.

## ii. Propagation of hexagonal patterns

In the previous sections we considered stationary patterns that fill uniformly a certain region. When Marangoni number is increased, the hexagonal pattern forms from some random fluctuations nonuniform in space, and it coexists with the equilibrium state. Thus *fronts* between the convective pattern and the equilibrium state must appear (Pomeau, 1986).

Near the onset of the pattern formation, the analysis of the pattern selection and propagation of fronts may be performed by means of the amplitude equations introduced in the previous subsection. We will not attempt to describe such calculations in detail here, but only summarize some significant results (Pismen and Nepomnyashchy, 1994; Hari and Nepomnyashchy, 2000).

In the variational case there exists a Lyapunov functional decreasing with time, thus prescribing the direction of front motion. If both the hexagonal pattern and the equilibrium state are locally stable (i.e. in the subcritical region of the hexagonal pattern), the velocity of the front is unique and depends on the difference between Lyapunov functional densities of both patterns and the orientation of the front with respect to the basic wavevectors of the pattern (Pomeau, 1986; Malomed *et al.*, 1990; Hari and Nepomnyashchy, 1994) Fig. 4.14 (left) illustrates the relationship between $\gamma$ and $C$. It vanishes at the point where the Lyapunov densities are equal for the hexagonal pattern and the equilibrium state, which occurs at

$$\gamma = \gamma_0 = -\frac{2}{9(1 + 2B)} \qquad (4.88)$$

In the vicinity of this point, $c \sim (\gamma - \gamma_0)$ but this linear relationship does not persist for a wide range of $\gamma$ as seen in Fig. 4.14.

When the Marangoni number increases, the equilibrium state becomes unstable (at $\gamma > 0$), and instead of a front between two locally

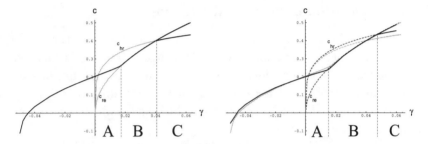

Figure 4.14: Benard-Marangoni convection. The velocity of the
hexagon-(mechanical) equilibrium front $c$ vs $\gamma$. Left (variational
case): the light lines correspond to the velocities $c_{hr}$ and $c_{re}$ given
by the marginal stability criterion. The dark lines correspond to
velocities calculated numerically. For large $\gamma$, the dark line splits
into two; these are the velocities of the two separate fronts (the
upper is $c_{re}$, and the lower is $c_{hr}$). These measured velocities
correspond with calculated ones ($B = 2$). Right (non-variational
case): the dashed lines correspond to the theory: $c_{hr}$ and $c_{re}$. The
velocity $c_{hr}$ was calculated in the potential case (light dashed line)
and in the non-potential case (dark dashed line). The velocity
$c_{re}$ is the same in both cases. The light continuous line gives the
velocity measured for the variational case (left plot), and the black
continuous line presents the velocity in the non variational case.
Regions A, B, and C are described in the main text.

stable structures, a front between a stable state and an unstable state appears. This case should be considered separately, also taking into account that at $\gamma > 0$, the roll pattern is also a possible solution. The propagation of a stable hexagonal pattern out of an unstable quiescent state and unstable roll patterns was investigated by Pismen and Nepomnyashchy (1994), for $K = K_0 = 0$ in Eqs (4.70). Both "generic" and "non-generic" fronts were found (van Saarloos, 1989; Powell et al., 1991).

Let us describe now how one can determine the velocity of the fronts analytically (Dee and van Saarloos, 1988). We focus on the case where the hexagonal pattern is stable, while both the (mechanical) equilibrium and a roll pattern are unstable. Hence, one can expect in general the existence of hexagon-equilibrium fronts (h-e), roll-equilibrium fronts (r-e), and hexagon-roll fronts (h-r). Kolmogorov et al. (1937) have proved (KPP criterion) that, even though the front velocity might not be unique when the front propagates from a stable state (hexagons in our case) into an unstable (quiescent) state, the system does prefer one velocity among of all available possibilities (Aronson and Weinberger, 1978). A linear stability principle yields (Dee and van Saarloos, 1988)

$$c - \left.\frac{d\omega(k)}{dk}\right|_{k=k_*} = 0, \qquad (4.89)$$

where $k_*$ satisfies

$$Im[ck_* - \omega(k_*)] = 0, \qquad (4.90)$$

and $\omega(k)$ is the dispersion relation for the disturbances of the linearly unstable state. In the case of a front propagating into the equilibrium (quiescent) state,

$$\omega = i(\gamma - k^2), \qquad (4.91)$$

and hence

$$k_* = ic/2, \quad c = c_{re} \equiv 2\sqrt{\gamma}. \qquad (4.92)$$

This velocity was found by Kolmogorov et al. (1937) for the case of a one-dimensional system with two stationary points, one stable and the other unstable.

Using the same method the velocity of propagation of the front into an unstable roll pattern can be calculated (Pismen and Nepomnyashchy, 1994). One can use the equations (4.82), (4.83) with $K = K_0 = 0$, assuming that $F = A_0$ and $G = A_1 = A_2$ are real, and linearizing them around the roll solution $F = \sqrt{\gamma}$, $G = 0$. Then, one obtains

$$\frac{\partial \tilde{G}}{\partial t} = \gamma \tilde{G} + \frac{1}{4}\frac{\partial^2 \tilde{G}}{\partial x^2} + \sqrt{\gamma}\tilde{G} - B\gamma\tilde{G}. \qquad (4.93)$$

Using the dispersion relation

$$\omega = i(\gamma - \frac{k^2}{4} + \sqrt{\gamma} - B\gamma), \qquad (4.94)$$

from (4.89), one finds $c = -ik_*/2$, and from (4.90),

$$c = c_{hr} \equiv \sqrt{\sqrt{\gamma} - (B-1)\gamma}. \qquad (4.95)$$

This velocity exceeds $c_{re}$ in the interval $0 < \gamma < (B+3)^{-2}$ (Fig. 4.14).

In some cases, however, the "generic" front with the velocity calculated above is replaced by the "non-generic" front (van Saarloos, 1989; Powell *et al.*, 1991). To examine this issue, Hari and Nepomnyashchy (2000) have numerically solved the system (4.82)-(4.83) with boundary conditions $F = G = A_h$ at $x = 0$ (hexagons), $F = G = 0$ at $x = L = 600$ (equilibrium), for different values of $\gamma$ in the region of $0 < \gamma < (B-1)^{-2} \equiv \gamma_2$ where the hexagons are stable and the mechanical equilibrium state and rolls are unstable, and have measured the steady state velocity. They used two kind of initial conditions: 1) $F = G = A_h$, $0 \leq x < L/2$; $F = G = 0$, $L/2 < x \leq L$; 2) $F = G = A_h$, $0 \leq x < (L - L_i)/2$; $F = \sqrt{\gamma}$, $G = 0$, $(L - L_i)/2 < x < (L + L_i)/2$; $F = G = 0$, $(L + L_i)/2 < x \leq$. Different behavior of the fronts was observed in the following three main regions:

*ii.1 Region A*

$\gamma > 0$ but almost vanishing, where the front is described by the non-generic solution. According to (4.92), the velocity $c_{re} = 0$ as $\gamma = 0$. However, in reality $c(\gamma)$ is continuous at the point $\gamma = 0$. Then, using $F(x, t) = \bar{F}(\xi)$ and $G(x, t) = \bar{G}(\xi)$ with $\xi = x - ct$, the system (4.82)-(4.83) allows to determine the asymptotic behavior

$$\bar{F} \sim e^{\kappa_F \xi}, \quad \bar{G} \sim e^{\kappa_G \xi} \text{ as } x \to \infty, \qquad (4.96)$$

with

$$\kappa_F^{\pm} = -\frac{c}{2} \pm \frac{1}{2}\sqrt{c^2 - 4\gamma}$$

$$\kappa_G^{\pm} = -2c \pm 2\sqrt{c^2 - \gamma}. \qquad (4.97)$$

For $\gamma < 0$, we have $\kappa_F^- < 0$ and $\kappa_F^+ > 0$, and hence the asymptotic behavior of the front tail for $F$ is prescribed by $\kappa_F = \kappa_F^-$. When $\gamma$ changes sign from negative to positive, both $\kappa_F$ become negative. Because, in general, $|\kappa_F^+| < |\kappa_F^-|$ the solution tending to zero behaves like $\sim \exp(\kappa_F^+ x)$ as $x \to \infty$, but there exist also a special trajectory approaching $\exp(\kappa_F^- x)$. In their computations, Hari and Nepomnyashchy (2000) obtained $\kappa_F(\gamma)$ in the tail of the function $F$ at large values of $x$, and found that the measured $\kappa_F$ agrees with the *non-generic* value $\kappa_F^-$ calculated using Eq. (4.97), with $c$ taken from the numerics. The results are shown in Fig. 4.15. Thus we see that the system prefers the non-generic solution, with $\kappa_F = \kappa_F^-$.

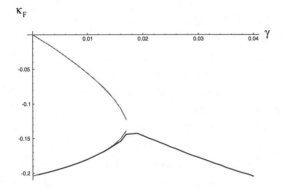

Figure 4.15: Generic and non-generic curves, $\kappa_F$, vs $\gamma$, calculated for the value of $c$ obtained numerically. The light lines are generic $\kappa_F^+$ (upper) and non-generic $\kappa_F^-$ (lower). The dark line is $\kappa_F$ obtained using the numerical results.

*ii.2 Region B*

$\gamma > 0$ and far from zero, where the KPP criterion may be used. The front moves with velocity $c_{re} = 2\sqrt{\gamma} < c_{hr}$. In this region an intermediate layer of unstable rolls is observed, between the synchronized roll-equilibrium and hexagon-roll fronts. The axes of the rolls are parallel to the front. In the steady state, the roll layer has a constant finite width, thus forming two fronts, both moving with the same velocity.

*ii.3 Region C*

$\gamma > (3 + \lambda)^{-2}$ which is the point where $c_e$ becomes higher than $c_r$: there, the h-r front starts lagging behind the r-e front, and they separate, each of them moving with its own velocity. Hence, in this region one observes front splitting, leading to an intermediate rolls layer which grows in time.

The velocities in each region are given in Figure 4.14. Further details and plots of the amplitude functions for the various solutions exist (Hari and Nepomnyashchy, 2000). These authors have also studied the influence of the non variational terms $(K, K_0 \neq 0)$, that lead to violating the analogy between non-equilibrium pattern formation and equilibrium phase transitions. Because the non variational terms contain spatial derivatives of amplitudes, they do not appear in the case of perfect patterns with all wavenumbers equal to $k_c$), and thus a Lyapunov functional can be constructed and calculated for different kinds of *perfect* patterns. However, in contradistinction to the variational case (Pomeau, 1986), the comparison of densities of the Lyapunov functional for different uniform patterns generally provides no information about the directions of front motion, because the nonuniformity of a system with a front, has no functional that tends to decrease. Numerical simulations and analytical results of Hari and Nepomnyashchy (2000) have shown that non variational terms induce a change of the normal velocity of the front (Fig. 4.14, (right), which however remains small for low enough values of $K$ and $K_0$. The marginal stability criterion may also be used in the non variational case, in regions B and C : while the hexagon/roll-equilibrium front velocity is unaffected, because non variational contributions disappear in the linearization process, the hexagon-roll front velocity does depend on $K$ and $K_0$, as also seen in Fig. 4.14 (right).

Another peculiarity of pattern formation processes governed by the non variational equations (4.70) is a nontrivial *wavenumber selection* in the presence of a front, in a way like that described in the previous section for lateral walls. In the variational case the critical wavenumber $k_c$ providing the minimal value of the Lyapunov functional is usually selected (Malomed *et al.*, 1990) while for the non variational case this is not so (Hari and Nepomnyashchy, 2000).

### iii. Pentagon-heptagon defects in hexagonal patterns

In addition to linear defects such as fronts, hexagonal patterns generally contain point defects, most often in the form of penta-hepta pairs. These are not only observed experimentally (Koschmieder, 1993; Cerisier *et al.*, 1987; Afenchenko *et al.*, 2001), but also in direct numerical integrations, either of the original system of equation (Bestehorn, 1993), or from the Knobloch equation (Fig. 4.4). These defects can also be studied using the amplitude equations (4.70), and correspond to a bound state of two dislocations on two different roll subsystems (or sublattices), such as seen in Fig. 4.16.

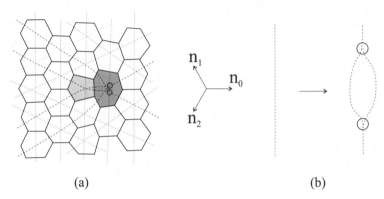

(a)                                                          (b)

Figure 4.16: Benard-Marangoni convection. (a): Sketch of a penta-hepta defect in an hexagonal pattern. The lighter and darker gray cells have 5 and 7 sides, respectively, while all other cells are (regular or irregular) hexagons. Dashed lines indicate basic roll subsystems (roll axes are orthogonal to directions $n_0$, $n_1$, and $n_2$). The thicker dashed lines correspond to rolls containing dislocations, where a single roll is forced to split into two rolls. The dislocations are encircled, and seen to occur in roll subsystems corresponding to orientations $n_1$ and $n_2$. (b): Mechanism of generation of a new pair of dislocations in one of the roll subsystems (corresponding to the orientation $n_0$).

The basic properties of dislocations in roll patterns are well known (Siggia and Zippelius, 1981; Pomeau et al., 1983; Cross and Hohenberg, 1993), while the behavior of penta-hepta pairs in convective hexagonal structures has been studied only recently (Pismen and Nepomnyashchy, 1993; Rabinovich and Tsimring, 1994), mostly for variational systems ($K = K_0 = 0$ in Eqs (4.70)). Note that a single dislocation, say in the roll subsystem $n_1$, corresponds to a point where both real and imaginary parts of the corresponding amplitude function, $A_1$, vanish. Accordingly, there must be a phase jump of $2\pi$ along a contour encircling the dislocation. If the phase of each amplitude function is denoted by $\varphi_j$, $j = 0, 1, 2$, and if $C$ denotes an arbitrary contour encircling the defect, one defines quantities like

$$Q_j = \frac{1}{2\pi} \oint_C \vec{\nabla} \varphi_j . \vec{dl}, \quad j = 0, 1, 2 \tag{4.98}$$

which are called the topological charges. In the case of Fig. 4.16(a), one sees that if $C$ is a contour encircling the PHD, $Q_j$ is zero for $j = 0$ (the phase $\varphi_0$ does not contain any dislocation and hence is analytic), while it is equal to $+1$ or $-1$ for $j = 1$ or $j = 2$. Thus, the PHD is characterized by these three topological charges, and can be denoted by $(0, +1, -1)$ in the case of Fig. 4.16(a). It is also clear that from the topological point of view, nothing precludes the possibility of two dislocations appearing with opposite topological charges in the same sublattice (Fig. 4.16b), an event which does not affect the total topological charge.

By suitably selecting the initial conditions, it is possible to create e.g. a $(0, 0, +1)$ defect (one dislocation in $A_2$). This type of defect is not motionless in the case of hexagonal patterns, as shown by Rabinovich and Tsimring (1994). However, the synchronization of the pattern ($\varphi = \varphi_0 + \varphi_1 + \varphi_2 = 0$) induced by quadratic terms in the amplitude equations (4.70) is not possible everywhere, and a thin region where $\varphi = \pi$ survives, originating at the core of the defect. For $K = K_0 = 0$, the system of equation is variational, and this energetically unfavorable situation (the region where $\varphi = \pi$ corresponds to down-hexagons) is eliminated by a relatively fast motion of the defect in the direction of this "corridor".

When two dislocations (in different sublattices) are initially located a certain distance apart from each other, the evolution depends on their relative topological charges. If their charges $Q_j$ have opposite signs, a region where $\varphi = \pi$ is created joining the two defects, therefore promoting their motion towards each other. When the two dislocations meet, the system cannot lower the total energy anymore (the phases are synchronized everywhere except at the defect), and a stable PHD is created for which the two defects are bound a small distance apart from each other. When the two dislocations have the same charge, they mutually repel each other, save their disappearance at the lateral walls. Finally, when the two dislocations are in the same sublattice, a similar evolution occurs. However, when they have opposite topological charge, their motion towards each other is followed by annihilation of the two dislocations, thereby leaving a perfect hexagonal planform.

This picture is complicated by the presence of non-variational effects introduced by quadratic derivative terms $(K, K_0 \neq 0)$. In particular, the PHDs are generally no longer motionless, but acquire a velocity parallel to the wavevector $\mathbf{n}_j$ of the dislocation-free sublattice (Colinet et al., 1997). This behavior is similar to the influence of the phase gradients studied by Tsimring (1996) for the potential case. Similarly to the behavior of dislocations predicted in roll structures (Siggia and Zippelius, 1981; Pomeau et al., 1983) for variational systems, Tsimring has shown that a PHD is motionless in an optimal hexagonal pattern (i.e., phase gradients vanish in the far-field of the defect, and the wavenumber of the pattern is $k_c$). If, however, such gradients exist (e.g. imposed externally), a constant velocity motion of the defect occurs, whose velocity and direction are determined by the phase gradients of the modes containing singularities, and by a certain mobility tensor. It can be shown that the quadratic non-variational terms can also produce a contribution to the force promoting defect motion, even in the absence of background phase gradients. Then the velocity of the defect depends on details of the core structure and cannot be calculated analytically.

Colinet et al. (1997) have studied the influence of the non variational terms with coefficient $K$ on the dynamics of PHDs in hexagonal patterns. They have shown that for a PHD in an hexagonal pattern with wavenumber $k_c$, increasing the control parameter $\gamma$ leads to a higher velocity of the defect. However, when $\gamma$ is increased past a critical value $\gamma_c$, the moving PHD undergoes a nonlinear instability and splits into two new PHDs with different orientations. For instance, starting with a PHD with charges $(0, +1, -1)$ moving along $n_0$, such as in Fig. 4.16(a), two new dislocations are created in the dislocation-free sublattice (such as in Fig. 4.16(b)), and each of the two new dislocations recombines with the dislocations already present in sublattices with orientation $n_1$ and $n_2$, therefore leading to PHDs $(+1, 0, -1)$ and $(-1, +1, 0)$. These PHDs then acquire motion in their respective directions, i.e. $60°$ away from the direction of motion of the initial PHD.

This individual event is then able to repeat, as the new PHDs, when sufficiently separated, actually undergo a similar evolution leading to four defects, which may then split again, and so on, leading eventually to a *multiplication of defects* and to a significant change of the whole hexagonal pattern. The evolution of the system in the long run possibly may be rather complicated, with the lateral walls

of the domain of integration always playing non negligable role in
the evolution of the system. Even if a large number of PHDs may be
generated, the final stages appear to be only weakly time-dependent,
because most of the defects ultimately move towards lateral walls
and disappear there. Recombination of defects by pairs may also be
occasionally observed.

In the course of defect multiplication and motion, strong longitudinal phase gradients are generally created, which correspond to a
shift of the wavenumber of the background pattern induced by the
process of multiplicating defects (wavenumber selection). On the
background of these phase gradients, it appears that the motion of
the PHDs is significantly slowed down, and PHDs eventually do not
multiply anymore. Colinet *et al.* (1997) have performed direct numerical computations for different initial wavenumbers (or winding
numbers $q$) and various values of $\gamma$. They have shown that PHD
multiplication can lead to an important reduction of the range of
stability of hexagons, when compared to the stability domain for
defect-free patterns (Sect. 4.4.1).

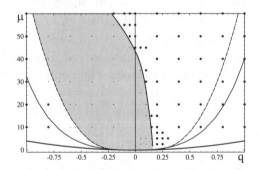

Figure 4.17: Benard-Marangoni convection. Stability region of
hexagonal patterns containing PHDs (light grey zone). The lower
thick solid line is the existence line of hexagons, while the thin
solid and dashed lines indicate vanishing of phase diffusion coefficients (Eckhaus instabilities). Bigger dots indicate values of the
winding number $q$ and supercriticality parameter $\mu \sim \gamma$ at which
a PHD has been observed to generate at least one new pair of
dislocations.

The phenomenon of PHD multiplication described in this section drastically depends on the non-variational terms and has not
been observed when $K = K_0 = 0$, where each defect provides a
positive contribution to the Lyapunov functional. A more complete

study of the role of the different coefficients in Eqs (4.70) has also
shown that multiplication of PHDs is not the only effect induced
by non variational terms. Indeed another interesting observation is
that non-variational terms can induce a non-zero distance between
the two dislocations of a PHD, in agreement with the experiment
(Afenchenko *et al.*, 2001). Furthermore, depending on the values of
$K$ and $K_0$, separation of dislocations has also been observed, quite
similar to what was observed for the hexagon-square transition (Eckert and Thess, 1999).

## 4.5   Strongly nonlinear patterns

Although Knobloch's equation (4.40) is a model of limited applicability when used at high values of the supercritical Marangoni number,
it, astonishingly, reproduces well basic features of highly nonlinear
patterns observed experimentally (Orell and Westwater, 1962; Linde
and Loeschcke, 1966; Schatz and Neitzel, 2001) or obtained from
direct numerical simulations of the original equations (4.15)–(4.21)
for Benard-Marangoni convection (Thess and Orszag, 1994, 1995).
Pontes *at al.* (1999) have performed direct numerical simulations
of Knobloch's equation at vanishing dissipation parameter $\alpha$, corresponding to $\epsilon = 1$ in Eq. (4.42), i.e. the maximal value allowed
(for $\epsilon > 1$, the homogeneous mode is amplified, which is unphysi-
cal). They have observed a growth of the pattern wavelength with
time, which tends to a single large convection cell filling the entire
simulation domain (Figure 4.18).

When, for otherwise equal conditions, the supercritical Marangoni
number, hence $\epsilon$ is progressively increased, the convective states
reached by the pattern after a sufficient time indeed correspond to
larger and larger wavelengths, as seen in Fig. 4.19. This "coars-
ening" mechanism is in qualitative agreement with experiments by
Schatz and Neitzel (2001).

Moreover, both in experiments (Schatz and Neitzel, 2001; Orell
and Westwater, 1962; Linde and Loeschcke, 1966) and in numerics
(Thess and Orszag, 1994, 1995), the boundaries of convection cells
are seen to be very sharp and straight at high supercritical Marangoni
number due to the strong nonlinear effect of surface advection of the
free surface temperature field. This produces "ripples", because the
surface flow, directed from hot to cold at the free surface, compresses

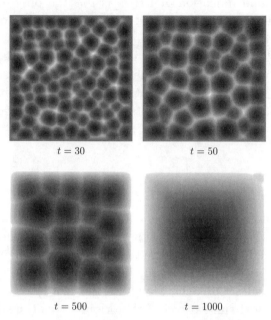

Figure 4.18: Evolution of the pattern predicted from Knobloch's equation (4.40) for $\alpha = 0$, $\mu = 2$, $\beta = -\frac{\sqrt{7}}{8}$, $\delta = 3\frac{\sqrt{7}}{4}$ and $\gamma = 0$ (after Pontes *et al.*, 1999).

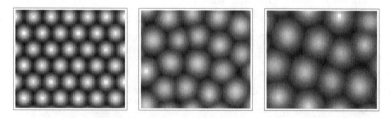

Figure 4.19: Quasi-steady patterns predicted after a time $t = 1000$ by direct numerical integration of Knobloch's equation (4.40) with $\mu = 2$, $\beta = -\sqrt{7}/8$, $\delta = -3\sqrt{7}/4$, $\gamma = 0$ and $\kappa = 1$. Left: $\epsilon^2 = 0.7$, center: $\epsilon^2 = 0.9$, right: $\epsilon^2 = 0.95$. The initial condition was random noise of small amplitude in each case.

the cold points, i.e. the periphery of cells, while the hot zones expand and finally fill most of the total surface area.

This behavior is well described by the infinite-Marangoni-number study made by Thess and Orszag (1994, 1995) with assumption that in this limit, the effect of thermal dissipation is negligible. Eq. (4.16) taken at the (flat) open surface, $z = 0$, becomes

$$\frac{\partial T_s}{\partial t} + (\mathbf{v_s} \cdot \nabla_s)T_s = 0, \qquad (4.99)$$

where $T_s = T(z = 0)$ is the free surface temperature field, $\nabla_s$ is the gradient along the free surface, and $\mathbf{v_s} = \mathbf{v}(z = 0)$ is the free surface velocity field, which in the limit of infinite Prandtl number liquids, can be obtained from the solution of the (zero Reynolds number) Stokes problem

$$\nabla^2\mathbf{v} - \nabla p = 0, \quad \nabla \cdot \mathbf{v} = 0, \qquad (4.100)$$

with boundary conditions at the open surface

$$v_z = 0, \quad \frac{d^2v_z}{dz^2} - M\nabla_s^2 T_s = 0 \qquad (4.101)$$

and corresponding boundary conditions at the bottom plate $z = -1$. Then $\mathbf{v_s}$ can be expressed as a function of $T_s$, hence reducing the three-dimensional problem to a single two-dimensional equation for the free surface temperature field $T_s$ (Thess and Orszag, 1994, 1995). The numerical solution of the latter equation showed that starting from random initial conditions, the surface temperature field develops shocks (cold ripples), resulting in blow-up after finite-time. Then, re-introducing *horizontal* dissipation in the form of a supplementary term $\nabla_s^2 T_s$ in the right-hand-side of Eq. (4.99) regularizes the thermal ripples, hence preventing blow-up.

In experiments conducted by Schatz, such highly supercritical patterns were also seen to acquire an intrinsic time-dependence, some (large) convection cells splitting, other (small) ones collapsing, eventually resulting in chaotic regimes. To account for these features goes, in principle, well beyond the range of validity of Knobloch's equation (4.40). However, one hopes to see significant nonlinear effects that could correspond to experiments with high Marangoni numbers. On the other hand the infinite-Marangoni-number model of Thess is not good either, unless one adds in Eq. (4.99) some external stirring force term (e.g. a white noise, as is often done in turbulence modeling). Another possibility, allowing self-sustained chaotic regimes to be generated, is to re-incorporate the effect of *vertical* heat diffusion in Eq. (4.99), i.e. the term $d^2T/dz^2$ evaluated at the free surface. This term is indeed essential as far as the energy input from the externally imposed gradient is concerned. Unfortunately, it appears that no rigorous method exists to express this contribution as a function of $T_s$ alone, i.e. to achieve a reduction of the full problem to two dimensions.

Presumably, the term $d^2T/dz^2(z = 0)$ plays an essential role even in the limit of infinite Marangoni number, because of the presence of an interfacial thermal boundary layer, where heat conduction cannot be neglected. Direct numerical simulations of the *steady* convective states of the original problem (4.15-4.21) either for very low Prandtl number liquids (Boeck and Thess, 1997) or at infinite Prandtl number have shown that the thickness of this thermal boundary layer decreases as $\delta \sim M^{-1/3}$ with the increase of the Marangoni number. Outside the boundary layer, i.e. in the liquid bulk, the temperature field is quite efficiently homogenized by the flow, while within it, the temperature gradient is unperturbed (if the Biot number is vanishingly small, the heat flux through the free surface is bound to remain constant). Then, in this steady regime, the maximal surface velocity, $V_s$, and the typical temperature difference along the free surface, $\Delta T_s$, are found to scale as $V_s \sim M^{2/3}$ and $\Delta T_s \sim M^{-1/3}$, respectively.

How about the stability of the thermal boundary layer? Considering that the externally imposed thermal gradient is not affected by thermocapillary convection within the thermal boundary layer of thickness $\delta \sim M^{-1/3}$, the linear stability analysis presented in Section 4.1 should remain applicable for normal modes with wavenumber $k \gg M^{1/3}$, i.e. provided the penetration depth of such modes does not extend outside the boundary layer. As mentioned above, the short-wave asymptotics ($k \to \infty$) of Pearson's theory (4.32) is $M_0 \to 8k^2$ (Scanlon and Segel, 1967), and corresponds to velocity and temperature perturbations

$$v_z \to -4k^3 z \exp(kz), \quad T \to (1 - kz + k^2 z^2) \exp(kz) \qquad (4.102)$$

indeed decaying exponentially below a depth $\sim k^{-1}$. Hence, modes with wavenumbers in the range $M^{1/3} \ll k \ll M^{1/2}$ can be expected to destabilize the steady power-law regime described above, and may lead to transitions to time-dependent or turbulent regimes. The threshold value of $M$ for these secondary instabilities has not been determined yet, due to the difficulties mentioned above, and to the fact that even direct numerical integration of the original problem (4.15)–(4.21) at such high Marangoni numbers is too demanding in computer resources, owing to the formation of very small scale structures in the temperature field (Thess and Orszag, 1995).

## 4.6 Extensions of Knobloch's equation and related model equations

### 4.6.1 Influence of the mean flow for low Prandtl number liquids

Three-dimensional cellular convection patterns can be obtained numerically directly from the original problem (4.15)–(4.21). However, as our approach in this chapter is to use simplified models capturing the essential features of patterns, flows, and instabilities, we shall now show how Knobloch's equation (4.40), valid for long-wave convection patterns at infinite Prandtl number, $P$, can be modified to account for the effect of vertical vorticity and mean flows at moderate or small $P$. Applying the curl operator to the horizontal component of Eq. (4.15), the vertical component of the result is

$$\Delta \omega_z = \frac{1}{P} \left[ \frac{\partial \omega_z}{\partial t} + \mathbf{e} \cdot \nabla \times [(\mathbf{v} \cdot \nabla) \mathbf{v}_\perp] \right], \qquad (4.103)$$

where $\omega_z = \mathbf{e} \cdot \nabla \times \mathbf{v}$ is the vertical component of the vorticity, $\mathbf{e}$ is the unit vector directed upwards, and $\mathbf{v}_\perp$ is the horizontal component of the velocity field. With generality it can be written as

$$\mathbf{v}_\perp = \nabla_\perp \varphi - \nabla \times (\Psi \mathbf{e}) + \mathbf{v}_{0\perp}(z, t) \qquad (4.104)$$

where $\nabla_\perp$ is the horizontal gradient, $\varphi$ is a potential function, $\Psi$ is the streamfunction of the horizontal flow field, and $\mathbf{v}_{0\perp}$ is the mean horizontal flow, for which an equation can be derived by averaging the horizontal component of Eq. (4.15) over horizontal coordinates. Assuming perturbations to remain bounded everywhere (e.g. periodic), we obtain

$$\frac{\partial^2 \mathbf{v}_{0\perp}(z, t)}{\partial z^2} = \frac{1}{P} \left[ \frac{\partial \mathbf{v}_{0\perp}(z, t)}{\partial t} + \frac{\partial}{\partial z} \left( \overline{v_z \mathbf{v}_\perp} \right) \right] \qquad (4.105)$$

where the bar over the product denotes its horizontally averaged value. Note that $\omega_z = \nabla^2_\perp \Psi$, and according to the incompressibility condition (4.17), we have

$$\partial v_z / \partial z + \nabla^2_\perp \varphi = 0 \qquad (4.106)$$

At the linear stage, Eqs (4.103) and (4.105) show that both the vertical vorticity and the mean flow are decoupled from the evolution of the potential part, $\varphi$, of the horizontal flow field (related to the vertical velocity by Eq. (4.106), and hence participates in the mechanism leading to instability). At finite values of $P$, mean flow and vertical vorticity can only be generated by nonlinear interactions, thus making the problem of pattern selection quite involved. However, for perfect patterns (e.g., defect-free rolls or hexagons in, say, infinite domains), vertical vorticity and mean flow do not affect amplitude equations, at the cubic order at which the latter are usually truncated. In contrast, mean flow and vertical vorticity can be generated in the vicinity of defects in the convective structure, a feature particularly important for low Prandtl number liquids.

In the case of large-scale convective modes (e.g., for convection between poorly conducting boundaries), Knobloch's equation (4.40) may be generalized to account for the effect of large-scale flows. Indeed, redefining appropriately the asymptotics, Eqs. (4.15)–(4.21) yields the equation (Knobloch, 1990; Golovin et al., 1995)

$$\frac{\partial \phi}{\partial t} = -\alpha\phi - 2\mu\Delta\phi - \Delta^2\phi + \nabla \cdot [(\nabla\phi)^2\nabla\phi] + \\ \beta\nabla \cdot [\Delta\phi\nabla\phi] + \delta\Delta\left[(\nabla\phi)^2\right] - \mathbf{e} \cdot (\nabla\phi \times \nabla\Psi) \qquad (4.107)$$

where, compared to Eq. (4.40), non-Boussinesq effects have been neglected ($\gamma = 0$), but an additional term accounts for the coupling between the order-parameter field, $\phi$, and the streamfunction, $\Psi$, of the horizontal flow field. The latter is itself slaved by $\phi$, according to

$$\Delta\Psi = p^{-1}\mathbf{e} \cdot (\nabla\Delta\phi \times \nabla\phi) \qquad (4.108)$$

where $p$ is a coefficient proportional to the Prandtl number $P$.

No systematic exploration of the system (4.107)–(4.108) has been carried out. Still, the effects of vertical vorticity have been studied using similar equations, rather based on the Swift-Hohenberg model (4.41) coupled to a second order-parameter equation similar to Eq. (4.108) (Manneville, 1983a,b; 1990; Bestehorn et al., 1993). Considering roll patterns, it has been shown that for low enough Prandtl number liquids the vertical vorticity generated by defects leads to much longer lasting transients than those observed at large $P$. In particular, turbulent bursts are generated by the nucleation of new dislocations in the roll structure, which might be at the origin of some permanent low-frequency regimes observed experimentally (Ahlers and Behringer, 1978; Libchaber and Maurer, 1978; Pocheau et al., 1985). Bestehorn et al. (1993) has shown using similar models that *spiral* and *target* patterns can be generated, as seen in Rayleigh–Benard experiments (Bodenschatz et al., 1991, 1992; Morris et al., 1993), and other direct numerical integration of the thermohydro-dynamic equations (Decker et al., 1994; Pesch, 1996). Experiments have also shown that a transition to *spiral-defect chaos* occurs above a certain value of the constraint depending on the system size (Hu et al., 1995).

To our knowledge, no experimental studies exist about Benard–Marangoni convection in low Prandtl number liquids in spatially extended systems. As a transition to squares rather than to rolls would generally be expected in this case, above a certain value of the supercritical Marangoni number, it would certainly be interesting to examine whether large scale flows play an important role in the dynamics of the transition, and whether there is an analog of spiral-defect chaos for such systems. Answers to these questions are coming from the work by Eckert et al. (1998), with direct numerical integrations of the original equations (4.15)–(4.21) for a wide range of Prandtl number values. Evidence appears about the role of the vertical vorticity as a "lubricant" during the transition, thus favoring the motion of dislocations against pinning effects, and generally leading to more perfect square patterns. This is the case for Prandtl numbers such as $P = 50$.

For low Prandtl number liquids (e.g., at $P = 0.5$), the transition to squares is found to occur at lower Marangoni numbers, and, eventually, squares may coexist with rolls. Then large scale flows are responsible for an intrinsic time-dependence in the form of oscillations between both types of cells.

### 4.6.2 Influence of surface deformation

When surface deformability cannot be neglected, i.e. the Galileo number, $G$, and capillary number, $C$, are not sufficiently large, one has to consider the more general system (1.28-1.34). In order to avoid inconsistencies related to the use of the Boussinesq approximation, as discussed in Chapter 1, we shall here neglect buoyancy, $R = 0$.

First of all, one has to consider the modification of Pearson's result (4.32) for finite $G$ and $C$. The linearized version of the system (1.28-1.34) yields the critical Marangoni number for cellular convection as (Scriven and Sternling, 1964; Smith, 1966)

$$M_0(k) = \tag{4.109}$$

$$\frac{16k(k\cosh[k] + Bi\sinh[k])(2k - \sinh[2k])}{4k^3\cosh[k] + 3\sinh[k] - \sinh[3k] - 32k^5\cosh[k](G + k^2C)^{-1}}$$

which indeed reduces to Pearson's relation (4.32) in the limit $G, C \to \infty$. For finite $G$ and $C$, a significant modification of the neutral stability result occurs in the region of long waves ($k \to 0$), as seen in Fig. 4.20.

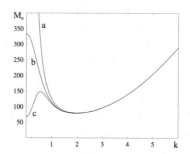

Figure 4.20:  Benard-Marangoni convection.  Neutral stability curves for a liquid lying on a rigid heat-conducting plate and having its deformable surface open to ambient air ($Bi = 0$). a: Pearson's result ($G, C \to \infty$); b: $G = 500$, $C = 10^3$; c : $G = 100$, $C = 10^3$.

In addition to the minimum of the neutral stability curve at $k \simeq 2$, the critical Marangoni number tends to a constant value $M_0 = \frac{2}{3}(1 + Bi)G$ for $k \to 0$. This value, $M_0$, may become lower than 80 if the Galileo number is small enough i.e. very thin, highly viscous liquid layers). Accordingly, the system will become unstable first to large-scale modes. This is another example of Goldstone mode. Here, the order-parameter whose uniform shift is neutrally stable is a change of the layer depth, hence leading to a supplementary mode of vanishing growth rate for $k \to 0$.

Let us now sketch the evolution of such large-scale deformational modes in the nonlinear regime. Several approaches exist, depending on the relative importance of various effects: thermocapillarity, gravity, interface curvature, etc. In what follows we shall use a smallness parameter $\varepsilon$, linked to the horizontal scale of convection cells. The wavenumber $k$, and the horizontal gradient $\nabla$, are assumed to be of order $\varepsilon \ll 1$. We must also distinguish between two cases, according to the thermal conductivity of the bottom plate.

### i. Heat-conducting bottom plate

In practice, $G$ and $C$ are often very large, and we shall follow Davis (1987) and VanHook *et al.* (1997), who use the scalings $k \sim \varepsilon \ll 1$, $G \sim \varepsilon^{-1}$, $C \sim \varepsilon^{-3}$, $M \sim \varepsilon^{-1}$, which allow a balance between hydrostatic pressure, Laplace overpressure and the Marangoni stress at the linear stage. Indeed, using these scalings, Eq. (4.109) yields, at the leading order,

$$M_0 \simeq \frac{2}{3}(G + k^2 C) \qquad (4.110)$$

Referring to VanHook *et al.* (1997) for details about the multiscale analysis used to derive the nonlinear evolution equation for the deformable surface displacement (here denoted by $h(\mathbf{x}_\perp, t)$), we focus attention on the features found. At the lowest order, one gets a pressure perturbation induced by gravity and surface curvature effects

$$p = G(h - 1) - C\Delta h \qquad (4.111)$$

which are independent of the vertical coordinate $z$. Hence, a horizontal (potential) flow field is generated, which is also influenced by thermocapillarity, according to

$$\mathbf{v}_\perp = M\, z\nabla h + \left(\frac{z^2}{2} - zh\right)(G\nabla h - C\nabla \Delta h) \qquad (4.112)$$

The first term of Eq. (4.112) accounts for the thermocapillary flow due to the change in surface temperature induced by the surface deformation (e.g. at local depressions, the interface is closer to the heated bottom, and therefore hotter), while the second term is the flow generated by the pressure perturbation (4.111).

Then, as done in lubrication theory (Chapter 3; Starov *et al.*, 1994, 1997; Oron *et al.*, 1997), the variation of the local film thickness is equal to minus the divergence of the total horizontal mass flux, that is

$$\frac{\partial h}{\partial t} = -\nabla \cdot \int_0^h dz \, \mathbf{v}_\perp \qquad (4.113)$$

Combining Eqs (4.112) and (4.113), one gets the evolution equation for the surface height $h(\mathbf{x}_\perp, t)$ as (Davis, 1987)

$$\frac{\partial h}{\partial t} = -\nabla \cdot \left\{ \frac{M}{2} h^2 \nabla h - \frac{G}{3} h^3 \nabla h + \frac{C}{3} h^3 \nabla \Delta h \right\} \qquad (4.114)$$

VanHook et al. (1997) have considered a more general situation, including conduction in the gas phase and hence two-layer model such that Eq. (4.114) argumented with new terms and parameters,

$$\frac{\partial h}{\partial t} = -\nabla \cdot \left\{ \frac{M}{2} \frac{1+F}{(1+F-Fh)^2} h^2 \nabla h - \right. \qquad (4.115)$$

$$\left. \frac{G}{3} h^3 \nabla h + \frac{C}{3} h^3 \nabla \Delta h \right\}$$

where $F = (1 - \lambda)/(a + \lambda)$ is, generally, positive ($a$ is the ratio of gas to liquid depths in the heat conducting, (mechanical) equilibrium state, and $\lambda$ is the ratio of gas to liquid thermal conductivities).

Equation (4.115) incorporates the effect of deformation on the surface temperature profile. Indeed, while both equations (4.114) and (4.115) predict the formation of local depressions (dry spots) seen in experiments, only the two-layer model (4.115) accounts for bumps, hence local elevations (high spots), which occur typically for $F$ higher than about 0.5. These features have been experimentally observed for increasing liquid depths (hence decreasing $a$ and increasing $F$). As it corresponds to $F = 0$, the one-layer model does not predict high spots. VanHook et al. (1997) have also shown that the bifurcation described by Eq. (4.115) is always subcritical, with an unstable solution branch which never turns over into a stable branch. Numerical integration has also been permitted to establish that no stable deformed states exist, but that the solution always blows up in finite time, which is coherent with the experimental observation of dry and high spots.

Golovin et al. (1994, 1997) and Kazhdan et al. (1995) have considered the nonlinear interaction of large-scale deformational modes with the cellular convective modes at $k \simeq 2$, when their thresholds nearly coincide. Depending on which mode is first unstable upon increasing the Marangoni number, they obtain a wide variety of dynamical regimes: steady surface reliefs, traveling and standing waves, blow-up in finite time, or (space-time) chaotic behavior.

## ii. Very poorly-conducting bottom plate

When both boundaries of the Benard set-up are very poor heat conductors (say, the bottom plate and free surface are characterized by very low Biot numbers $Bi_0$ and $Bi_1$, respectively), one expects the cellular convection mode to be of large-scale. Let us comment now on approaches to describe its interaction with the surface deformational mode described above.

### ii.1 Weak surface deformation

First, let us consider $C, G = O(1)$, a peculiar situation corresponding to a very thin highly viscous liquid layer, very low surface tension, or, on the other hand, low effective-gravity. It can be shown from the linear stability analysis at $k \sim \varepsilon \ll 1$, $(Bi_0, Bi_1) = (a, b)\varepsilon^4$ that the critical Marangoni number (up to a correction of order $\varepsilon^2$) is given by

$$M_c = \frac{48\, G}{72 + G} \qquad (4.116)$$

which, on the one hand, leads to $M_c = 48$ for $G \gg 1$ [see Eq. (4.35) for $Bi \to 0$] and $M_c = 2G/3$ for $G \ll 1$ [see Eq. (4.110) for $k \to 0$]. Considering the surface displacement to be given by $h(\mathbf{x}_\perp, t) = 1 + \xi(\mathbf{x}_\perp, t)$ with $\xi \sim \varepsilon^2$, it can be shown that at the leading order, $T = \varphi - G\xi/72$ couples temperature and surface deformation fields, with $\varphi$ being a *spatially uniform* integration constant. Expressing the total volume conservation $< \xi >= 0$ (where $< . >$ denotes the horizontal average over the thin liquid layer), it follows that $\varphi =< T >$. Note that both $T$ and $\xi$ (and $\varphi$) evolve on the slow time scale $\tau = \varepsilon^4 t$. If one consider that $M - M_c \sim \varepsilon^2$, then using an appropriate multiscale analysis, the result obtained at order $\varepsilon^6$ is a nonlinear evolution equation for the coupled surface deformation and temperature fields, namely

$$\frac{\partial \xi}{\partial t} = -\alpha\xi - \mu\Delta\xi - \nu\Delta^2\xi - \gamma\nabla\cdot[\xi\nabla\xi] + \gamma'\left\{(\nabla\xi)^2 - \left\langle(\nabla\xi)^2\right\rangle\right\} \quad (4.117)$$

which is a two-dimensional generalization of the Kuramoto-Velarde (KV) equation (Garcia-Ybarra, Castillo and Velarde, 1987a,b; Castillo, Garcia-Ybarra and Velarde, 1988; Funada, 1987). Eq. (4.117) corresponds to a low-order truncation of Knobloch's equation (4.40), save the last term. The coefficients, not needed here, are identical to those found by Garcia-Ybarra *et al.* (1987b). The mean value of the temperature perturbation, i.e. $\varphi = <T>$, is found to evolve according to

$$\frac{\partial \varphi}{\partial t} = -(Bi_0 + Bi_1)\varphi - \frac{1}{2}\left\langle (\nabla \xi)^2 \right\rangle \qquad (4.118)$$

by expressing the volume conservation $<\xi> = 0$ (if $<\xi> = 0$ at $t = 0$, Eq. (4.117) shows that this remains the case for all $t > 0$, e.g. assuming periodic boundary conditions).

Although the nonlinear term $\nabla \cdot [\xi \nabla \xi]$ cannot saturate the linear instability (Funada and Kotani, 1986; Hyman and Nicolaenko, 1987; Rodriguez-Bernal, 1992; Christov and Velarde, 1992), the second nonlinear term $(\nabla \xi)^2$ not only prevents blow-up in finite time, but also generates quite complex (space-time) chaotic patterns (even in dimension one). Such features, as in the case of Knobloch's equation (4.40), go beyond the range of validity of Eq. (4.117). It is beyond our scope here to discuss chaotic regimes of Eq. (4.117). The reader is also referred to the related results found by Manneville (1990), Chate and Manneville (1987), Misbah and Valance (1994), Elder *et al.* (1997), and Paniconi and Elder (1997), for the damped Kuramoto–Sivashinsky equation, a particular case of Eq. (4.117) corresponding to $\gamma = 0$ ($G = 36$). Note that steady cellular structures and in particular hexagonal patterns may be found right above threshold, increasing the constraint further (or decreasing the dissipation $\alpha$) leads to transitions to chaotic regimes, both for one-dimensional and two-dimensional situations. Note also that the one-dimensional version of Eq. (4.117) with $\alpha, \gamma = 0$ is the equation describing the phase turbulence of traveling waves (Manneville, 1990; Mori and Kuramoto, 1998).

*ii.2 Very weak surface deformation*

Let us consider now a different asymptotics, tailored to the case $G$ and $C$ very high, specifically $G \sim \varepsilon^{-2}$ and $C \sim \varepsilon^{-4}$ (Golovin *et al.*, 1995). These authors have shown that for a two-layer system Benard-Marangoni convection is indeed long-wave only when the gas phase is shallow enough. For large gas layer depths, there is a short-wave mode of instability.

Golovin *et al.*, (1995) have also shown that surface deformation is smaller ($\xi \sim \varepsilon^2 T$), than in the case of the previous subsection, and does not affect the linear stability threshold $M_c = 48$ (hence the same, as for one-layer convection). Actually, the evolution of the surface deformation is slaved to the dynamics of the temperature field, here denoted by $\phi$, according to the non-local equation

$$\Delta(g\xi - c\Delta\xi) = -\Delta\phi \qquad (4.119)$$

where $g \sim G$, and $c \sim C$.

The analysis of Golovin *et al.* (1995) is not restricted to infinite Prandtl number liquids, such that vertical vorticity effects (Sect. 4.6.1) are included. The streamfunction, $\Psi$, of the non-potential part of the horizontal velocity is also slaved by the temperature, according to

$$\Delta\Psi = p^{-1}\mathbf{e} \cdot (\nabla\Delta\phi \times \nabla\phi) - q\mathbf{e} \cdot (\nabla\xi \times \nabla\phi) \qquad (4.120)$$

which, compared with Eq. (4.108), contains a new term linked to surface deformation. Here, $p \sim Pr$ and $q = O(1)$. Accordingly, vertical vorticity is generated by surface deformations *even at infinite Prandtl number*, due to the non-zero vertical component of the velocity at the interface. Finally, Knobloch's equation (4.40) describing the evolution of the thermal order-parameter, $\phi$, is found to be completed by the coupling with surface deformation and vertical vorticity, namely

$$\frac{\partial\phi}{\partial t} = -\alpha\phi - 2\mu\Delta\phi + 2\Delta\xi - \Delta^2\phi + \nabla \cdot [(\nabla\phi)^2\nabla\phi] + \beta\nabla \cdot [\Delta\phi\nabla\phi] + \\ \delta\Delta[(\nabla\phi)^2] - \nu\nabla \cdot [\xi\nabla\phi] - \mathbf{e} \cdot (\nabla\phi \times \nabla\Psi)$$

$$(4.121)$$

where $\mu \sim (M - M_c)/M_c$ and $\alpha$ is proportional to the bottom plate Biot number.

Golovin *et al.* (1995) have also analyzed pattern selection in the system (4.119-4.121), varying $g \sim G$ and $c \sim C$ over a wide range of values. Fig. 4.21 gives an idea of the regimes found for a water–air system.

Finally recall that for perfect patterns, vertical vorticity does not influence pattern selection, as mentioned in Sect. 4.6.1. For large $G$ and $C$ (very weak surface deformation), squares and up-hexagons are recovered. When decreasing $G$ and $C$, the effect of surface deformation increases and the competition between squares and rolls turns in favor of the latter. Up-hexagons near the threshold are then replaced by down-hexagons, when further lowering the values of $G$ and $C$. Then, squares may again become preferred over rolls, and finally become subcritical at small $G$ and $C$ (rolls become also

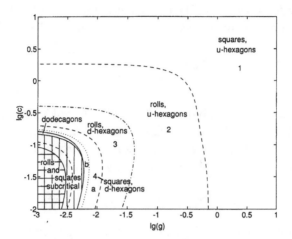

Figure 4.21: Log-log plot indicating where Benard-Marangoni convection patterns in the $(G, C)$-plane are expected for a two-layer water–air system at atmospheric pressure, 20 °C, and gas to liquid depth ratio $a = 0.5$. Vertical shading – squares subcritical; horizontal shading – rolls subcritical. Regions 4a and 4b differ by the intervals of bistability of squares and hexagons.

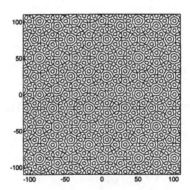

Figure 4.22: Quasi-crystalline dodecagonal convection pattern.

subcritical at another threshold). Just before this transition, an interesting dodecagonal pattern (Fig. 4.22), i.e. a superposition of 6 modes with equal amplitude and with wavevectors each separated by 30°, is found to be stable in a small region, though it does not correspond to an absolute minimum of the Lyapunov function of perfect patterns and hence is metastable.

# Chapter 5

# Interfacial oscillations and waves

In the preceding chapters we have considered stationary flows and patterns that are developed due to a *monotonic* instability of the static equilibrium. However, systems with an interface like liquid layers with an open surface are prone to both monotonic and oscillatory instability. These, and hence wavy interfacial motions, shall be discussed in the present chapter, in particular when the Marangoni effect acts at the interface.

## 5.1 Classification of oscillatory instabilities

There are different ways to perform the classification of oscillatory instabilities. One way is to determine the physical nature of waves generated by the instability. As we shall see in following sections, from this point of view the oscillatory instabilities observed in systems with an interface can be separated into three classes, because they are related to the excitation of three types of waves: capillary-gravity (transverse), dilational (longitudinal), and internal waves. An alternative taken in the present section is to discuss their "mathematical" origin which can be the same for instabilities of different physical nature.

Typically, oscillatory instabilities occur when *negative feedback* exists and hence a disturbance switches on some compensating mechanisms that try to *diminish* this disturbance. However, these mechanisms do not always suppress the disturbance, but sometimes they

159

lead to *overshooting oscillations (overstability)*. Roughly speaking, factors suppressing stationary (monotonic) instability can produce in some cases an oscillatory instability. In the following sections we shall repeatedly find such situations. The physical nature of the stabilizing factor may be different: an anomalous thermocapillary effect suppressing a monotonic buoyancy instability (subsection 5.3.2), a redistribution of a surfactant suppressing a buoyancy- or a thermocapillary-driven instability (subsection 5.5), etc. It is essential that the destabilizing and stabilizing factors and the stabilizing factor have different time scales, so that their counteraction is characterized by a certain effective *time delay*. The asynchronous changes of fields of different physical variables lead to the appearance of oscillations instead of a monotonic growth or decay of the disturbance. Typical shapes of the regions of oscillatory and monotonic instabilities in this case are schematically shown in Fig. 5.1(a) (the numbers characterize the dimension of the unstable manifold of the equilibrium, base state). The behavior of the growth rate as a function of the characteristic driving parameter is shown in Fig. 5.1(b) (dashed line corresponds to an oscillatory mode, solid lines correspond to a monotonic mode). As it can be seen from the latter figure, the instability is related to a pair of complex eigenvalues which are produced by the merging of two real ones. Note that the fragment of the monotonic instability boundary situated above the oscillatory instability boundary, is actually the boundary of a *restabilization* of an unstable monotonic mode.

The transition from a monotonic instability to an oscillatory one can take place also when there are two monotonic instability mechanisms which *compete* with each other, i.e. each instability mode tends to act as a suppressing factor on another mode, and the competition leads to overshooting. A typical example of such an instability is the appearance of the oscillatory (buoyancy-driven) Rayleigh-Benard convection in two-layer systems (Gershuni and Zhukhovitsky, 1982; Rasenat *et al.*, 1989). In the latter case, there are two monotonic instability modes, one of which is related to the onset of convection mainly in the top layer and another one mainly in the bottom layer. For some pairs of fluids these modes tend to suppress each other, which leads to the appearance of an oscillatory instability in a certain interval of values of the thicknesses' ratio (Fig. 5.2[a]). The dependence of the growth rate on the Rayleigh number for such oscillations is shown schematically in Fig. 5.2(b). A similar mechanism

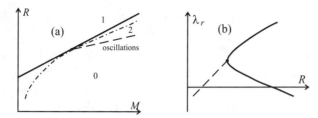

Figure 5.1: Oscillatory instability caused by the suppression of a monotonic instability. (a) Monotonic instability (solid line); and oscillatory instability boundaries (dashed line); the dashed-dotted line is the boundary of the region where oscillatory disturbances exist. The numbers (0, 1, 2) characterize the dimension of the unstable manifold in the corresponding region of parameters. (b) Dependence of the instability growth rate ($\lambda_r$) on the driving parameter $R$. Solid lines correspond to real eigenvalues while, the dashed line corresponds to a pair of complex eigenvalues.

of generation of an oscillatory instability but based on the competition of surface tension gradient - driven (Marangoni) *monotonic instabilities* originated at two opposing interfaces is characteristic for three-layer fluid systems (see Sect. 5.6).

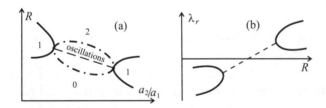

Figure 5.2: Oscillatory instability caused by the competition of two monotonic instability modes. (a) Monotonic instability (solid lines); and oscillatory instability boundaries (dashed line); dashed-dotted lines are the boundaries of the region where oscillatory disturbances exist. The numbers (0, 1, 2) characterize the dimension of the unstable manifold in the corresponding region of parameters. (b) Dependence of the growth rate ($\lambda_r$) on the driving parameter $R$. Solid lines correspond to real eigenvalues while, the dashed line corresponds to a pair of complex eigenvalues.

Another type of oscillatory instabilities can be produced by the interaction of two *decaying* oscillatory modes. Indeed if the frequencies of two different oscillatory modes approach each other, the phenomenon of *mixing* can take place (Fig. 5.3[a]), so that along each branch of the dispersion relation the physical nature of oscillations is changed (see, e.g., Lifshitz and Pitaevskii, 1981). It is much less known that because of a "repulsion" between complex eigenvalues which approach each other one of the eigenvalues can cross the neutral stability line $\mathrm{Re}\sigma = 0$, and then an instability takes place (Fig. 5.3[b]). The above mentioned repulsion of the eigenvalues is the manifestation of the phenomenon of *avoided crossing* observed in many physical systems (see, e.g., Triantafyllou and Triantafyllou, 1991). Examples of such oscillatory instabilities will be described in subsections 5.2.3 and 5.3.1.

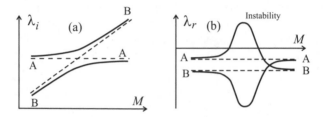

Figure 5.3: Oscillatory instability caused by mixing of two decaying oscillatory modes. (a) Dashed lines show the frequencies for modes A and B in the absence of their interaction; solid lines show the actual behavior. (b) The same as in (a) but for growth rates.

Another way of classification of oscillatory instabilities (as well as of monotonic instabilities) is connected with the location of the minimum of the neutral curve. If the wavenumber, $k_c$, of the fastest growing disturbance is non-zero, the oscillatory instability is, generally, a *short-wavelength* one, otherwise (if $k_c = 0$) it is a *long-wavelength* instability. In the latter case, it is necessary to distinguish between two cases.

In the first case, the eigenvalue $\lambda(k, M)$, which determines the growth rate, $\lambda_r$, and the frequency, $-\lambda_i$, of the oscillations, can be expanded into a Taylor series near the critical point $(k_c, M_c)$ as

$$\lambda_r = \left(\frac{\partial \lambda_r}{\partial R}\right)_c (M - M_c) + \frac{1}{2}\left(\frac{\partial^2 \lambda_r}{\partial k^2}\right)_c k^2 + \ldots, \qquad (5.1)$$

$$\lambda_i = (\lambda_i)_c + \left(\frac{\partial \lambda_i}{\partial k}\right)_c k + \left(\frac{\partial \lambda_i}{\partial M}\right)_c (M - M_c) + \frac{1}{2}\left(\frac{\partial^2 \lambda_i}{\partial k^2}\right)_c k^2 + \ldots \quad (5.2)$$

This type of oscillatory instability is known in reaction-diffusion systems (Kuramoto and Tsuzuki, 1976). In this case, the spatially homogeneous disturbance with $k = 0$ oscillates and grows with the largest growth rate.

However, there is a wide class of problems where the growth of a spatially homogeneous disturbance is forbidden by a *conservation law*. For instance, in the case of the surface waves in a layer of an incompressible liquid, a homogeneous change of the depth of the layer is impossible because of the conservation of the fluid volume. In this case, besides thee eventual role played by the confinement of the liquid (lateral boundaries), there exists the *Goldstone mode* with

$$\lambda(0, M) = 0 \tag{5.3}$$

for any $M$, which can produce a long-wavelength instability for small but non-zero $k$. In the latter case, not $\lambda_r(0, M)$ but $\partial^2 \lambda_r / \partial k^2 (0, M)$ changes its sign at the threshold of instability, $M = M_c$:

$$\lambda_r = \left(\frac{\partial^3 \lambda_r}{\partial k^2 M}\right)_c k^2 (M - M_c) + \frac{1}{24} \left(\frac{\partial^4 \lambda_r}{\partial k^4}\right)_c k^4 + \dots \tag{5.4}$$

$$\lambda_i = \left(\frac{\partial \lambda_i}{\partial k}\right)_c k + \left(\frac{\partial \lambda_i}{\partial k \partial M}\right)_c k(M - M_c) + \frac{1}{6} \left(\frac{\partial^3 \lambda_i}{\partial k^3}\right)_c k^3 + \dots. \tag{5.5}$$

In the supercritical region, the growth rate is proportional to $k^2$ at small $k$ ("negative" viscosity).

The behavior of the eigenvalue in the vicinity of the point $(k_c, M_c)$ is crucial for the evolution of the weakly nonlinear waves generated by the oscillatory instability. In the case of a *short-wavelength* where the fastest growing disturbance has a finite wavenumber, $k_c \neq 0$, the complex Ginzburg-Landau equation

$$\frac{\partial a}{\partial t} = \lambda_0 a + \lambda_2 \frac{\partial^2 a}{\partial x^2} - \delta |a|^2 a \tag{5.6}$$

is a universal equation for an envelope function, $a$, describing the modulation of waves (Stewartson and Stuart, 1971); here $\lambda_0$, $\lambda_2$ and $\delta$ are complex constant coefficients.

In the case of *long-wavelength* instabilities ($k_c = 0$), the situation is much more intricate, and there is a variety of different amplitude equations describing the weakly nonlinear evolution of disturbances (Nepomnyashchy, 1995). If the dependence $\lambda(k, M)$ is determined by Eqs. (5.1) and (5.2), the Ginzburg-Landau equation is again the generic amplitude equation, but this time the amplitude function, $a$, describes the wave itself rather than the wave envelope (Kuramoto and Tsuzuki, 1976).

Let us consider now the case of a "negative" viscosity [dispersion relation (5.4), (5.5)] caused by a conservation law. The structure of an amplitude equation compatible with a conservation law is

$$\frac{\partial a}{\partial t} = \hat{L}a + \frac{\partial}{\partial x}Q(a), \tag{5.7}$$

where the linear operator $\hat{L}$ has the following structure:

$$\hat{L} = \lambda_1 \frac{\partial}{\partial x} + \lambda_2 (M - M_c)\frac{\partial^2}{\partial x^2} + \lambda_3 \frac{\partial^3}{\partial x^3} + \lambda_4 \frac{\partial^4}{\partial x^4} + \ldots, \tag{5.8}$$

$$\lambda_1 = \left(\frac{\partial \lambda_i}{\partial k}\right)_c + \left(\frac{\partial^2 \lambda_i}{\partial k \partial M}\right)_c (M - M_c), \quad \lambda_2 = -\frac{1}{2}\left(\frac{\partial^3 \lambda_r}{\partial k^2 \partial M}\right)_c,$$

$$\lambda_3 = -\frac{1}{6}\left(\frac{\partial^3 \lambda_i}{\partial k^3}\right)_c, \quad \lambda_4 = \frac{1}{24}\left(\frac{\partial^4 \lambda_r}{\partial k^4}\right)_c, \ldots,$$

while

$$Q(a) = \delta_1 a^2 + \delta_2 \frac{\partial}{\partial x}(a^2) + \ldots. \tag{5.9}$$

It should be noted that if the conservation law is an approximate one, the amplitude equation can contain additional small terms. The term containing $\lambda_1$ is eliminated by a suitable Galilean transformation to a moving reference frame.

First, let us consider the generic case where all the coefficients $\lambda_n$, $n = 1, 2, \ldots$ in Eq. (5.8) are of order of unity. Taking $M - M_c = O(\epsilon^2)$, $\partial/\partial x = O(\epsilon)$, where $\epsilon \ll 1$, and assuming that the largest nonlinear term $\delta_1 \partial(a^2)/\partial x^2$ is balanced by the dispersion term $\lambda_3 \partial^3 a/\partial x^3$, we find that $a = O(\epsilon^2)$, and the amplitude equation is

$$\frac{\partial a}{\partial t} = \lambda_3 \frac{\partial^3 a}{\partial x^3} + \delta_1 \frac{\partial(a^2)}{\partial x} + \delta_2(M - M_c)\frac{\partial^2 a}{\partial x^2} + \delta_4 \frac{\partial^4 a}{\partial x^4} + \delta_2 \frac{\partial^2(a^2)}{\partial x^2}. \tag{5.10}$$

The first two terms in the right-hand side are of order $O(\epsilon^5)$, and the last three terms describing an instability at long wavelengths, the stabilization at short wavelengths, and a nonlinear correction to the "negative" viscosity coefficient, are of order $O(\epsilon^6)$. Thus, we find that the generic amplitude equation (5.10) is a *dissipation-modified Boussinesq-Korteweg-de Vries equation.*[1] The equation with $\delta_1 = \delta_4 = \delta_2 = 0$ is the BKdV-Burgers equation (Jonhson, 1972; Whitham, 1974; Canosa and Grazdag, 1977; Chu and Velarde, 1991; Gonzalez and Castellanos, 1994).

$$(283 \; bis) \qquad \frac{dh'}{dt} + \omega_\circ \frac{d}{ds}\left[ h' + \frac{k''}{2}\left( \frac{2+k}{2}\frac{h'^2}{H} + \frac{k'H^2}{3}\frac{d^2h'}{ds^2} \right) \right] = 0 \, ,$$

Figure 5.4: The Korteweg–de Vries (1895) equation as earlier written by Boussinesq (1877).

This equation has been derived in several physical problems, e.g. for the oscillatory Marangoni convection (Garazo and Velarde, 1991), instability of a liquid film flowing down a slightly inclined plane (Nepomnyashchy, 1976), Eckhaus instability of periodic waves (Janiaud et al.,, 1992) etc. We shall consider this equation in more detail in subsection 5.2.2 in the context of the nonlinear dynamics of transverse (capillary-gravity) waves sustained by the Marangoni effect. It will be shown that contrary to the conservative BKdV equation (Whitham, 1974; Lamb, 1980; Grimshaw, 1986; Drazin and Johnson, 1989; Shen 1993), the amplitudes of steady solitary and periodic waves governed by the equation (5.10) are not arbitrary albeit depending on the initial conditions but take some prescribed values due to an energy-dissipation, input-output dynamic balance that ignores initial conditions.

If the dispersion is small ($\delta_3 = O(\epsilon)$) (for instance, this feature is characteristic for the modulational instability of periodic waves with small wave-numbers; see Gershuni et al., 1989), it is balanced by the nonlinearity if $a = O(\epsilon^3)$, and one obtains the *Kawahara equation* (Topper and Kawahara, 1978; Kawahara, 1983; Kawahara and Toh,

---

[1]We shall use Boussinesq-Korteweg-de Vries equation (BKdV), instead of the standard terminology Korteweg-de Vries equation (KdV), because the latter equation (Fig. 5.4) was first obtained by Boussinesq (Boussinesq, 1877, p. 360; Korteweg and de Vries, 1895; see also Christov et al., 1996). For the history and, mathematical questions, the soliton concept and applications of the BKdV equation see e.g. Boussinesq, 1871, 1872, 1877; Rayleigh, 1876; Favre, 1935; Zabusky and Kruskal, 1965; Zabusky and Galvin, 1971; Benjamin, 1972, 1982; Scott et al., 1973; van der Blij, 1978; Miles, 1980; Ursell, 1983; Kivshar and Malomed, 1989; Sander and Hutter, 1991; Crighton, 1995; Zeytounian, 1995; and Wu, 1998.

1985):

$$\frac{\partial a}{\partial t} = \lambda_2(M - M_c)\frac{\partial^2 a}{\partial x^2} + \lambda_3\frac{\partial^3 a}{\partial x^3} + \lambda_4\frac{\partial^4 a}{\partial x^4} + \delta_1\frac{\partial(a^2)}{\partial x}. \qquad (5.11)$$

Solutions of this equation and their stability were studied by several authors (Toh and Kawahara, 1985; Kawahara and Toh, 1985, 1988; Elphick et al., 1991; Chang et al., 1993).

If the dispersion vanishes ($\lambda_3 = 0$), the Kawahara equation is reduced to the *Kuramoto-Sivashinsky equation* (KS)

$$\frac{\partial a}{\partial t} = \lambda_2(M - M_c)\frac{\partial^2 a}{\partial x^2} + \lambda_4\frac{\partial^4 a}{\partial x^4} + \delta_1\frac{\partial(a^2)}{\partial x}. \qquad (5.12)$$

typical for instabilities of oscillations in reaction-diffusion systems (Kuramoto and Tsuzuki, 1976), instabilities of flame fronts (Sivashinsky, 1977) etc. Also, this equation can be obtained in the case where the instability region is narrow because of large $\lambda_4$: $M - M_c = O(1)$, $\lambda_4 = O(\epsilon^{-2})$. This kind of scaling is used in the case of a film flow instability in a liquid with large surface tension (Homsy, 1974; Nepomnyashchy, 1974), where the KS equation is obtained as a weakly-nonlinear limit of the strongly-nonlinear longwave equation of Gjevik (1970).

The Kuramoto-Sivashinsky equation, which has both locally stable regular wavy solutions and spatiotemporal chaotic regimes, provides a paradigmatic example of the transition between regular and chaotic patterns (Nepomnyashchy, 1974; Cohen et al., 1976; Yamada and Kuramoto, 1976; Chang, 1986; Hyman and Nicolaenko, 1986; Hyman et al., 1986; Frisch et al., 1986; Demekhin et al., 1991; Feudel et al., 1993; Oron and Rosenau, 1997 and references therein), as well as a paradigmatic model for the application of the dynamical systems approach to turbulence (Bohr et al., 1998. For its three-dimensional generalizations, see Lin and Krishna, 1977; Frenkel and Indireshkumar, 1996).

It should be noted that the Kuramoto-Sivashinsky equation is formally equally valid for both oscillatory and monotonic instabilities. The only "sign" of the "wavy" origin of this equation is the lack of the invariance of the nonlinear term to the parity transformation $x \to -x$. However, the transformation $a = h_x$ provides another form of the Kuramoto-Sivashinsky equation,

$$\frac{\partial h}{\partial t} = \lambda_2(M - M_c)\frac{\partial^2 h}{\partial x^2} + \lambda_4\frac{\partial^4 h}{\partial x^4} + \frac{1}{2}\delta_1\left(\frac{\partial h}{\partial x}\right)^2, \qquad (5.13)$$

which is invariant to the parity transformation. Eq. (5.13) and its extensions describe the nonlinear development of stationary instabilities of interfaces in different physical systems (Benney, 1966; Nekorkin and Velarde, 1994; Koulago and Parseghian, 1996). As noted in Chapter 4, a generalization of the KS equation is the *Kuramoto-Velarde* (KV) equation

$$\frac{\partial h}{\partial t} + \nu \frac{\partial^4 h}{\partial x^4} + \frac{\partial^2 h}{\partial x^2} + \beta h + \gamma \left(\frac{\partial h}{\partial x}\right)^2 + \frac{\delta}{2} \frac{\partial^2 (h^2)}{\partial x^2} - \frac{\gamma}{L} \int_0^L \left(\frac{\partial h}{\partial x}\right)^2 dx = 0$$

(5.14)

which describes the deformation of the surface in Benard-Marangoni convection in the limit of vanishingly small Biot number (Garcia-Ybarra, Castillo and Velarde, 1987a,b, Castillo, Garcia-Ybarra and Velarde, 1988). The last term in the left-hand side of the equation (5.14) ensures volume conservation. The propagation of solitary waves and kinks governed by the KV equation was studied by Christov and Velarde (1992).

The analysis of long-wavelength instabilities given in the present section was based on the assumption of the *analyticity* of the function $\lambda(k, M)$ in the point $k = 0$. If the analyticity condition does not hold, the amplitude equation can contain some non-local integral terms (see, e.g., Sivashinsky (1977)). We shall find such a situation in subsection 5.2.2 when considering the dynamics of transverse Marangoni waves in the presence of a bottom friction.

## 5.2 Transverse and longitudinal oscillatory instabilities

### 5.2.1 Two interfacial wave modes in the absence and in the presence of the Marangoni effect

At the surface of a liquid, capillary-gravity waves are possible if surface deformations are allowed. However, other wavy motions can appear also by *compressions-dilations* of the surface, if the surface manifests some *elastic* properties.

Lucassen (1968) and Lucassen-Reynders and Lucassen (1969) were the first to investigate, theoretically and experimentally, the dispersion relation for surface waves assuming that the surface tension $\sigma$ may be changed by compressions-dilations of the surface. They characterized the variation of the surface tension, i.e. the surface elasticity, by a *surface dilational modulus*

$$\epsilon = \frac{d\sigma}{d\ln A_e}, \tag{5.15}$$

where $A_e$ is the area of a surface element. $\epsilon$ or $E$ have been used in the literature to account for (5.15) as a solutal or elasticity Marangoni number. For a one-dimensional wave, the surface tension gradient $\partial\sigma/\partial x$ in the boundary condition for tangential stresses, Eq. (1.25), is replaced by $\epsilon\partial^2\xi/\partial x^2$, where $\xi$ is the horizontal displacement of the point at the surface (i.e. $\partial\xi/\partial t = v_x$). For an infinitely deep liquid layer, the analysis shows that in addition to the capillary-gravity (*transverse*) wave characterized by the dispersion relation

$$\omega^2 = gk + \frac{\sigma}{\rho}k^3, \tag{5.16}$$

there exists another branch of oscillations, a *dilational* (longitudinal) wave which in the high frequency limit $\omega \gg \nu k^2$ is governed by the dispersion relation

$$\eta\omega(i\omega/\nu)^{1/2} = i\epsilon k^2. \tag{5.17}$$

The physical origin of the surface elasticity may be diverse. For instance, it can be caused by a redistribution of a surfactant concentrated in an insoluble monolayer (see, e.g. Cini *et al*, 1987; Myers, 1999). In this case

$$\epsilon = -\frac{d\sigma}{d\ln\Gamma}, \tag{5.18}$$

where $\Gamma$ is the excess surface concentration of the adsorbed surfactant. If the surfactant is soluble, the parameter $\epsilon$ is neither frequency-independent nor real:

$$\epsilon = -\frac{d\sigma}{d\ln\Gamma} \cdot \frac{1}{1 - \frac{inD}{\omega}\left(\frac{d\Gamma}{dC}\right)^{-1}}, \quad n^2 = k^2 + \frac{i\omega}{D}, \tag{5.19}$$

where $d\Gamma/dC$ characterizes the relation between $\Gamma$ and the bulk concentration, $C$, of the surfactant, and $\omega$ is the frequency of oscillations. The imaginary part of $\epsilon$ characterizes the effective *dilational viscosity* of the surface.

It should be emphasized that in any case the longitudinal surface waves described by the dispersion relation (5.17) decay rapidly: their decay rate is of the order of the time period.

Mass or heat *transfer* through the interface may also lead to a surface elasticity effect. Assume, for instance, that an infinitely deep liquid layer is heated *from above*; the equilibrium temperature gradient is equal to $A$. In this case, the boundary condition for tangential stresses (1.25) is

$$\eta \left( \frac{\partial u}{\partial z} + \frac{\partial w}{\partial x} \right) = \frac{d\sigma}{dT} \frac{\partial}{\partial x} (\theta - A\zeta), \qquad (5.20)$$

where $\theta$ is a disturbance of the temperature, and $\zeta$ is the vertical deviation of the interface from the initial level reference state. Disregarding $\theta$ for sake of simplicity (that corresponds to the limit of high heat diffusivity, i.e. $P \to 0$) and taking into account the relation

$$ik\xi_k + \left( \frac{i\omega}{\nu} \right)^{1/2} \zeta_k = 0 \qquad (5.21)$$

for Fourier components of the surface displacements, which is obtained from the continuity equation in the high-frequency limit, we find that the effective surface elasticity due to the Marangoni effect can be estimated as

$$\epsilon = \left( -\frac{d\sigma}{dT} \right) A \left( \frac{i\omega}{\nu} \right)^{-1/2}. \qquad (5.22)$$

Substituting (5.22) into (5.17), we find

$$\omega^2 = \frac{1}{\rho} \left( -\frac{d\sigma}{dT} \right) A k^2. \qquad (5.23)$$

Thus, the analysis of the dispersion relation reveals (as we shall study in detail further below) the existence of a sound-like, non-dispersive dilational mode of oscillations with the squared frequency proportional to the temperature gradient. The origin of such a mode can be understood in the following way. Assume that a liquid element rises to the free surface. Because its temperature is lower than that of the interface, it creates a cold spot. The surface tension gradient acts toward this spot and generates an incoming surface flow. Because of the mass conservation, this surface flow produces a flow in the bulk directed downwards. The latter flow pushes the liquid element back to the bulk. The theory of dilational waves will be discussed in subsection 5.2.3.

## 5.2.2   Transverse (capillary-gravity) waves in the presence of the Marangoni effect

The above mentioned waves may be observed as fluctuations in equilibrium, for instance, using low angle (light, x-ray) scattering, but they are damped by viscosity. In the *non-equilibrium* case, when the fluid layer is subject to a temperature or concentration gradient, the wavy modes described in the previous subsection can be *spontaneously* generated because of an oscillatory surface tension gradient-driven (Marangoni) instability. That phenomenon leads to the appearance of complicated *wavy* patterns.

The problem of the theoretical description of wavy patterns produced by an instability is very difficult, and we are far from its complete solution. In the present subsection we shall present some attempts of a simplified description of the problem based on the long-wave approximation.

### i. Stress-free lower boundary

First, we shall consider the analytically tractable case of the transverse wave generation in a layer with a stress-free lower boundary by heating the liquid *from above*, i.e. from the ambient air. Assuming that the minimum of the neutral stability curve is situated at $k = 0$, the problem under consideration belongs to the class of problems where the growth of a spatially homogeneous disturbances is precluded by a *conservation law*. Indeed, the homogeneous change of the layer thickness is impossible because of the conservation of the fluid volume. As it was explained in Sect. 5.1, in this case there exists the *Goldstone mode* with the growth rate $\lambda(k, M) = \lambda_r(k, M) + i\lambda_i(k, M)$ which satisfies the condition

$$\lambda(0, M) = 0. \tag{5.24}$$

This mode generates an instability when the quantity

$$\frac{\partial^2 \lambda_r}{\partial k^2}(0, M)$$

becomes positive, which happens as $M > M_c = 12$, as shown by Garazo and Velarde (1991) and Nepomnyashchy and Velarde (1994). Near the threshold point $k = 0$, $M = M_c$, the growth rate can be expanded into the Taylor series:

$$\lambda_r(k, M) = \lambda_{21}k^2(M - M_c) + \lambda_{40}k^4 + \dots, \tag{5.25}$$

$$\lambda_i(k, M) = \lambda_{10}k + \lambda_{11}k(M - M_c) + \lambda_{30}k^3 + \dots, \qquad (5.26)$$

where

$$\lambda_{21} = \frac{P}{6}, \; \lambda_{40} = -\frac{2}{105}(17GP^2 + 204P^2 + 134P + 22),$$

$$\lambda_{10} = \sqrt{G + 12}, \; \lambda_{11} = \frac{\sqrt{P}}{2\sqrt{G + 12}},$$

$$\lambda_{30} = \frac{C - G(8P/5 + 1/3) - (96P/5 + 56/5)}{2\sqrt{G + 12}}.$$

Recall that $G = ga^3/\nu\kappa$ is the (modified) Galileo number and $C = \eta\kappa/\sigma_0 a$ is the capillary number. The instability interval is $0 < k < k_m$, where

$$k_m = \sqrt{\frac{\lambda_{21}(M - M_c)}{-\lambda_{40}}}.$$

Extending to the surface tension gradient-driven problem of a shallow liquid layer the scaling and longwave asymptotics used by Gardner et al. (1967) Velarde and collaborators (Chu and Velarde, 1991; Garazo and Velarde, 1991, 1992; Nepomnyashchy and Velarde, 1994) have derived Eq. (5.10) and its three-dimensional counterpart. The theory targeted the natural extension of the soliton-bearing BKdV equation to the case where the Marangoni effect leads to the transverse waves just described. We shall not describe the methodology here but rather we shall recall salient features of those studies.

### i.1. One-dimensional waves

In the simplest case of a one-dimensional wave, the longwave expansions give rise to an evolution equation for the surface deformation $h(x, t)$ ("order parameter equation") which can be written in the form (see subsection 5.1):

$$\frac{\partial h}{\partial t} = Lh + \frac{\partial}{\partial x}Q(h) \qquad (5.27)$$

where

$$L = [\lambda_{10} + \lambda_{11}(M - M_c)]\frac{\partial}{\partial x} - \lambda_{21}(M - M_c)\frac{\partial^2}{\partial x^2} - \qquad (5.28)$$

$$\lambda_{30}\frac{\partial^3}{\partial x^3} + \lambda_{40}\frac{\partial^4}{\partial x^4}$$

is the linear operator corresponding to the expressions (5.25) and (5.26),

$$Q(h) = \delta_1 a^2 + \delta_2 \frac{\partial}{\partial x}(a^2) + \ldots, \tag{5.29}$$

with

$$\delta_1 = -\frac{3(G+8)\sqrt{P}}{4\sqrt{G+12}}, \ \delta_2 = -2P.$$

As earlier noted, Eq. (5.27) is a *dissipation-modified Boussinesq-Korteweg-de Vries* equation. By means of a suitable Galilean transformation and scale transformation of variables, it can be reduced to the standard form

$$H_T + H_{XXX} + 3(H^2)_X + \delta[H_{XX} + H_{XXXX} + D(H^2)_{XX}] = 0, \tag{5.30}$$

with

$$\delta = \frac{\sqrt{-\lambda_{21}\lambda_{40}(M - M_c)}}{|\lambda_{30}|}, \ D = -\frac{3\lambda_{30}\delta_2}{\lambda_{40}\delta_1}. \tag{5.31}$$

When $\delta$ vanishes, the ideal BKdV equation, obtained from the Euler equation for an inviscid shallow layer, is recovered. It is known that this equation has an infinite number of conservation laws. The BKdV equation possesses a family of traveling wave solutions

$$H(X,T) = H(\xi), \xi = X - cT \tag{5.32}$$

with arbitrary spatial period $L = 2\pi/k$:

$$H(\xi + 2\pi/k) = H(\xi). \tag{5.33}$$

We take also

$$\langle H \rangle = \int_0^{2\pi/k} H(\xi)d\xi = 0, \tag{5.34}$$

because the mean surface deformation is equal to zero by definition. The corresponding solutions (cnoidal waves) are known: (Nepomnyashchy, 1976; Bar and Nepomnyashchy, 1995, 1999; Rednikov *et al.*, 1995)

$$H(\xi) = A \left[ \text{dn}^2 \left( \left( \frac{A}{2} \right)^{1/2} (\xi - \xi_0) \right) - \frac{E}{K} \right], \tag{5.35}$$

$$c = 2A \left( 2 - \kappa^2 - \frac{3E}{K} \right) \tag{5.36}$$

where

$$A = \frac{2k^2 K^2}{\pi^2}, \tag{5.37}$$

dn is Jacobi's delta amplitude function with the modulus $\kappa$, and $E = E(\kappa)$ and $K = K(\kappa)$ are complete elliptic integrals. The limit $\kappa \to 1$ corresponds to solitary wave solutions. Clearly, in the case $\delta = 0$, the wavelength $L = 2\pi/k$ and the wave amplitude

$$Amp(k) = H_{max} - H_{min} = \frac{2k^2 K^2 \kappa^2}{\pi^2} \tag{5.38}$$

are independent parameters which can take arbitrary values.

When $\delta \neq 0$, the governing equation (5.30) contains three additional terms. The first two terms, which are linear, account for long-wave instability ("negative" viscosity, hence an energy input term) and shortwave dissipation (energy output), respectively. The last term, which is nonlinear, describes an energy redistributions between Fourier modes and nonlinear dissipation. The various conservation laws characteristic for the ideal BKdV equation, are not fulfilled anymore, except that of the conservation of the fluid volume. For instance, in the case where $H(X,T)$ is spatially periodic with the period $L = 2\pi/k$, the time evolution of the squared deflection of the interface ("momentum") is governed by the equation

$$\frac{d}{dT} \int_0^L H^2 dX = \delta \left( \int_0^L H_X^2 dX - \int_0^L H_{XX}^2 dX + 2D \int_0^L H H_X^2 dX \right). \tag{5.39}$$

Eq. (5.39) is an energy-like input-output balance and its right-hand side vanishes only for some definite values of the wave amplitude $Amp(k)$.

In the limit of small $\delta$, the stationary values of the wave amplitude have been calculated analytically (Nepomnyashchy, 1976; Kawahara and Toh, 1985; Bar and Nepomnyashchy, 1995). There are three different types of behavior of $Amp(k)$, depending on $D$.

(i) If $D \leq 5/4$, the function $Amp(k)$ is uniquely defined in the whole region $0 < k < 1$. However, the spatially periodic solutions with the stationary value of the amplitude $Amp(k)$ are stable only inside a certain sub-interval $k_- \leq k \leq k_+$.

(ii) If $5/4 < D < 2$, there are two solutions for $Amp(k)$ in a certain region $k_{min}(D) < k < 1$ (the lower branch and the upper branch), one solution for $k > 1$ and no solutions for $0 < k < k_{min}(D)$.

Only the solutions on the lower branch in the interval $k_{min}(D) < k < 1$ are stable with respect to strictly periodic disturbances with the same spatial period $2\pi/k$. However, even these solutions are unstable with respect to disturbances violating the periodicity of the solution.

(iii) If $D \geq 2$, there is a unique solution for $k > 1$, which is unstable.

For finite values of $\delta$, the traveling wave solutions of Eq. (5.30) were studied qualitatively by Nekorkin and Velarde (1994) (see also Velarde *et al.* (1995)). Significant numerical results for Eq. (5.30) were obtained by Christov and Velarde (1995), Oron and Rosenau (1997), and Kliakhandler *et al.* (2000).

Noteworthy results are depicted in Figures 5.5 – 5.8. Fig. 5.5 shows the forms predicted by the qualitative analysis of Nekorkin and Velarde (1994). The elevation wave shows asymmetry in its central peak, a wavy head/tail and the possibility of forming bound states of waves, all of equal amplitude but variable crest-to crest separation, hence including chaotic wave trains. Figures 5.6 - 5.8 provide experimental numerical support (Christov and Velarde, 1995) to the qualitative analysis and, in view of their collision properties, the consideration of such waves as dissipative solitons (see also Argentina *et al.*, 2000 and various contributions to Velarde and Christov, 1995).

Fig. 5.6 shows the expected *aging* of a solitary wave, born as an arbitrary initial perturbation here chosen already as a Sech$^2$ not a solution of Eq. (5.30) and terminating with the amplitude imposed by the steady input-output energy balance. Figs. 5.7 and Fig. 5.8 illustrate suitably prepared collision events.

*i.2 Two-dimensional waves*

In the nearly one-dimensional case, when some transverse modulations of one-dimensional waves are taken into account, the problem is governed by the *dissipation-modified Kadomtsev-Petviashvilii* equation (Garazo and Velarde, 1992). By means of a scaling transformation, this equation is reduced to the following form:

$$\{H_T + H_{XXX} + 3(H^2)_X + \delta[H_{XX} + H_{XXXX} + D(H^2)_{XX}]\}_X - 3sH_{YY} = 0 \tag{5.40}$$

where $\delta$ and $D$ are determined by (5.31),

$$s = \text{sign}\left(\frac{\lambda_{30}}{\lambda_{10}}\right). \tag{5.41}$$

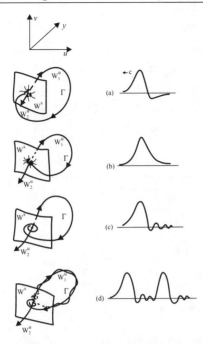

Figure 5.5: Homoclinic trajectories and dissipative localized structures. (a) - (c) "Solitons" and (d) "bound solitons" (after Nekorkin and Velarde, 1994).

In the framework of Eq. (5.41), it has been shown (Bar and Nepomnyashchy, 1999) that all one-dimensional waves are unstable with respect to transverse modulations.

In the general case, the waves generated by the instability can propagate in arbitrary directions simultaneously. The full system of longwave equations which describes such waves is rather complicated (Nepomnyashchy and Velarde, 1994). This system was used for the investigation of collisions of dissipative solitary waves.

In the case of two solitary waves that move obliquely (the angle between their normal vectors is $2\psi$), it was found that in the lowest order in $\epsilon$ their interaction leads to a mere *shift* in their trajectories (called a phase shift) whose value is proportional to $\Psi A(\lambda_{30}/\delta_{11})$, where $A$ is the amplitude of the solitary wave, and

$$\Psi = \frac{\sin^2 \psi_*}{\sin^2 \psi} - 1, \tag{5.42}$$

$$\sin^2 \psi_* = \frac{1}{4} \left( 3 - \frac{M_c}{G + M_c} \right). \tag{5.43}$$

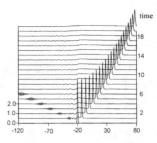

Figure 5.6: Evolution or "aging" of a Sech towards the soliton solution obeying the input-output energy balance of the dissipation-modified BKdV equation (5.10) (after Christov and Velarde, 1995).

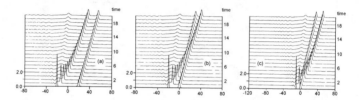

Figure 5.7: Numerical experiment of suitably prepared overtaking interactions of dissipative solitary waves (5.10): (a) Large initial separation. (b) Moderate initial separation. (c) Small initial separation (after Christov and Velarde, 1995).

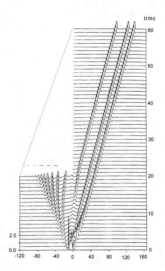

Figure 5.8: Overtaking interaction of Sech pulses with different amplitudes leading to a three-crest (all equal amplitude but unequal spacing between crests) soliton bound state (after Christov and Velarde, 1995).

For a given experiment, the change of sign of the phase shift takes place at the "critical" angle $\psi = \psi_*$ (for *neutral* collisions; later on we shall use either $\psi_*$ or $\psi_c$). Wider angles than the critical value yield a Mach-Russell third wave or stem. In the limit $G \gg M_c = 12$ the result of Miles (1977) is recovered for interacting solitary waves in ideal viscous-free shallow liquid layers: $\psi_* \to \pi/3$. In the opposite limit, $\psi_* \to \pi/4$. Thus the critical angle $2\psi_*$ between normal vectors is between $2\pi/3$ and $\pi/2$, and the corresponding critical angle between wave fronts, $\pi - 2\psi_*$, is between $\pi/3$ and $\pi/2$.

Incidentally, the critical value, $\psi_*$, for *neutral* collisions (or wall reflections considered as collisions of waves with their mirror images) with no phase shift appears as a geometric or kinematic property of asymptotic trajectories for both solitary waves and shocks or hydraulic pumps, both in theory and experiment (Russell, 1844, 1885; Bazin, 1865; Mach and Wosyka, 1875; Bouasse, 1924; Courant and Friedrichs, 1948; Crossley, 1949; Wiegel, 1964a,b; Van Dyke, 1975; Maxworthy, 1976; Newell and Redekopp, 1977; Weidman and Massworthy, 1978; Melville, 1980; Su and Mirie, 1980; Mirie and Su, 1982; Lugt, 1983; Renouard *et al*, 1985; Hornung, 1988; Krehl and van der Geest, 1991; Weidman *et al.*, 1991; Linde *et al.*, 1993a-c; Cooker *et al.*, 1997; Santiago-Rosanne *et al.*, 1997; Linde *et al.*, 2001a,b; Krehl, 2001). We shall return with details to this question further below in this chapter.

In the case of a head-on collision of two solitary waves, the phase shift is proportional to $A^{1/2}(\lambda_{30}/\delta_{11})$. One can conclude that the phase shift is negative for "humps", which are called also bumps, elevation waves or *bright* solitons (in optics, Bose condensates, and other fields) and positive for "hollows" (depression waves, *dark* solitons; Vanden-Broeck, 1991; Aoosey *et al.*, 1992; Nepomnyashchy and Velarde, 1994; Huang *et al.*, 2001).

Huang *et al.* (1998) have extended the theory to the case of cylindrical solitary waves $H(\epsilon(r - ct), \epsilon^3 r, \epsilon\phi)$. A *dissipation-modified cylindrical Kadomtsev-Petviashvili equation* has been derived, which after rescaling can be written in the form (cf. Eq. (5.40))

$$\{H_T + \frac{1}{2T}H + H_{XXX} + 3(H^2)_X + \delta[H_{XX} + H_{XXXX} + D(H^2)_{XX}]\}_X$$
(5.44)
$$+\frac{1}{2T^2}H_{\Phi\Phi} = 0.$$

They have also provided some details of the time evolution of a cylin-

drical solitary wave and head-on collisions between concentric cylindrical solitary waves.

## ii. Solid lower boundary

Eq. (5.27) and hence the analysis given above comes from a study
that completely ignores the friction of the fluid at the bottom. Such
a drastic simplification has permitted, however, a theoretical study
that provides significant predictions for nonlinear dissipative waves
due to an instability triggered by the Marangoni effect. However, the
friction at the solid support and hence the no-slip boundary condition
seems to play quite a significant role. In the case of a layer with finite
thickness, $a$, and a solid lower boundary, the friction at the solid
lower boundary shifts the critical wavenumber from the $k \to 0$ limit
to a finite $k_c \neq 0$. The existence of transverse Marangoni waves in a
layer with finite thickness by heating from the gas side was justified
by Takashima (1981b). A remarkable phenomenon found by the
numerical solution of the eigenvalue problem is the non-monotonic
behavior of the critical Marangoni number, $M_c$, and the non-smooth
behavior of the critical wave number, $k_c$, and the critical frequency,
$\omega_c$, as functions of the capillary number. Note that the appearance
of the oscillatory instability was investigated for both positive and
negative values of the corresponding (static) Bond number, i.e. for a
liquid resting on solid support or hanging from it. Some additional
data concerning the transverse Marangoni waves have recently been
found (Zimmerman, 2001).

One can phenomenologically incorporate *ad hoc* the friction at
the solid bottom boundary and to study the influence of the non-
zero critical wave number by adding to Eq. (5.30) a term $\delta \alpha H$ with
$\alpha > 0$

$$H_T + H_{XXX} + 3(H^2)_X + \delta[H_{XX} + H_{XXXX} + D(H^2)_{XX} + \alpha H] = 0.,$$
$$(5.45)$$

In the framework of Eq. (5.45), the equilibrium solution $H = 0$
is unstable in the interval of wavenumbers $k_-(\alpha) < k < k_+(\alpha)$,
$k_\pm^2 = (1 \pm \sqrt{1 - 4\alpha})/2$, if $\alpha < 1/4$, and linearly stable as $\alpha > 1/4$.
The additional term with a constant coefficient $\alpha$ may approximate
the influence of the bottom friction reasonably well for wavenumbers
in the instability interval, but it is not satisfactory in the long-wave
limit, because it violates the basic condition (5.24). The cnoidal wave
solutions of the problem (5.44) are described by the same formulas

(5.35), (5.36), but the amplitude $A$ satisfies now a certain quadratic equation (Rednikov *et al.*, 1995b).

If $\alpha$ is relatively small ($0 < \alpha < 0.13$), there are no branches of solutions different from those bifurcating at $k_+(\alpha)$ and $k_-(\alpha)$. Both branches merge if $D < D_* = 5/4 - 7/80\alpha$ (Fig. 5.9(a), and they are separated otherwise (Figs. 5.9(b)-(c)). Here the stability is considered only with respect to disturbances with the same period as the basic cnoidal wave. If $0.13 < \alpha < 0.236$, a new configuration, which is intermediate between those of Fig. 5.9(a) and Fig. 5.9(b), exists in a certain interval of $D$ (see Fig. 5.10). It is interesting that there exists a stable branch of solutions (including the solitary waves) which is not related to the linear instability of the solution $H = 0$. If $0.236 < \alpha < 0.25$, the configuration similar to Fig. 5.10 is also obtained in a certain interval of $D$. As $D$ passes $5/4$, the stable solitary solution disappears at infinity; for a certain larger values of $D$ the branches reconnect. If $\alpha > 1/4$, the solution $H = 0$ is linearly stable; nevertheless, some nonlinear cnoidal wave solutions exist (Fig. 5.11) if $D > D_*$. As $D$ increases, first the branch with solitary waves appears (Fig. 5.11(a)). Further increasing of $D$ results in Fig. 5.11(b) and Fig. 5.11(c).

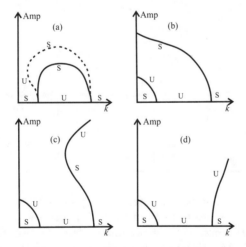

Figure 5.9: Cnoidal wave trains: amplitude versus wavenumber for $0 < \alpha < 0.13$. Transitions from (a) to (d) correspond to increasing values of the nonlinearity parameter $D$. Marked 'S' are stable states and 'U' - unstable ones. (Redrawn after Rednikov *et al.*, 1995b).

The theory described above oversimplifies the dispersion relation

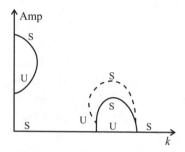

Figure 5.10: Cnoidal wave trains: amplitude versus wavenumber for $0.13 < \alpha < 0.236$. Marked 'S' are stable states and 'U' - unstable ones. (Redrawn after Rednikov *et al.*, 1995b).

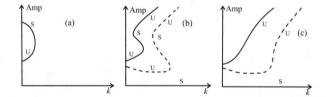

Figure 5.11: Cnoidal wave trains: amplitude versus wavenumber for $\alpha > 1/4$. Transitions from (a) to (c) correspond to increasing values of the nonlinearity parameter $D$. Marked 'S' are stable states and 'U' - unstable ones. (Redrawn after Rednikov *et al.*, 1995b).

for linear disturbances in the region of small wavenumbers. Another of its deficiencies is connected with the fact that the phenomena in the gas phase adjacent to liquid are ignored. In two-layer systems, the interaction between the transverse mode and the longitudinal mode is very important. The development of a complete theory in the general case is a formidable task. However, the problem can be drastically simplified in the realistic limit $G \gg 1$, because in this case the critical wavenumber $k_c$ is small, and the longwave approach can be applied. Also, one can take into account the fact that the *kinematic* viscosity and thermal *diffusivity* of a gas are typically much larger than corresponding parameters of a standard liquid.

An attempt to use the above-mentioned circumstances for the construction of the nonlinear theory of long transverse Marangoni waves was made by Velarde and Rednikov (2000). They considered the limit where the characteristic wavenumber $k \sim \epsilon \ll 1$ but $\nu \sim \kappa \sim G^{1/2}\epsilon \gg 1$ (actually, the relation $\epsilon \sim G^{-1/10}$ was selected). In this limit, another more general *dissipation-modified* BKdV equation was obtained, which after rescaling of variables can be written as

$$H_T + H_{XXX} + 3(H^2)_X + \int_{-\infty}^{+\infty} Q(X - X')H(X')dX' = 0, \quad (5.46)$$

The Fourier transform, $\hat{Q}(k)$, of the kernel $Q(X - X')$ is a cumbersome function of $k$, $m = M/G$, $P$, $a$ and $\nu_*$, where $\nu_* = \nu/(\epsilon G^{1/2}) = O(1)$. Eq. (5.46) extends to the surface tension gradient-Marangoni-driven waves an earlier result of Miles (1976) for BKdV waves with bottom friction.

Let us conclude this subsection by noting that Zimmerman (2001) has performed a direct simulation of the nonlinear thermohydrodynamic equations (1.28)–(1.34) which describe the finite-amplitude surface tension gradient-driven convection in a layer with a deformable free surface and a solid bottom surface. The flow was assumed to be two-dimensional. The Galerkin finite element method with the grid adaptation procedure developed formerly by Goodwin and Schowalter (1995) was used. In the first series of simulations, the free surface is initially flat up to numerical noise. The latter turned out to be sufficient to produce some Marangoni flow near the surface. In the subcritical region, it was found that the decay of this flow is essentially related to the its deep penetration into the layer and friction losses in the vicinity of the bottom. Thus, the prediction of the linear stability theory (Takashima, 1981a,b; Zimmerman, 2001) on the increase of the critical Marangoni number due to the bottom friction has been justified (see also susection 5.7.1). In the second series of simulations, the motion was produced by an initial hump disturbance. In this case some growth of disturbances was observed even below the threshold predicted by the linear stability theory.

### 5.2.3  Mixing of transverse and longitudinal waves

#### i. Infinitely deep layer

As it was explained in subsection 5.1.1, there exists another type of interfacial oscillations, *dilational (longitudinal)*. In order to understand the physical nature of this kind of waves in the presence of a vertical temperature gradient, let us consider an infinitely deep liquid layer heated from above (Levchenko and Chernyakov, 1981; Garcia-Ybarra and Velarde, 1987; Chu and Velarde, 1988; Rednikov *et al.*, 2000a,b); the temperature gradient is equal to $A$. In this case, we cannot use the thickness of the layer as the length scale. Instead of that, we use the *capillary length*

$$l = (\sigma_0/\rho g)^{1/2} \tag{5.47}$$

and define with the scale set by (5.47) the Marangoni number

$$M = \frac{\alpha A l^2}{\eta \kappa}, \tag{5.48}$$

as well as the Galileo number

$$Ga = \frac{g l^3}{\nu^2}. \tag{5.49}$$

As the unit of time, we take $l^2/\nu$.

In view of earlier given argument, two kinds of wavy motions are possible in such a system. One is the dilational, or longitudinal wave. A qualitative explanation of the physical origin of dilational waves has been done in subsection 5.2.1. This kind of wave is *genuinely* caused by the thermocapillary effect and hence its existence is only possible in viscous liquids. It can be shown that the dependence of the non-dimensional frequency $\omega$ on the non-dimensional wavenumber $k$ is determined by the relation (Levchenko and Chernyakov, 1981)

$$\omega^2 = \omega_{long}^2 = |M|Pk^2/(P^{1/2} + 1) \tag{5.50}$$

and hence exhibits no dispersion. Note that when writing the dispersion relations we shall use the absolute value of the Marangoni or Rayleigh numbers. However, in a one-layer system the purely longitudinal waves *decay* for any values of $M$, and they do not produce an instability.

The other type of waves possible in a system with a deformable surface is the *transverse*, or capillary-gravity wave. It is caused by the joint action of gravity and surface tension. The corresponding dispersion relation (5.16) is

$$\omega^2 = \omega_{tr}^2 = Gak(1 + k^2/Bo). \tag{5.51}$$

In the high-frequency limit, when the decay of waves is neglected, the full dispersion relation of the system can be written as (Lucassen, 1968; Levchenko and Cherniakov, 1981; Chu and Velarde, 1988)

$$(\omega^2 - \omega_{long}^2)(\omega^2 - \omega_{tr}^2) = 0. \tag{5.52}$$

However, the next order corrections to the dispersion relation (5.52) describe a certain *mixing* between two kinds of waves, which is especially strong in the resonant case where the frequencies of both waves are close. Also, if the next order corrections are taken into account, any solution $\omega(k)$ has a non-vanishing *imaginary part* which corresponds to decay or growth of waves in time due to the Marangoni effect. The neutral curve is

$$M = 2^{5/2}(GaP^2)^{3/4}\frac{(1 + k^2)^{3/4}}{k^{1/4}}. \tag{5.53}$$

It has the minimum at

$$k = k_c = \sqrt{5}/5. \tag{5.54}$$

### ii. System with two layers of finite thicknesses

Let us now consider the possible oscillatory instabilities in the general case of two-layer systems with a deformable interface and hence incorporating the ambient gas or liquid phase as an active rather than passive layer, no longer using the large disparity of (dynamic) shear viscosities and heat conductivities.

Let us consider a system of two horizontal layers of immiscible fluids between solid boundaries $z = a_1$ and $z = -a_2$. In the absence of deformations, the interface is located at $z = 0$. The governing equations and boundary conditions have been described in subsection 1.2.1. As the units of length, time, velocity, pressure and temperature we use $a_1$, $a_1^2/\nu_1$, $\nu_1/a_1$, $\rho_1\nu_1^2/a_1^2$ and $\theta$, the total temperature drop across both layers. We shall define $\eta = \eta_1/\eta_2$, $\nu = \nu_1/\nu_2$, $\lambda = \lambda_1/\lambda_2$, $\kappa = \kappa_1/\kappa_2$, $a = a_2/a_1$ (1 and 2 refer to upper and lower layers, respectively).

We shall consider the realistic but analytically tractable limit of large Galileo number, $Ga$, small capillary number, C, and Marangoni number, $M$ (Rednikov et al., 1998). It will be assumed that $m = M/G = O(1)$ and $Bo = GaC = O(1)$. In this limit the frequencies of both modes, the dilational one and the capillary-gravity one, are large (except the case of small wavenumbers), because $\omega_{long} \sim M^{1/2}$, $\omega_{cg} \sim Ga^{1/2}$. Because of that, there appear boundary layers ("skin-layers") at the interface and near the rigid boundaries, while the bulk of each layer can be well approximated by inviscid flow. The thickness of boundary layers is proportional to $\omega^{-1/2} \sim Ga^{-1/4}$. Thus,

$$\epsilon \equiv Ga^{-1/4} \tag{5.55}$$

is the natural smallness parameter of the problem. All the variables are expanded as power series in $\epsilon$ and take $\omega = Ga^{1/2}\Omega$. For the description of the variables in boundary layers, we shall define stretched coordinates $Z_0 = z/\epsilon$ in the boundary layer near the interface between fluids, $Z_1 = (z - 1)/\epsilon$ and $Z_2 = (z + a)/\epsilon$ in the boundary layers near the top and bottom rigid boundaries, correspondingly.

In the leading order, the dilational (longitudinal) and capillary-gravity (transverse) interfacial modes are non-dissipative and non-mixed. The dispersion relation (5.52) has the following form:

$$(\Omega^{(0)2} - \Omega_{long}^2)(\Omega^{(0)2} - \Omega_{tr}^2) = 0, \tag{5.56}$$

where

$$\Omega_{long}^2 = \frac{sm(\sqrt{\nu} - \sqrt{\chi})\kappa\eta k^2}{P(1 + \sqrt{P})(\sqrt{\nu} + \sqrt{\kappa P})(\eta + \sqrt{\nu})(\lambda + \sqrt{\kappa})(1 + \lambda a)} \tag{5.57}$$

and

$$\Omega_{tr}^2 = \frac{(\rho^{-1} - 1)\, k + Bo^{-1}k^3}{\coth k + \rho^{-1}\coth ka}. \tag{5.58}$$

Later on, we assume that the condition of existence of longitudinal waves for arbitrary large $M$

$$s(\sqrt{\nu} - \sqrt{\kappa}) > 0 \tag{5.59}$$

is satisfied. In this case, $\Omega_{long}^2 = \Omega_{tr}^2$ when $m = m_{res}$,

$$m_{res} = s\Omega_{tr}^2 \frac{P(1+\sqrt{P})(\sqrt{\nu}+\sqrt{\chi P})(\eta+\sqrt{\nu})(\kappa+\sqrt{\chi})(1+\kappa a)}{(\sqrt{\nu}-\sqrt{\kappa})\lambda\eta k^2}.$$

(5.60)

In the next order, one obtains complex corrections to the modes' frequencies which correspond to the decay or growth of waves, as well as to the shift of frequencies. If $m$ is not too close to $m = m_{res}$, the growth rate $\lambda^{(1)}$ (the imaginary part of the above-mentioned correction) satisfies the following equations:

$$2\lambda^{(1)}\left(1 - \frac{m_{res}}{m}\right)(\coth k + \rho^{-1}\coth ka) +$$

$$k\sqrt{\frac{\Omega_{long}^{(0)}}{2}}\left[\frac{1}{1+\rho\sqrt{\nu}} + \frac{\sqrt{\nu}+\sqrt{\kappa P}}{\rho\sqrt{\nu P}(\sqrt{\nu}-\sqrt{\kappa})}\right] \times$$ (5.61)

$$[(\rho-1)(1+\nu^{1/2})\coth k \coth ka +$$

$$\frac{m_{res}}{m}(\coth ka - \sqrt{\nu}\coth k)(\coth ka + \rho\coth k)] = 0$$

for the longitudinal mode and

$$2\lambda^{(1)}\left(1 - \frac{m}{m_{res}}\right)(\coth k + \rho^{-1}\coth ka) + k\sqrt{\frac{\Omega_{tr}^{(0)}}{2}}\left(1 - \frac{m}{m_{res}}\right) \times$$

$$[\coth k(\coth k + \coth ka) + \csc^2 k + 1/(\rho\sqrt{\nu})\csc^2 ka]$$ (5.62)

$$+k\sqrt{\frac{\Omega_{tr}^{(0)}}{2}}\left[1 + \frac{m}{m_{res}} \cdot \frac{(1+\rho\sqrt{\nu})(\sqrt{\nu}+\sqrt{\kappa P})}{\sqrt{P}(\sqrt{\nu}-\sqrt{\kappa})}\right] \times$$

$$\frac{(\coth ka - \rho\sqrt{\nu}\coth k)(\coth k + \coth ka)}{1 + \rho\sqrt{\nu}} = 0$$

for the transverse mode. The formulas (5.61) and (5.62) determine the regions of the excitation and the decay for both wave modes. Note that the formula (5.61) is correct only if $\Omega_{long}^2 > 0$, i.e. in the region where the signs of $s$ and $\sqrt{\nu} - \sqrt{\chi}$ coincide. Also, the expressions for $\lambda^{(1)}$ are correct if $k \gg G^{-1/2} \sim M^{-1/2}$, because in their derivation it was assumed that the thicknesses of the boundary layers are small with respect to the thicknesses of the bulk fluid layers. Expressions (5.61) and (5.62) diverge as $m \to m_{res}$. In that limit, the *mixing* of both types of modes should be taken into account.

### iii. Oscillatory instabilities generated by the Marangoni effect with surfactants

In the previous sections, we considered in detail the case where the Marangoni instability was caused by heat transfer through the interface. Now we shall study the case when the Marangoni instability is *caused* by the mass transfer of a surfactant through the interface. The temperature disturbances will be neglected.

There are similarities and differences between the cases of heat and mass transfer. First, the field of temperature is continuous at the interface, while the concentration of a surfactant has a jump across the interface. However, this difference is easily eliminated by a certain transformation of variables. That is why the basic results of Sternling and Scriven (1959) are equally valid for the longitudinal instabilities generated by heat or mass transfer (see also Hennenberg *et al.*, 1979).

The most important difference between the mass transfer of a surfactant and the heat transfer is the adsorption-desorption process with eventual accumulation of the surfactant at the interface, (Sect. 1.2.2) which has no analogue in the case of the heat transfer. The distribution of a surfactant on the interface is governed by an *additional* evolution equation, Eq. (1.42), which significantly influences the development of instabilities.

An extensive investigation of instabilities in a system of two semi-infinite fluid layers was carried out by Chu and Velarde (1988, 1989) where the problem was studied by means of the one-layer approach, and that by Hennenberg *et al.* (1992) where the finite time ratio between the adsorption and diffusion processes was taken into account. The crucial parameters which determine the type of instability are the ratio of kinematic viscosities $\nu$ and the ratio of diffusion coefficients $D = D_1/D_2$. These works generalized in proper context the earlier theory by Sternling and Scriven (1959).

In the case $\nu < 1$, both longitudinal and transverse instabilities appear in the system. Typical neutral curves are shown in Fig. 5.12. The low-frequency branch is associated with longitudinal waves, while the high-frequency branch corresponds to transverse waves. In the case $\nu > 1$, $D > 1$, only one oscillatory instability mode is typically observed (Fig. 5.13). If $\nu > 1$ and $D < 1$, the system can be monotonically unstable, but no oscillatory instability was found.

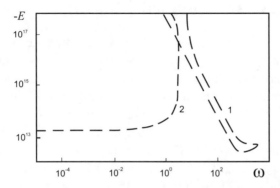

Figure 5.12: Neutral curve for overstability providing the critical (surfactant/solutal) Marangoni number – also called elasticity number and denoted $E$ - for capillary-gravity or transverse (line 1) and dilational or longitudinal (line 2) waves as a function of frequency, $\omega$, in a two-layer system. $\nu = 3/8$. (Redrawn after Chu and Velarde, 1989).

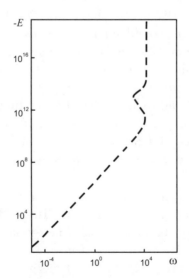

Figure 5.13: Neutral curve as in Fig. 5.12 for $\nu = 15/4$. (Redrawn after Chu and Velarde, 1989).

# 5.3   Oscillatory instabilities in the mixed Rayleigh-Benard-Marangoni convection

In the previous section we studied the generation of oscillatory instability only due to the Marangoni effect. However, surface tension gradients are only dominant in very thin layers. In not too thin layers, the stability of the mechanical equilibrium state is determined by the combined action of thermocapillarity (Marangoni effect) and buoyancy (Rayleigh effect). The joint action of both effects may lead to the appearance of some specific types of oscillations which will be described in the present section. In subsection 5.3.1 we shall consider the case of heating from above where the instability is produced by the mixing of longitudinal interfacial waves and internal waves caused by a stable density stratification of the fluid (Rednikov *et al.*, 2000b). In subsection 5.3.2 the case of heating from below is studied. In the latter case, the oscillations are generated by the competition of Marangoni and Rayleigh effects leading to instability (Simanovskii and Nepomnyashchy, 1993; Boeck *et al.*, 2001).

## 5.3.1   Mode mixing of interfacial and internal waves

In this subsection, we consider the stability of the mechanical equilibrium state under the combined action of the Marangoni stresses and (negative) buoyancy when a horizontal liquid layer is heated *from above*. It will be shown that even though both buoyancy and surface tension gradients are expected to play a stabilizing role, their combined action may lead to an oscillatory instability. This instability is caused by the resonant interaction and mode mixing for two physically different types of *decaying* waves, the above discussed dilational (longitudinal) waves and internal waves in the stably stratified liquid layer.

The nature of the dilational (longitudinal) interfacial waves in the absence of buoyancy ($R = 0$) was discussed in detail in subsection 5.2.3. Here we shall use a non-dimensionalization different from that of subsection 5.2.3: the layer thickness $a$ is chosen as the length unit, and the quantity $a^2/\nu$ as the time unit. In this case the dependence of the real part of the dimensionless frequency, $\omega$, on the non-dimensional wavenumber, $k$, is, recalling (5.50), determined by the relation

$$\omega^2 = \omega_{long}^2 = |M|Pk^2/(P^{1/2} + 1), \qquad (5.63)$$

where $M = \alpha A a^2 / \eta \kappa$. The imaginary part of the frequency is negative which corresponds to decay of oscillations.

Another type of oscillation is produced by (negative) buoyancy as the heating helps stabilizing the stratified liquid layer. Assume $M = 0$. By heating from above, there exists a density gradient which generates *internal waves* with the Brunt-Väisälä frequency

$$\omega^2 = \omega_{int}^2 = |R| P k^2 / (k^2 + k_z^2), \tag{5.64}$$

where $k$ is the horizontal component of the wavevector, and $k_z$ is the vertical component of the wavevector (Landau and Lifshitz, 1987; Normand *et al.*, 1977). The non-deformability of the surface and of the rigid bottom leads to a discrete spectrum of vertical wavenumbers, $k_z = n\pi$, $n = 1, 2, \ldots$. Both viscosity and thermal diffusivity tend to dampen the internal waves.

Let us consider now the case of the combined action of Marangoni stresses and buoyancy, and hence $M$ and $R$ are not equal to zero. The relative strength of both factors is determined by the dynamic Bond number that we recall is $Bd = R/M$. The evolution of disturbances is governed by a general dispersion relation which contains $M$ and $R$ and describes both branches of waves, longitudinal interfacial waves and internal waves. To avoid confusion here we shall not use the word dilational. However, if the values of the frequencies $\omega_{long}$ and $\omega_{int}$ given by the expressions (5.63) and (5.64) are drastically different from each other, then both kinds of waves are clearly distinguishable, and, in principle, both of them decay. The situation is different near the points of *resonance* where $\omega_{long} = \omega_{int}$. The resonance between both kinds of waves takes place for the values of the dynamic Bond number

$$Bd_n = \frac{k^2 + \pi^2 n^2}{P^{1/2} + 1}, \quad n = 1, 2, \ldots. \tag{5.65}$$

The asymptotic analysis performed by Rednikov *et al.* (2000b) in the limit $\epsilon = (P/R)^{1/4} \ll 1$ has shown that near the resonance point there is a continuous transition, as a function of $Bd - Bd_n$, from the *longitudinal* one, and vice versa, and hence there is a *mode mixing*. Moreover, the analysis of the dispersion relation near the point $Bd = Bd_n$ shows that one of the eigenvalues (corresponding to a complex frequency) has a positive imaginary part and the other has a negative imaginary part. Thus, near the resonance point one of the modes is unstable, while the other one is stable (Fig. 5.3(b)).

The results of the asymptotic analysis have been verified by numerical analysis of the full linear stability problem (Rednikov *et al.*, 2000b). The stability diagram in the plane $(M, R)$ obtained by minimization of neutral curves with respect to wavenumbers is shown in Fig. 5.14. Fig. 5.14(a) corresponds to the case $n = 1$. For a given $Bd$, the intersection of the straight line $R = BdM$ with the line 1 of Fig. 5.14(a) yields the critical temperature gradient for the onset of instability. The lower branch has an oblique asymptote with slope 0.715, thus there is no instability as $Bd < 0.715$. The critical wavenumber tends to zero along the lower branch and to infinity along the upper branch. Fig. 5.14(b) also contains the results for $n = 2$ and $n = 3$.

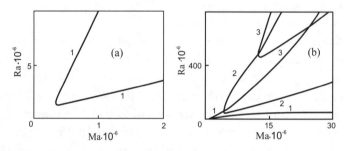

Figure 5.14: Stability boundaries in the plane $(M, R)$ at $P = 6$. (a) Fundamental mode $(n = 1)$, (b) modes $n = 1$ (line 1), $n = 2$ (line 2) and $n = 3$ (line 3). (Redrawn after Rednikov *et al.*, 2000b).

A weakly nonlinear analysis has been performed for several values of $P$ and $Bd$ by Rednikov *et al.* (2000b). In all the cases, a standing wave was found to be stable, and traveling waves turned out to be unstable. This result coincides with that of the direct numerical simulation of finite-amplitude waves obtained by Shklyaev (2001).

### 5.3.2    Competition between Marangoni and Rayleigh instability mechanisms

When the liquid layers are heated *from below*, the competition between surface tension gradients and buoyancy may lead to the appearance of a specific type of oscillations which will be described in the present subsection using a two-layer system. The cases of normal thermocapillarity $(d\sigma/dT < 0)$ and anomalous thermocapillarity $(d\sigma/dT > 0)$ will be considered separately.

## i. Normal thermocapillarity

In each layer, the buoyancy may be characterized by the corresponding "local" Rayleigh numbers $R_m = g\beta_m |A_m| a_m^4 / \nu_m \kappa_m$, $m = 1, 2$ which are not independent (1 and 2 refer to upper and lower layers, respectively). Both Rayleigh numbers are proportional to the Grashof number, $Gr = R/P$, which will be defined using the temperature difference between the bottom rigid boundary and the top surface, $\theta$, and the parameters of the upper layer, $g\beta_1 |\theta| a_1^3 / \nu_1^2$. If the ratio $R_2(Gr)/R_1(Gr)$ differs much from unity, buoyancy–driven natural convection appears first in the layer with the higher value of its corresponding "local" Rayleigh number, and then a weak induced flow arises in the other layer.

Let us consider the case $R_1(Gr) < R_2(Gr)$ where natural convection takes place mainly in the lower layer. As one can see in Fig. 5.15(a), a temperature disturbance on the interface generates buoyancy bulk forces with thermocapillary tangential stresses acting *in the same direction*. In this case, the action of the Marangoni effect leads to a *decrease* of the critical Grashof number.

In the opposite case $R_1(Gr) > R_2(Gr)$, natural convection arises first in the upper layer (Fig. 5.15(b)). Then, buoyancy and thermocapillary tangential stresses oppose each other. Indeed, let us assume that there is a local negative temperature fluctuation ("cold spot") on the interface. Buoyancy generates a downstream flow above the cold spot, and an outgoing flow on the interface near the cold spot. At the same time, the normal thermocapillary effect produces tangential stresses that tend to form an incoming flow on the interface near the cold spot and an upstream flow in the upper layer. Thus, in this case the thermocapillary stresses tend to suppress natural convection. The competition between buoyancy and the Marangoni effect leads to stabilization of the possible stationary, monotonic instability. However, the asynchronous action of two factors working in opposite directions can produce *overstability* and hence oscillatory motions.

Note that if $R_1(Gr) > R_2(Gr)$, but $R_1(Gr)$ and $R_2(Gr)$ are close, the buoyancy acting in the lower layer should be also taken into account. The buoyancy in the lower layer, which acts in the same direction as the thermocapillarity, can contribute to the appearance of oscillations (Fig. 5.15(c)).

The most important factors, which influence the appearance and

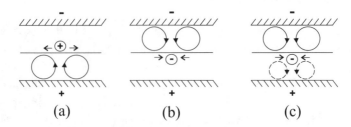

Figure 5.15: Interaction between buoyancy and thermocapillarity in the standard case when surface tension decreases with increasing temperature. (a) $R_1 \ll R_2$; (b) $R_1 \gg R_2$; (c)$R_1 \sim R_2$.

disappearance of oscillations, are the ratio $M/Gr$ and the ratio of thicknesses of the layers, $a$. We shall discuss these factors separately. *i.1 Influence of $M/Gr$ for a fixed ratio of the layers thicknesses.* As an example, let us consider the appearance of convection in the system water-silicone oil (200) at $a = 1.6$ (in this case, $R_2/R_1 \approx 0.23$). If $M = 0$, the neutral curve consists of two separate monotonic fragments. The lower neutral curve has the minimum at $Gr = 270$; it corresponds to convection arising in the upper layer. The convection in the lower layer is excited at the upper neutral curve which has the minimum at $Gr = 2860$.

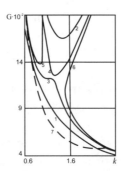

Figure 5.16: The neutral curves for a water-silicone oil (200) system ($a = 1.6$). (Redrawn after Simanovskii and Nepomnyashchy, 1993).

As $M$ increases, the neutral curve for the disturbances in the upper layer slowly goes upward (see lines 1 and 3 in Fig. 5.16), and that for the disturbances in the lower layer quickly goes downward (see lines 2 and 4). For the higher value of $M$ the monotonic curves coincide in a certain point, and then they are separated into a "long-wave" (line 5) and a "shortwave" (line 6) parts. The dependence of the critical value of the Grashof number, $Gr_c$, on the Marangoni number, $M$, for the monotonic neutral curve is shown in Fig. 5.17 (line 1).

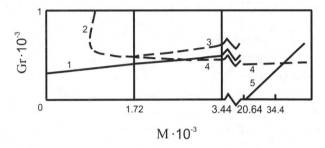

Figure 5.17: Dependence of $Gr_c$ on $M$ for a water-silicone oil (200) system ($a = 1.6$). (Redrawn after Simanovskii and Nepom-nyashchy, 1993).

Even earlier a fragment of an oscillatory neutral curve appears. It is situated below the monotonic neutral curve for the disturbances in the upper layer in a finite interval of the wavenumbers, $k$. At the extreme boundaries of this interval, the frequency of oscillations tends to zero. The location of these points in the plane $(Gr, M)$ is shown in Fig. 5.17 (lines 2, 3). After the separation of the monotonic neutral curves into longwave and shortwave fragments, the oscillatory neutral curve connects both "monotonic" fragments. The minimum of the oscillatory neutral curve (line 4 in Fig. 5.17) becomes lower than that of the shortwave monotonic neutral curve (line 1 in Fig. 5.17) as $M$ is sufficiently large.

In the case $Gr = 0$, stationary surface tension gradient-driven convection appears as $M > 2.2 \cdot 10^4$. When $Gr$ increases, the critical Marangoni number grows (line 5 in Fig. 5.17). The lines 5 and 4 cross at $M \approx 3.6 \cdot 10^4$. For $M > 3.6 \cdot 10^4$ the equilibrium state is unstable for any value of the Grashof number.

Thus, there is a large interval of Marangoni number values, where the oscillatory instability is the most "dangerous". Let us empha-size, that in the case of normal thermocapillary effect the oscillations appear when the natural convection is excited mainly in the upper layer.

*i.2 Influence of a.* The influence of the thicknesses' ratio on the appearance and disappearance of oscillations was studied by Juel *et al.* (2000). Calculations have been performed for the system *n*-hexane-acetonitrile. The total thickness of both fluid layers was kept constant, while the ratio of layer thicknesses was varied. Oscillations have been found by heating the two-layer system from below in the interval $0.5 < a < 1.2$ (the width of this interval slightly depends on the total two-layer thickness). The ratio of the "local" Rayleigh numbers is $R_2/R_1 \sim 0.456a^4$. Oscillations are observed in the interval $0.03 < R_2/R_1 < 0.95$. One can expect the appearance of oscillations caused by the competition of buoyancy in the upper layer, from one side, and thermocapillarity and buoyancy in the lower layer, from the other side (Fig. 5.15(c)). Note that the ratio of thermal diffusivities $\kappa = 0.743 < 1$. It can be shown that the stationary surface tension gradient-driven convective instability appears only by heating from above if $a < a_* = 1/\sqrt{\kappa} \sim 1.16$ (Simanovskii and Nepomnyashchy, 1993). This circumstance is favourable for the onset of oscillatory convection by heating the system from below. In the case $a > a_*$ the stationary surface tension gradient–driven (Marangoni) instability with small wavenumbers becomes possible and hence can compete with the oscillatory instability. Indeed, when $a$ increases, one observes a transition to relatively longwave monotonic instability. The transition value of $a$ is rather close to $a_*$, despite the influence of buoyancy which is not taken into account when calculating $a_*$. It is found that the oscillations disappear also by diminishing $a$.

## ii. Anomalous thermocapillarity

Let us now discuss the interaction between buoyancy and thermo-capillary instability mechanisms in a two-layer system in the case of *anomalous* thermocapillarity $(d\sigma/dT > 0)$ and hence when the interfacial tension increases with increasing temperature. The possibility of the anomalous thermocapillary effect should be taken into account in realistic models of multilayer convection, because there are indications that the occurence of an anomalous thermocapillary effect might be a typical property of various liquid-liquid systems. Specifically, it was observed in aqueous alcohol solutions, nematic liquid crystals, binary metallic alloys etc. (see, e.g., Legros (1986); Eckert and Thess, 2001).

In the case of the anomalous thermocapillary effect, the oscillatory instability caused by the competition between buoyancy and thermocapillary tangential stresses will appear when natural convection is generated mainly in the lower layer.

As an example, let us recall results obtained for the 10cS silicone oil-ethylenglycol system with the the thickness ratio of $a = 1.8$ corresponding to a ratio of local Rayleigh numbers $R_2/R_1 = 3.19$. Some typical neutral curves are shown in Fig. 5.18. One can see that in the region $15 < Gr < 15.5$ the monotonic neutral curves (solid lines) change rather slowly with $Gr$. As $Gr < Gr_1 = 15.1$, the oscillatory neutral curve does not appear. At $Gr > Gr_1$, a closed region of oscillatory instability (dashed line) appears. It grows rapidly with increasing $Gr$ and at last touches the monotonic neutral curve as $Gr = Gr_2 = 15.3$. The stability boundaries in the $(M\text{-}Gr)$-plane for the monotonic and oscillatory instabilities are shown in Fig. 5.19. One can conclude that when $K = |M|/Gr < K_* = 31.7$ only the former will be observed. For the specific parameter values of the system, one finds that the latter case will take place if the total thickness of the two-layer system is larger than 5.9 mm. If the thickness of the two-layer system is smaller than the critical one, some slow oscillations appear near the threshold. The *dimensional* frequency is $\Omega = \omega \nu_1/a_1^2$. For instance, if the total thickness $a_1 + a_2 = 4$ mm ($K = 70.3$) we find $\Omega = 0.21$ sec$^{-1}$.

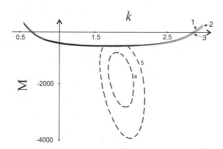

Figure 5.18: Neutral curves in the case $a = 1.8$: $Gr = 15$ (line 1), $Gr = 15.2$ (lines 2, 4), $Gr = 15.5$ (lines 3, 5). (Redrawn after Braverman et al., 2000).

In order to describe the main stages of oscillations, let us recall results of 2D simulations performed in the case $Gr = 18$ and $M = -3933$ (Braverman et al., 2000).

Figs. 5.20(a1), 5.20(a2) provide an illustration of the structure of the convective motion (let us call it structure A). The upward

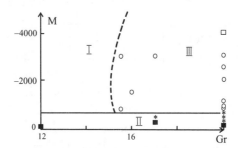

Figure 5.19:  Diagram of stability regions: I - stability, II - monotonic instability, III - oscillatory instability. (Redrawn after Boeck
*et al.*, 2001).

motion in the lower layer generates the temperature field on the
interface which has a maximum in the middle of the interface. In the
present case $(d\sigma/dT < 0)$, the tangential stresses are directed toward
this maximum. These stresses produce a four-vortex motion near
the interface, so that a three-storey structure is produced (see Figs.
5.20(b1), 5.20(b2)). Because the Prandtl numbers of both fluids are
rather high, the field of temperature is much more inertial than that
of the stream function. That is why the temperature field generated
by the structure A exists during some time and supports both the
buoyancy-induced motion in the lower part of the second layer and
the thermocapillarity-induced motion around the interface. Finally
the former is completely washed out by the latter (Figs.  5.20(c1),
5.20(c2)). Consequently, the temperature maximum in the middle
of the interface disappears. The thermocapillary motion near the
interface decays, while in the lower layer another type of convection
appears (Figs.  5.20(d1), 5.20(d2)); it will be called structure B. The
transition between the structures A and B takes place during the
first half of the period. The subsequent evolution can be understood
in similar way: the temperature field generated by the structure B
produces a thermocapillary motion near the interface (Figs.  5.20(e1),
5.20(e2)) which replaces the buoyancy-induced motion in the lower
layer (Figs.  5.20(f1), 5.20(f2)), but, subsequently, the temperature
field in the lower layer reorganizes itself and the structure A (Figs.
5.20(a1), 5.20(a2)) is restored.

In reality, the patterns generated by the instability are typically
three-dimensional. Let us describe results of three-dimensional nonlinear simulations obtained by means of a pseudospectral numerical

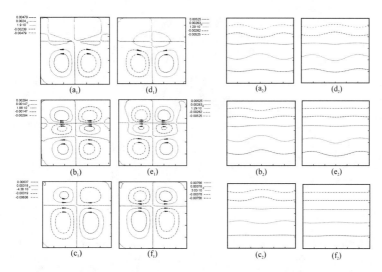

Figure 5.20: (a1)-(f1) Streamlines, and (a2)-(f2) isotherms for the oscillatory motion in the system with $a = 1.8$; $Gr = 18$, $M = -3933$. (Redrawn after Braverman et al., 2000).

method (Boeck et al., 2001).

First, we shall describe the results for convective regimes generated by the monotonic instability (region II, Fig. 5.19). Typical stationary patterns are shown in Fig. 5.21. Generally, two types of stationary patterns appear. For relatively large values of $Gr$, a roll pattern is observed, while for relatively high values of the Marangoni number $M$, the typical pattern is hexagonal. This numerical result agrees with the results of a weakly nonlinear theory (Parmentier et al., 1996) and direct numerical integration (Tomita and Abe, 2000) for the combined buoyancy and surface tension gradient–driven (Rayleigh-Benard-Marangoni) convection in one-layer systems.

An increase of the Marangoni number leads to the appearance of an alternating roll pattern (see Fig. 5.22) which turns out to be the typical kind of oscillatory pattern in region III. This pattern is a nonlinear superposition of two systems of standing waves with orthogonal wave vectors. The temporal phase shift between standing waves of different spatial orientations is equal to $T/4$, where $T$ is the period of oscillations. Accordingly, one observes some kind of roll patterns that change their orientation with the time interval $T/4$.

In a certain intermediate region, starting from stationary hexagons,

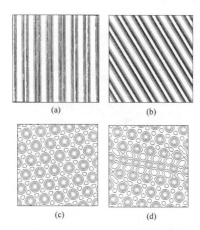

Figure 5.21:   Contour plots of interfacial temperature for (a) $Gr = 20$, $M = -200$; (b) $Gr = 17$, $M = -400$; (c) $Gr = 15.5$, $M = -400$; (d) $Gr = 20$, $M = -400$. (Redrawn after Boeck *et al.*, 2001).

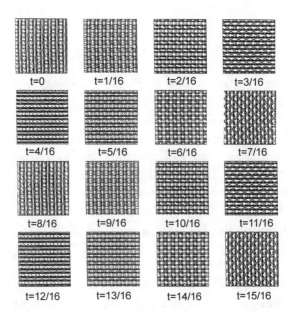

Figure 5.22: Snapshots of the interfacial temperature field during one period of oscillations of alternating rolls for $Gr = 20$, $M = -1000$. (Redrawn after Boeck *et al.*, 2001).

periodic pulsations of the hexagonal pattern are observed. The hexagonal structure is conserved, but some periodic changes of the isotherms take place. In addition to the oscillation of these modes, the amplitudes of the basic modes of the hexagonal pattern also oscillate slightly about a nonzero mean value. This regular pattern was not observed when a simulation was started from random initial conditions. In this case, an irregular hexagonal pattern with chaotic time dependence persists.

## 5.4 Longitudinal instability in two-layer systems

In two-layer systems, longitudinal waves can become unstable without mixing with transverse waves.

Longitudinal oscillatory instability in two-layer fluid systems was reported by Sternling and Scriven (1959) almost simultaneously with the appearance of Pearson's theory (1958) for monotonic instability leading to steady cellular (Benard) patterns, both due to the Marangoni effect. Because of the thermocapillary effect, the *hot* fluid in the lower layer goes up towards the hot spot, while the *cold* fluid in the upper layer goes down towards the hot spot. The crucial point is that the two-layer system is characterized by four different time scales (viscous and heat diffusion time scales for each fluid). Accordingly, the counteraction of convective heat fluxes generated by motions in the upper and lower layers has some *time delay*. Since the fields of velocity and temperature in two fluids do not change synchronously, some *oscillations* can arise instead of a monotonic growth or decay of the disturbance. One can expect that if the temperature gradient is large enough, these oscillations may grow. The theory of Sternling and Scriven (1959) dealt with a system of two layers of *infinite thicknesses* ($d \to \infty$) but disturbances were of finite wavelengths $\lambda$. Thus, it actually described the *short-wave asymptotics* of the neutral curve ($\lambda \ll d$). However, as well as in the case of the stationary instability, the most dangerous disturbances have a wavelength comparable with the thicknesses of the layers. In the case $\lambda \sim d$ the criteria for instability significantly changes relative to the result in the limit $\lambda \ll d$ (Reichenbach and Linde, 1981; Nepomnyashchy and Simanovskii, 1983a). However, the physical nature of the oscillatory instability is not changed: it is the competition between processes that take place in both fluids, as thoroughly discussed by Chu and Velarde (1989).

For a fixed value of the wavenumber $k$, the monotonic surface tension gradient–driven (Marangoni) instability of the motionless state can appear only for a definite way of heating, either from below or from above, but not for both of them simultaneously. It takes place, e.g., for $s > 0$, as $k < k_d$, and for $s < 0$, as $k > k_d$; at $k = k_d$ the monotonic neutral curve has a discontinuity. The absence of the monotonic instability for a definite value of $k$ is a circumstance favourable for the appearance of the oscillatory instability. In general, there exists a codimension-2 point $(k_*, sM_*)$ where the frequency of the oscillations tends to zero, and the oscillatory neutral curve (dashed line) terminates in the monotonic one. The behaviour of the growth rate as a function of the Marangoni number, $M$, for $k > k_*$ (monotonic instability) and for $k < k_*$ (oscillatory instability) is as shown in Fig. 5.1(b) (the dashed line corresponds to an oscillatory mode, while the solid lines correspond to monotonic modes). Note that on the monotonic neutral curve $sM = sM_m(k)$, where the growth rate $\lambda(M_m, k)$ is equal to zero, the quantity

$$\lambda'_M(k) \equiv \left( \frac{\partial \lambda}{\partial M} \right)_{M=M_m(k)} \tag{5.66}$$

is positive in the case $k > k_*$ and negative in the case $k < k_*$. Thus, the role of the monotonic neutral curve is different in both cases. In the case $k > k_*$, it is indeed an instability boundary. In the case $k < k_*$, the monotonic neutral curve is the boundary of the *stabilization* for one of the two monotonic modes. We can see that the inequality $\lambda'_M(k) > 0$ held on the monotonic neutral curve is the sign of a true monotonic instability and of the absence of oscillatory instability. On the contrary, the inequality $\lambda'_M(k) < 0$ is an indication of the existence of an oscillatory instability for the same value of $k$ and a lower value of $M$.

The expression for the $\lambda'_M(k)$ on the monotonic neutral curve can be obtained analytically, but it is rather cumbersome (Simanovskii and Nepomnyashchy, 1993). It has been shown that $\lambda'_M(k)$ changes its sign at the value of $k = k_d$ corresponding to the discontinuity of the monotonic neutral curve. Accordingly, *for any two-fluid system* there exists an oscillatory instability near the point $k = k_d$ for a certain way of heating. The opposite way of heating does not produce an oscillatory instability. Recall that the discontinuity of the monotonic neutral curve at a certain $k = k_d$ takes place if $\kappa > 1$, $a < a_* = \kappa^{-1/2}$ or $\kappa < 1$, $a > a_*$.

Some relatively simple criteria for the appearance of the oscillatory instability can be obtained in the limits of small $k$ (long-wave disturbances) and large $k$ (short-wave disturbances). The long-wave disturbances are especially important in the case where $a$ is close to $a_*$. It has been shown in this case (Simanovskii and Nepomnyashchy, 1993) that the condition $\lambda'_M(k) < 0$ of the oscillatory instability is fulfilled in the cases $\nu > \kappa$, $s = -1$ and $\nu < \kappa$, $s = 1$. In other words, oscillations arise in a long-wave region if the heating is performed from the side of the fluid with larger value of the Prandtl number.

In the short-wave limit ($k \to \infty$), the analysis of the expression $\lambda'_M(k)$ shows that there is no short-wave oscillatory instability in the case $\kappa < 1$, $s = -1$, while in the case $s = 1$ the short-wave oscillatory instability arises as $1 < \kappa < \kappa_m$, where

$$\kappa_m = 1 + (1 + \nu\kappa)\left(\frac{\eta + \nu}{1 + \eta} + 4P\right)^{-1} \tag{5.67}$$

(Simanovskii and Nepomnyashchy, 1993).

One more criterion can be obtained by the consideration of the short-wave asymptotics of the oscillatory neutral curve, which corresponds to the limit of semi-infinite fluid layers (Sternling and Scriven, 1959). The short-wave oscillatory instability appears if the quantities $sM_0 f_1$ and $f_2$ have opposite signs, where

$$sM_0 = \frac{8(1 + \lambda a)(1 + \lambda)(1 + \eta)}{\lambda\eta(\kappa - 1)} \tag{5.68}$$

determines the leading order short-wave asymptotics of the neutral curve $(sM \sim sM_0 k^2)$, and

$$f_1 = \frac{P}{2}\left(\frac{\lambda + \kappa}{1 + \lambda} + 1 + \kappa\right) + \frac{\eta + \nu}{4(\eta + 1)} + \frac{\kappa\nu - 1}{4(\kappa - 1)}, \tag{5.69}$$

$$f_2 = \frac{1 + \lambda a}{\lambda\sqrt{2}}\kappa^{1/2}(\kappa^{1/2} + \lambda)(\nu^{1/2} + \eta)(\nu^{1/2} + \kappa^{1/2}P^{1/2})(1 + P^{1/2})\times$$

$$\frac{[(\eta + 1)(\eta + \nu^{1/2})^{-1} + P^{-1/2}](\kappa^{1/2} - \nu^{1/2}) + (1 - \kappa^{1/2})(1 + P^{-1/2})}{(\kappa^{1/2} - \nu^{1/2})^2}. \tag{5.70}$$

Note, that in the case $\kappa = 1$, $a = 1$ there is no monotonic instability at all. The oscillatory instability was found in this case numerically for $\eta = \nu = 0.5$, $\lambda = P = s = 1$ (Nepomnyashchy and Simanovskii, 1983a). The latter case was studied also by means of numerical nonlinear computations (Nepomnyashchy and Simanovskii, 1983b) in a closed cavity with aspect ratio $L = 2.5$. The instability appears as $M > M_1 \approx 3.14 \cdot 10^4$ and leads to the appearance of an oscillatory four-vortex flow. With the increase of the Marangoni number the oscillations become essentially nonlinear, and their period $\tau$ grows. In the region $4.5 \cdot 10^4 < M < M_2 \approx 5.0 \cdot 10^4$ the period of oscillations is well approximated by the formula

$$\tau^{-2} = 0.31(M_2 - M). \tag{5.71}$$

As $M = M_2$, the limit cycle corresponding to the oscillatory motion becomes the separatrix of a saddle-node fixed point. In the region $M > M_2$, the limit cycle does not exist, and a steady convective flow arises. A more complicated sequence of nonlinear oscillations was observed in the case $L = 2$; it includes a period doubling bifurcation with the violation of the flow symmetry.

Note that in the case of a closed cavity, the steady deformation of the interface caused by the contact angle on the lateral boundaries can drastically influence the flow regimes. This deformation generates a thermocapillary flow below the threshold of the instability, which suppress the oscillatory instability mode. The numerical simulations done for the above-mentioned model system by Nepomnyashchy and Simanovskii (1984) showed that the oscillations are completely suppressed as the contact angle is less than 84°.

For a certain ratio $a = a_0$ of the layer thicknesses $a$, the monotonic instability threshold and the oscillatory instability threshold coincide. The case where $a$ is close to $a_0$ was studied by Colinet et al. (1996) for the system methanol-octane ($\nu = 1.14$, $\eta = 1.02$, $\kappa = 0.934$, $\lambda = 0.698$; $a = a_0 \approx 0.726$; $k_o = 2.67$; $k_m = 7.03$) by means of a weakly-nonlinear analysis. The investigation of the nonlinear interaction of disturbances with $k = k_m$ and $k = k_o$ (non-resonant case, $k_m \neq 2k_o$), led to no stable regimes were found when the oscillatory instability is the most "dangerous" one (as $a < a_0$). In the opposite case (as $a > a_0$) it was found that the first bifurcation generates a spatially periodic steady flow, and the second bifurcation produces a flow which is quasiperiodic in space but periodic in time. For a specific choice of wavenumbers $k_1 = 3$ and $k_2 = 2k_1 = 6$ of the interacting monotonic and oscillatory disturbances (resonant case; the monotonic and oscillatory instability boundaries coincide as $a_0 = 0.7488$), a much wider variety of dynamical behaviors, including quasiperiodic relaxational oscillations and temporal chaos, have been found. Chaos appears with the growth of the Marangoni number after an infinite sequence of "loop-doubling" homoclinic gluing bifurcations (Lyubimov and Zaks, 1983). Further increasing the Marangoni number, a reverse cascade is observed.

## 5.5 Oscillatory instability in the presence of both thermal gradient and surfactant transport

### 5.5.1 Nondeformable interface

In this subsection, we discuss the influence of surfactants on the thermal convection with heat transfer in systems with an interface. Recall that if the amount of surfactant absorbed at the interface is large enough, it forms a monolayer too "rigid" and suppresses completely any motion at the interface. However, if the excess surface concentration of the surfactant, $\Gamma$, is small, its molecules form a "surface gas". As discussed in subsection 1.2.2, the transfer of the surfactant at the interface is governed by the equation

$$\frac{\partial \Gamma}{\partial t} + \nabla_s \cdot (\mathbf{v}_s \Gamma) = D_s \nabla_s^2 \Gamma + j, \qquad (5.72)$$

where we use here $\mathbf{v}_s = \mathbf{v} - v_n \mathbf{n}$ to denote the tangential component of the fluid velocity at the interface, $\nabla_s = \nabla - \mathbf{n}(\mathbf{n} \cdot \nabla)$, $D_s$ is the surface diffusion coefficient, and $j$ is the mass flux of the surfactant from the bulk to the interface.

Generally, the addition of surfactants diminishes the surface tension, and that may lead to the suppression of stationary convective flows and to the generation of oscillations and waves, as already discussed (see also Berg and Acrivos, 1965; Palmer and Berg, 1972). This effect can be explained in the following manner. Let us consider a flow in the bulk that generates at the interface a velocity field with a non-vanishing interfacial divergence $q = \nabla_\perp \cdot \mathbf{v}_\perp$. Assume that the surface diffusion coefficient is small, so that the diffusion of the surfactant is negligible as compared to the advection of the surfactant by the flow, and the surfactant is insoluble ($j_n = 0$). Assume also, that the initial distribution of the surfactant is homogeneous.

Due to the advection of the surfactant, the surface concentration decreases in the region of a divergent interfacial flow where $q > 0$, and it increases in the region of a convergent flow, $q < 0$. The tangential stresses generated by the inhomogeneity of the surfactant concentration are directed opposite to the fluid motion. Thus, they will prevent the development of a monotonic instability and suppress stationary flows. At the same time, such a "negative feedback" can lead to an oscillatory instability ("overstability"), as a particular case of those discussed in Sect. 5.1.

Note that the arguments given above do not depend on the physical nature of flow. Thus, we can expect that for both buoyancy-driven convection and surface-tension-gradient-driven convection the monotonic instability will be suppressed and replaced by oscillatory instability.

A thorough study of the influence of surfactants on the convective stability in two-layer systems was given by Nepomnyashchy and Simanovsky (1986, 1988, 1989a,b) and by Gilev *et al.* (1986).

In the case of surface tension gradient–driven (Marangoni) instability (Nepomnyashchy and Simanovsky, 1986), the linear stability of the mechanical equilibrium state with the constant surface concentration $\Gamma_0$ of the insoluble surfactant has been developed. An analytical expression was obtained for the monotonic neutral stability curve which shows that the increase of the monotonic instability threshold is proportional to $B/D_s$, where

$$B = \frac{\alpha_s \Gamma_0 a_1}{\eta_1 \nu_1}, \quad \alpha_s = -\left(\frac{\partial \sigma}{\partial \Gamma}\right)_{\Gamma = \Gamma_0}; \quad D_s = \frac{D_0}{\nu_1}. \tag{5.73}$$

Thus, the monotonic instability is strongly suppressed if $D_s$ is large enough.

It has been shown also that, for arbitrary small $B$, an oscillatory neutral curve appears in the longwave region $(0 < k < k_*(B))$. In that region, this curve is always below the monotonic one. It appears that the longwave asymptotics of the oscillatory neutral curve does not depend on $B$ in the leading order, and it coincides exactly with the longwave asymptotics of the monotonic neutral curve taken with $B = 0$:

$$sM = sM_o = -\frac{1}{1 - \kappa a^2} \frac{80(1 + \lambda a)^2 (1 + \eta a)}{\eta a^2 \lambda} k^{-2}. \tag{5.74}$$

Thus, in the longwave region the oscillatory instability just replaces the monotonic one. The longwave asymptotics of the frequency of oscillations $\omega$ is

$$\omega \sim \pm \omega^{(1)} k,$$

$$\omega^{(1)} = \pm B^{1/2} a \left\{ \frac{2}{15} (\eta a^2 + \nu^3) + \frac{1}{315} \frac{1 + a\eta}{1 - \kappa a^2} \times \right.$$

$$\left[ 10 P \frac{11(1 - \lambda \kappa^2 a^5) + 53a(\lambda - \kappa^2 a^3) + 42\kappa a^2 (1 - \lambda a)}{1 + \lambda a} + \quad (5.75) \right.$$

$$\left. \left. 19(1 - \nu \kappa a^4) \right] \right\}.$$

The range of parameter values for oscillatory instability grows with $B$, and, eventually, the oscillatory instability becomes the most "dangerous" one.

Let us discuss now the case of buoyancy-driven (Rayleigh) instability (Gilev $et$ $al.$, 1986). For small values of $B$, the behaviour of the oscillatory neutral curve is quite similar to that in the case of the surface tension gradient-driven (Marangoni) instability. The oscillations appear in the longwave region $0 < k < k_*(B)$, $k(B)$ grows with $B$, and the oscillatory instability becomes the most "dangerous" one when $B$ exceeds a certain value $B_1$. However, for large enough values of $B$, when the surfactant film becomes "rigid", there is an essential difference between the cases of Marangoni and Rayleigh instabilities. In contradistinction to the monotonic Marangoni instability, the monotonic Rayleigh instability does not disappear when the boundary between two layers is rigid. The oscillatory instability, which is essentially connected with the fluid motion at the interface, cannot exist in the limit of large $B$. That is why as $B$ increases, $k(B)$ eventually begins to decrease, and for some $B > B_2$ the monotonic mode again becomes the most dangerous (see Fig. 5.23).

The transition from the monotonic to the oscillatory instability in the presence of surfactants takes place also in the case of the so-called "anticonvective" buoyancy instability found by heating from above (Simanovskii and Nepomnyashchy, 1993). Because the monotonic anticonvective instability is essentially connected with the hydrodynamic interaction of two fluid layers and cannot exist when the boundary between fluid layers is rigid, it is not restored at large $B$, similarly to the case of Benard–Marangoni convection.

The influence of the solubility of the surfactant was also investigated by Nepomnyashchy and Simanovskii (1988, 1989a), but we shall not deal with that case here.

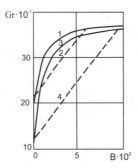

Figure 5.23: Grashof number versus parameter $B$ for $M = 0$ (lines 1, 2) and $M = 416$ (lines 3, 4). (Redrawn after Simanovskii and Nepomnyashchy, 1993).

### 5.5.2 Deformable interface

Let us now consider the influence of surfactants on the Marangoni convection in systems with a deformable interface. The stability is influenced now by the (modified) Galileo number that here is $Ga = ga_1^3/\nu_1\kappa_1$, and also by the capillary number here taken as, $C = \rho_1\nu_1\kappa_1$, and the ratio of fluid densities, $\rho = \rho_1/\rho_2$.

### i. Analytical results

The monotonic instability boundary can be determined analytically. As in the case of the non-deformable interface, the presence of a surfactant always leads to the increase of the monotonic instability threshold proportional to $B/D_s$, which is large as $D_s$ is small. In the limit $C \ll 1$ (strong interfacial tension) the deformation of the boundary is significant only in the long-wave region, $k \ll 1$. For $k = O(C^{1/2})$, the monotonic neutral curve is determined by the expression

$$sM = sM_m(k) =$$

$$\frac{2[Ga(\rho^{-1} - 1) + C^{-1}k^2](1 + \eta a + Ba/4D_s)(1 + \kappa)^2 a}{(1 + a)(1 - \eta a^2)}. \quad (5.76)$$

In the presence of surfactants on the interface the monotonic mode is replaced by an oscillatory one, as in the cases studied in subsection 5.5.1. Analytical results can be obtained in the limit $C \ll 1$, $k \ll 1$. The cases $k = O(C^{1/2})$ and $k = O(C^{1/4})$ should be considered separately.

In the region $k = O(C^{1/2})$ one obtains the following expressions for the oscillatory neutral curve $sM = sM_o(k)$ and for the frequency $\omega$ of neutral disturbances:

$$sM = sM_o(k) =$$

$$\frac{(1 + \kappa a)^2}{\eta a^2 \kappa (1 + a)(1 - \eta a^2)} \left\{ \frac{2}{3} \eta a^3 (1 + \eta a)[Ga(\rho^{-1} - 1) + C^{-1}k^2] \right. \quad (5.77)$$

$$\left. + P[2aB(1 + \eta a^3) + 2D_s(1 + 4\eta a + 6\eta a^2 + 4\eta a^3 + \eta^2 a^4)] \right\};$$

$$\omega^2 = B \frac{a^2 k^4}{1 + 4\eta a + 6\eta a^2 + 4\eta a^3 + \eta^2 a^4} \left\{ \frac{\eta a^2}{6}[Ga(\rho^{-1} - 1) + Wk^2]P^{-1} - \right.$$

$$\quad (5.78)$$

$$\left. - 2D_s(1 + \eta a^3) \right\} - D_s^2.$$

The comparison of the expressions (5.76) and (5.77) shows that $M_o < M_m$ in the whole region of the existence of the oscillatory instability, $\omega^2 > 0$.

In order to match the expressions (5.77) and (5.78) with the expressions (5.74) and (5.75), it is necessary to investigate the intermediate region $k = O(W^{-1/4})$, where one has

$$sM = \frac{80(1 + \lambda a)^2(1 + \eta a}{\eta a^2 \lambda (\kappa a^2 - 1)} \frac{k^2}{k^4 + k_*^4 S}; \quad (5.79)$$

$$\omega^2 = Ba^2 k^2 [L_1 + (L_2 + L_3 k^4)(k^4 + k_*^4 S)^{-1}]^{-1}. \quad (5.80)$$

with

$$k_*^4 = \frac{120(1 + a)C}{\eta a^3} \left| \frac{1 - \eta a^2}{\kappa a^2 - 1} \right|; \quad S = \text{sign} \left( \frac{1 - \eta a^2}{\kappa a^2 - 1} \right);$$

$$L_1 = \frac{a^2}{15} \left[ (\eta + a\nu) + \eta(1 + a)\frac{1 - \nu a^2}{1 - \eta a^2} \right];$$

$$L_2 = \frac{160(1 + \eta)(1 - \lambda)(1 - \eta a^2)PC}{\eta a(1 + \lambda a)};$$

$$L_3 = -\frac{2aP(1 + \eta a^2)}{63(\kappa a^2 - 1)(1 + \lambda a)} \left\{ P[11(1 - \lambda \kappa a^5) \right.$$

$$+ 53a(\lambda - \kappa a^3) + 42\kappa a^2(1 - \lambda a^2)] +$$

$$(1 + \lambda a) \left[ 4(1 - \kappa \nu a^4) - \frac{21(1 - \nu a^2)(1 - \kappa \eta a^4)}{10(1 - \eta a^2)} \right] \right\}.$$

If $S > 0$, the oscillatory neutral curve is continuous, and the function $M_o(k)$ has a maximum as $k = k_*$. If $S < 0$, the oscillatory neutral curve $sM = sM_o(k)$ has a discontinuity as $k = k_*$.

The expressions (5.79) and (5.80) tend to (5.77) and (5.78) as $k \ll C^{1/4}$, and to (5.74) and (5.75) as $k \ll C^{1/2}$.

## ii. Numerical results

Let us describe now some results of the numerical calculation of neutral curves for a few particular, albeit significant cases.

### ii.1 Air-water system heated from below

Let us consider now the influence of a surfactant on the instability in the air-water system by heating from below. Fix $a = 1$ and $W = 10^6$ (that corresponds to the thickness of each layer about 3 mm) and vary the parameter $Ga$ (which means varying and, eventually, the effective–gravity conditions).

The neutral curves obtained in the case of heating from below are shown in Fig. 5.24(a) ($Ga = 10$; the lines 1 - 4 correspond to $B = 0; 1; 5; 10$). In the absence of the surfactant ($B = 0$) the instability is monotonic. The neutral curve has two minima, the long-wave one and the short-wave one. When $B \neq 0$, the neutral curve splits into two neutral curves, monotonic and oscillatory, respectively. The monotonic neutral curve, which replaces the monotonic one, has two minima. The short-wave oscillations are mainly longitudinal, while the long-wave oscillations are essentially influenced by the deformation of the interface.

When $B$ grows, the critical $M$ grows for both minima. However, the stabilization of the short-wave oscillations is stronger than that of long-wave oscillations. Accordingly, the long-wave mode provides a lower maximum as $B$ is larger than a certain value $B_*$ (Fig. 5.24(a)). The frequency of oscillations grows monotonically with $B$ (Fig. 5.24(b); lines 1 – 3 for $B = 1; 5; 10$).

### ii.2 Air-water system heated from above

When the air-water system is heated from above, an oscillatory instability appears even *in the absence of surfactant* due to the interaction of transverse and longitudinal instability modes (subsection

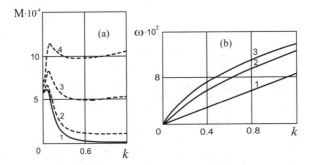

Figure 5.24: Influence of the surfactant concentration (a) on the stability curve, and (b) on frequency of oscillations in the case of heating from below.

5.2). Thus, in the latter case the physical nature of oscillations is quite different from that considered until now in the present section. The addition of a surfactant leads to the growth of the threshold value of $M$ (Fig. 5.25(a)) and to the growth of the oscillation frequency (Fig. 5.25(b)).

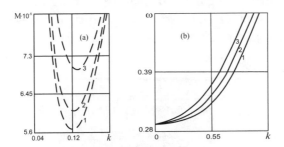

Figure 5.25: Influence of the surfactant concentration (a) on the stability curve, and (b) on the frequency of oscillations in the case of heating from below.

## 5.6 Oscillations and waves in multilayer systems

In the present section, we discuss the phenomena which appear in heated *multilayer* fluid systems due to the Marangoni effect. There is more than academic interest in this problem. Indeed, it has been shown that such systems underlie the liquid encapsulation crystal growth technique (Johnson, 1975; Campbell and Koster, 1995) used in variable or low effective–gravity facilities like space missions, which may allow to reach high quality of the growing crystals by putting the melt between the fluid layers. Then one expects that the simultaneous interaction of *several* interfaces with their bulks and eventually through bulk phases with each other would lead to rather complex dynamics and subtle effects.

Let the space between two rigid horizontal plates be filled by three immiscible fluids with different physical properties (Fig. 5.26). The equilibrium thicknesses of the layers are $a_i$, $i = 1, 2, 3$. The deformable interfaces are described by equations $z = h(x, y, t)$ and $z = -a_2 + h_*(x, y, t)$. The $i$-th fluid has density $\rho_i$, kinematic viscosity $\nu_i$, (dynamic) shear viscosity $\eta_i = \rho_i \nu_i$, thermal diffusivity $\kappa_i$ and heat conductivity $\lambda_i$. For simplicity let us consider the case of normal capillarity at both interfaces, hence the surface tension on the upper and lower interfaces, $\sigma$ and $\sigma_*$, are, as in most cases treated in this book, linear functions of temperature $T$: $\sigma = \sigma_0 - \alpha T$, $\sigma_* = \sigma_{*0} - \alpha_* T$ and normal thermocapillarity. The acceleration due to gravity is again $g$. To further simplify the problem, we do not take into account buoyancy which is negligible in the case of thin layers or under very low effective–gravity (microgravity) conditions. However, we shall retain hydrostatic contributions at the interfaces.

Figure 5.26: Side view of the three-layer configuration.

The horizontal plates are kept at different constant temperatures. The temperature difference can be positive or negative and the overall temperature drop is $\theta$.

Let us define

$$\rho = \frac{\rho_1}{\rho_2}, \; \nu = \frac{\nu_1}{\nu_2}, \; \eta = \frac{\eta_1}{\eta_2} = \rho\nu, \; \kappa = \frac{\kappa_1}{\kappa_2}, \; \lambda = \frac{\lambda_1}{\lambda_2}, \; a = \frac{a_2}{a_1},$$

$$\rho_* = \frac{\rho_1}{\rho_3}, \; \nu_* = \frac{\nu_1}{\nu_3}, \; \eta_* = \frac{\eta_1}{\eta_3} = \rho_*\nu_*, \; \kappa_* = \frac{\kappa_1}{\kappa_3}, \; \lambda_* = \frac{\lambda_1}{\lambda_3}, \; a_* = \frac{a_3}{a_1}.$$

As the units of length, time, velocity, pressure and temperature we use $a_1$, $a_1^2/\nu_1$, $\nu_1/a_1$, $\rho_1\nu_1^2/a_1^2$ and $\theta$. Then recalling the earlier used nonlinear equations governing surface tension gradient–driven convection we now have the following dimensionless expressions:

$$\frac{\partial \mathbf{v}_i}{\partial t} + (\mathbf{v}_i \cdot \nabla)\mathbf{v}_i = -e_i \nabla p_i + c_i \nabla^2 \mathbf{v}_i, \tag{5.81}$$

$$\frac{\partial T_i}{\partial t} + \mathbf{v}_i \cdot \nabla T_i = \frac{d_i}{P} \nabla^2 T_i, \tag{5.82}$$

$$\nabla \cdot \mathbf{v}_i = 0, \ \ i = 1, 2, 3, \tag{5.83}$$

with $e_1 = c_1 = d_1 = 1$, $e_2 = \rho$, $c_2 = 1/\nu$, $d_2 = 1/\kappa$, $e_3 = \rho_*$, $c_3 = 1/\nu_*$, $d_3 = 1/\kappa_*$; $\Delta = \nabla^2$, and $P = \nu_1/\kappa_1$ is the Prandtl number.

The boundary conditions on the rigid boundaries are:

$$\mathbf{v}_1 = 0, \ T_1 = 0 \ \text{ at } \ z = 1, \tag{5.84}$$

$$\mathbf{v}_3 = 0, \ T_3 = s \ \text{ at } \ z = -a - a_*, \tag{5.85}$$

with $s = 1$ for heating from below and $s = -1$ for heating from above. The boundary conditions on the deformable interfaces at $z = h$ can be written in the form

$$p_1 - p_2 + \frac{W_0}{R}(1 - \delta_\alpha T_1) + Ga\,\delta h$$

$$= \left[ \left( \frac{\partial v_{1i}}{\partial x_k} + \frac{\partial v_{1k}}{\partial x_i} \right) - \eta^{-1} \left( \frac{\partial v_{2i}}{\partial x_k} + \frac{\partial v_{2k}}{\partial x_i} \right) \right] n_i n_k, \tag{5.86}$$

$$\left[ \left( \frac{\partial v_{1i}}{\partial x_k} + \frac{\partial v_{1k}}{\partial x_i} \right) - \eta^{-1} \left( \frac{\partial v_{2i}}{\partial x_k} + \frac{\partial v_{2k}}{\partial x_i} \right) \right] \tau_i^{(l)} n_k - \frac{M}{P} \tau_i^{(l)} \frac{\partial T_1}{\partial x_i} = 0, \tag{5.87}$$

$$l = 1, 2, \tag{}$$

$$\mathbf{v}_1 = \mathbf{v}_2, \tag{5.88}$$

$$\frac{\partial h}{\partial t} + v_{1x} \frac{\partial h}{\partial x} + v_{1y} \frac{\partial h}{\partial y} = v_{1z}, \tag{5.89}$$

$$T_1 = T_2, \tag{5.90}$$

$$\left( \frac{\partial T_1}{\partial x_i} - \lambda^{-1} \frac{\partial T_2}{\partial x_i} \right) n_i = 0, \tag{5.91}$$

and at $z = -a + h_*$

$$p_2 - p_3 + \frac{W_{*0}}{R_*}(1 - \delta_{\alpha*}T_1) + Ga\,\delta_* h_*$$

$$= \left[\eta^{-1}\left(\frac{\partial v_{2i}}{\partial x_k} + \frac{\partial v_{2k}}{\partial x_i}\right) - \eta_*^{-1}\left(\frac{\partial v_{3i}}{\partial x_k} + \frac{\partial v_{3k}}{\partial x_i}\right)\right] n_{*i}n_{*k}, \qquad (5.92)$$

$$\left[\eta^{-1}\left(\frac{\partial v_{2i}}{\partial x_k} + \frac{\partial v_{2k}}{\partial x_i}\right) - \eta_*^{-1}\left(\frac{\partial v_{3i}}{\partial x_k} + \frac{\partial v_{3k}}{\partial x_i}\right)\right] \tau_{*i}^{(l)} n_{*k} \qquad (5.93)$$

$$-\frac{\bar{\alpha}M}{P}\tau_{*i}^{(l)}\frac{\partial T_3}{\partial x_i} = 0,\; l = 1, 2,$$

$$\mathbf{v}_2 = \mathbf{v}_3, \qquad (5.94)$$

$$\frac{\partial h_*}{\partial t} + v_{3x}\frac{\partial h_*}{\partial x} + v_{3y}\frac{\partial h_*}{\partial y} = v_{3z}, \qquad (5.95)$$

$$T_2 = T_3, \qquad (5.96)$$

$$\left(\lambda^{-1}\frac{\partial T_2}{\partial x_i} - \lambda_*^{-1}\frac{\partial T_3}{\partial x_i}\right) n_{*i} = 0, \qquad (5.97)$$

where now $M = \alpha\theta a_1/\eta_1\kappa_1$ is the Marangoni number, $Ga = ga_1^3/\nu_1^2$ is the Galileo number, $W_0 = \sigma_0 a_1/\eta_1\nu_1$, $\delta_\alpha = \alpha\theta/\sigma_0$, $\delta = \rho^{-1} - 1$, $W_{*0} = \sigma_{*0}a_1/\eta_1\nu_1$, $\delta_{\alpha*} = \alpha_*\theta/\sigma_{*0}$, $\delta_* = \rho_*^{-1} - \rho^{-1}$, $\bar{\alpha} = \alpha_*/\alpha$; $R$ and $R_*$ are radii of curvature, $\mathbf{n}$ and $\mathbf{n}_*$ are normal vectors and $\boldsymbol{\tau}^{(l)}$ and $\boldsymbol{\tau}_*^{(l)}$ are tangential vectors of the upper and lower interfaces; $p_i$ is the difference between the overall pressure and the hydrostatic pressure in the corresponding fluid.

The boundary value problem (5.81)-(5.97) has as multilayer base solution

$$\mathbf{v}_i = 0,\; p_i = 0,\; i = 1, 2, 3;\; h = 0;\; h_* = 0, \qquad (5.98)$$

$$T_1 = T_1^0 = -\frac{s(z - 1)}{1 + \lambda a + \lambda_* a_*}, \qquad (5.99)$$

$$T_2 = T_2^0 = -\frac{s(\lambda z - 1)}{1 + \lambda a + \lambda_* a_*}, \qquad (5.100)$$

$$T_3 = T_3^0 = -s\frac{\lambda_* z - 1 + (\lambda_* - \lambda)a}{1 + \lambda a + \lambda_* a_*} \qquad (5.101)$$

corresponding to the mechanical equilibrium state. Depending on physical parameters of fluids, the mechanical equilibrium state may become unstable with respect to different instability modes.

Recall that in a system with one interface, two basic instability mechanisms are known (see Chapter 4). The first one is caused by the positive feedback between the spatially inhomogeneous disturbances of temperature and velocity, with no deformation of the interface (Pearson, 1958). This surface tension gradient–driven (Marangoni) instability generates a short-wave cellular structure (Benard cells). The second mechanism generates long-wavelength interfacial deformations, which typically destroy the fluid layer (Scriven and Sternling, 1964; Smith, 1966). Most investigations of instability with the Marangoni effect have been carried out for the former case. The second instability mechanism can be observed only under conditions of low effective gravity or using very thin layers (VanHook et al., 1997).

## 5.6.1 The case of undeformable interfaces

The case of undeformable interfaces is better explored also for multilayer systems, and we shall start with its description. Note that all the experiments carried out until now, in practice, belong to this case (Georis and Legros, 1996; Georis et al., 1999; Legros and Georis, 2001).

The physical nature of the instability is essentially the same as in the case of a single interface. The new phenomena appear due to the *coupling* of instability modes generated on different boundaries.

The most remarkable feature of three-layer fluid systems is the appearance of an *oscillatory* instability due to the negative feedback characterizing the interaction of interfaces. Georis et al. (1993) have carried out a detailed investigation, both linear and nonlinear, of stationary and oscillatory instability modes in the case of identical top and bottom layers. Specifically, it was shown that the nonlinear development of the oscillatory instability leads to a so-called "flip-flop" oscilatory regime. The extension of the analysis to case of three-layer systems when all the three fluids were different was performed albeit in linear theory by Kats-Demianets et al. (1997a,b). It was shown that for the same system of fluids, depending on the ratio of thicknesses of the layers and the way of heating (from below and from above), various kinds of stationary and oscillatory instabilities can take place. Some results obtained for the system silicone oil (10 cSt) - ethylene glycol -fluorinert FC75 are shown in Figs. 5.27 - 5.29. In figures, $k^*$ and $M^*$ are respectively the wavenumber of disturbance and the Marangoni number normalized both with respect to the total width of the system,

$$k^* = k(1 + a + a_*), \ M^* = M(1 + a + a_*).$$

The flows are characterized by rather various typical wavenumbers and various structures of disturbances (including "multi-store" flow structures, Fig. 5.30).

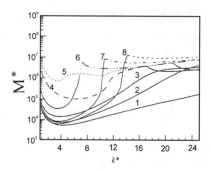

Figure 5.27: Neutral stability curves in the case of $a_* = 1$. Solid (broken) lines define monotonic (oscillatory) instability neutral curves when heating from above. 1: $a = 0.2$; 2: $a = 0.5$; 3: $a = 0.7$; 4: $a = 1.0$; 5: $a = 2.0$; 6: $a = 3.0$; 7: $a = 5.0$; 8: $a = 7.0$ (Redrawn after Kats-Demianets et al., 1997b).

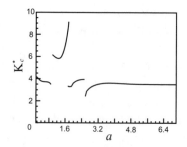

Figure 5.28: Critical number in the case $a_* = 1$ when heating from above. (Redrawn after Kats-Demianets et al., 1997b).

Under normal gravity conditions, the Marangoni effect acts simultaneously with buoyancy. The investigation of the combined action of both instability mechanisms started by Simanovskii et al. (1992), Liu and Roux (1992) and Liu et al. (1998), was continued by Georis, who carried out the linear analysis of different kinds of instabilities

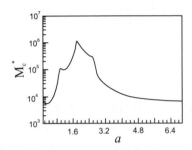

Figure 5.29: Critical Marangoni number in the case $a_* = 1$ when heating from above. (Redrawn after Kats-Demianets $et$ $al.$, 1997b).

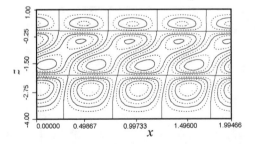

Figure 5.30: Streamlines of the flow field in the case of $a = a_* = 2$ when heating from above. The domain shown is twice larger than the wavelength of the critical perturbation (Redrawn after Kats-Demianets $et$ $al.$, 1997b).

together with numerical simulations of stationary and non-stationary flow regimes. An extensive analysis of nonlinear convection regimes under the action of different instability mechanisms was done by Nepomnyashchy and Simanovskii (1999).

Recently, the experimental discovery of oscillatory convection regimes, which formerly were predicted by Georis *et al.* (1993), has been reported (Legros and Georis, 2001).

### 5.6.2   Deformable interfaces

Let us consider now instabilities with interfacial deformation. The linear theory of the long-wavelength surface tension gradient-driven (Marangoni) instability in three-layer systems was developed by Nepomnyashchy and Simanovskii (1997). It was shown that the existence of two deformable boundaries yields in the limit of small wavenumbers, $k$, two critical Marangoni numbers $M = M_1$ and $M = M_2$ for *monotonic* instability. Depending on the parameters of the system, the signs of the critical Marangoni numbers can coincide, but they can also be different, which corresponds to the appearance of the monotonic instability by heating either from below or from above. Also, an *oscillatory* instability boundary may arise, unlike the case of a single interface. The long-wavelength oscillatory instability is characterized by a $\omega \sim k^2$ dispersion relation.

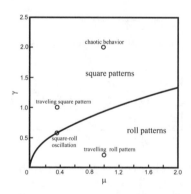

Figure 5.31: Parameter regions corresponding to wave patterns with different symmetries. Open circles correspond to various patterns found in numerical computation. (Redrawn after Golovin and Nepomnyashchy, 1999).

The nonlinear evolution of the stationary deformational mode in the case of a three-layer system is similar to that in the case of a two-layer system and leads to a destruction of the layers or to the appearance of a stationary relief of interface deformations. A more interesting problem is the nonlinear development of the oscillatory deformational instability. The weakly nonlinear theory of waves generated by this instability was developed by Kliakhandler *et al.* (1998) in the case where the thickness of the bottom layer is small in comparison with the thicknesses of the top and the middle layer. The interfacial deformations of both interfaces appear evolving coupled together. After appropriate rescaling the corresponding coupled equations are

$$\frac{\partial H}{\partial \tau} + \Delta(\Delta H + \mu H + \gamma H^2 - H^3 + H_*) = 0, \qquad (5.102)$$

$$\frac{\partial H_*}{\partial \tau} - \Delta H = 0. \qquad (5.103)$$

It was shown that two main wavy patterns can appear in the system: traveling one-dimensional waves (rolls) and a traveling square pattern; the latter type of pattern appears due to a certain resonant effect. The boundary between both flow regimes in the parameter space was calculated by a suitable perturbative method. Direct numerical computations of the coupled equations for deformations, which were carried out by Golovin and Nepomnyashchy (1999) revealed some new patterns of wavy motions (Fig. 5.31). Near the boundary between regions of traveling rolls (Fig. 5.32) and traveling squares (Fig. 5.33), a new type of motion was found in the form of regular oscillations between nearly standing roll waves of different orientations and square wavy patterns (so called "alternating rolls"; Fig. 5.34). Also, some spatially chaotic patterns were found. In the latter flow regime, the deformation of the upper interface displays irregular "spots" of nearly flat interface which split and merge in a chaotic manner (Fig. 5.35).

## 5.7  Experiments on surface tension gradient-driven waves

For over three decades much experimental material on mostly nonstationary surface tension gradient-driven (Marangoni) convection has been accumulated. Various phenomena (ripples and solitary waves,

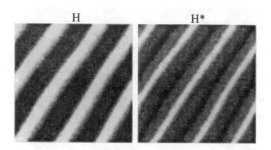

Figure 5.32: Traveling roll pattern, $\mu = 1.0$, $\gamma = 0.2$. (Redrawn after Golovin and Nepomnyashchy, 1999).

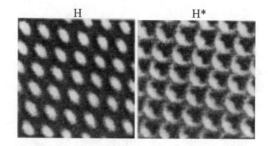

Figure 5.33: Traveling square pattern, $\mu = 0.368$, $\gamma = 1.0$. (Redrawn after Golovin and Nepomnyashchy, 1999).

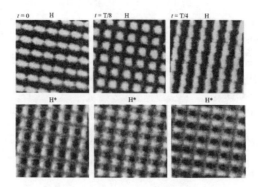

Figure 5.34: Alternating rolls, $\mu = 0.368$, $\gamma = 0.5415$. (Redrawn after Golovin and Nepomnyashchy, 1999).

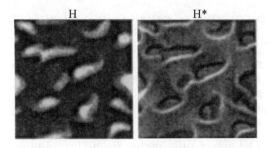

Figure 5.35: Chaotic pattern, $\mu = 1.0$, $\gamma = 2.0$. (Redrawn after Golovin and Nepomnyashchy, 1999).

propagating and mutually interacting nonlinear waves, chaotic motions, eruptions, dry-out and so on) have been observed. However, until recently, the experimental studies have provided qualitative rather than quantitative data. Besides, the nonstationary character of experiments with surfactants and the Marangoni effect makes difficult the comparison between theory and experiment. The available information refers to the three types of waves earlier described in this chapter.

Restricting to a reasonably succinct account, in the present section we shall describe a series of experimental results which illustrate some remarkable phenomena related to the excitation and interaction of waves during the process of mass transfer from a gas to a liquid or absorption of a surfactant by a liquid due to drop dissolution (Weidman *et al.*, 1992; Linde *et al.*, 1993a-c, 1997, 2001a,b; Wierschem *et al.*, 1999, 2000, 2001; Santiago-Rosanne *et al.*, 1997). These phenomena are: different kinds of collisions of solitary waves, formation of Mach-Russell waves by collisions and wall reflections, formation and interaction of periodic wave trains of interfacial and internal waves. The experiments have been carried out for different gas-liquid pairs, so that one can conclude that there are some general laws characterizing the behavior of waves driven by the Marangoni effect.

### 5.7.1   Typical mass-transfer experimental set-up and experimental runs

#### i. Set-up and methodologies

As an illustration let us consider a set-up described by Wierschem *et al.* (2000) prepared for the study of mostly one-dimensional waves and wavetrains, either interfacial (capillary-gravity and dilational) or internal. Other set-ups, including plain cylindrical, square and rectangular containers, have been used by Linde and collaborators (Linde *et al.*, 1993, 1997, 2001a,b; Wierschem *et al.*, 1999, 2000, 2001) to observe the various wave features and wave kinematics we shall describe below.

Fig. 5.36 shows a side view of a set-up. It consists of a cylindrical glass container in which two quartz rings are placed concentrically. The two rings were concentric within 1%. Besides that shown in Fig. 5.36, another container with an inner ring of $54.25 \pm 0.05$ mm diameter was used. Its outer diameter was $74.60 \pm 0.05$ mm while the height of the quartz-glass rings was $8.85 \pm 0.1$ mm. The rings were concentric within 3%. They were fixed with the two-component adhesive $Araldit^R$ from $Ciba - Geigy$. Contamination of the liquid by the adhesive or dissolution and aging of the adhesive points was not observed. The annular gap between the two quartz rings was filled with e.g. liquid toluene. The inner and outer cuvettes were filled with e.g. liquid pentane, which has a high vapor pressure at room temperature and a surface tension much lower than liquid toluene. Because the surface level was expected to rise due to the absorbed pentane the annular container was not filled brimful. The experiment began by covering the reservoir with a quartz-glass plate. To maintain the atmospheric pressure this plate was not tightly fixed and the enclosure was thus slightly open to the ambient air. The experimental runs were carried out at ambient temperature between 297.65 and 300.1 K.

The choice of toluene as the liquid and pentane as the absorbable and completely miscible surfactant was done as one of many other possibilities. Property values for the pure substances at atmospheric pressure are given in Table 5.1. The surface tension and viscosity of liquid pentane are much lower than those of liquid toluene. The same holds for the density. Note that this is a key feature for the possible generation of internal waves as the liquid layer becomes stably stratified due to the absorption of pentane. The refractive index

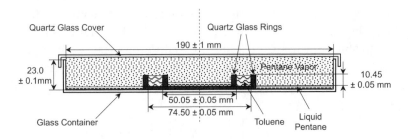

Figure 5.36: Typical set-up to observe surface tension gradient–driven–Marangoni–waves in mass transfer experiments (after Wierschem et al., 2000).

of pentane vapor is almost the same as that of air. Accordingly, one expects that optical rays traversing the vapor phase are not deflected. Furthermore, the boiling point of pentane is about 11 K above room temperature at atmospheric pressure. For toluene the corresponding value is higher. Within the range of the experimental runs the effect of temperature on the parameters is low and appeared negligible.

**Table 5.1. Properties of liquid toluene and liquid pentane**

| Property | Liquid toluene | Liquid pentane | Pentane gas |
|---|---|---|---|
| Molecular weight $u$ (g/mol) | 92.15 | 72.15 | 72.15 |
| Boiling point $T_B$ (K) | 383.75 | 309.22 | |
| Density $\rho$ at 293.15 K (g/cm$^3$) | 0.8669 | 0.6262 | |
| Refraction index $n$ at sodium $D$ line | 1.4961 | 1.3575 | 1.001711 |
| Surface tension $\sigma$ at 298.15 K (mN/m) | 27.93 | 15.49 | |
| Viscosity $\eta$ at 298.15 K (m Pas) | 0.560 | 0.224 | 0.00676 |

To illustrate the difficulties with quantitative estimates let us recall how Wierschem et al. (1999, 2000) estimated material properties. As a literature value for the diffusion of pentane in air and in liquid toluene was not found, an estimate of its magnitude was obtained from other data in the literature. The diffusion coefficient for a system of two simple gases is usually about $10^{-5}$ m$^2$s$^{-1}$. For pentane in air, the estimate was $4.2 \cdot 10^{-6}$ m$^2$s$^{-1}$ with a factor 2 as the expected error. The diffusivity of pentane in toluene can be approximated from the known values of the toluene-heptane and toluene-hexane systems. These systems have diffusion coefficients of $3.72 \cdot 10^{-9}$ and $4.21 \cdot 10^{-9}$ m$^2$s$^{-1}$, respectively. Thus, by extrapolation the diffusion coefficient of pentane in toluene was taken to be about $5 \cdot 10^{-9}$ m$^2$s$^{-1}$. In the temperature range 296 to 301 K the vapor pressure of pentane increases linearly from 63.0 to 76.1 kPa, taking the form $C_{PV} = -7.065$ (mol) $+ 0.025968$ (mol K$^{-1}$) $T$, while that of toluene grows from 3.36 to 4.37 kPa. As the two components do not react, the density changes linearly with concentration. For the (dynamic) shear viscosity, $\eta$, the following equation was used

$$\eta(C_P) = \left[ \eta_P^{1/3} + \eta_T^{1/3}(1 - C_P) \right]^3, \qquad (5.104)$$

where $C_P$ denotes the pentane concentration, and $\eta_P$ and $\eta_T$ are the viscosities of pure pentane and pure toluene, respectively. The surface tension, as a function of the concentration at $T = 298.15$ K, was measured semiautomatically with a *Lauda TE 1 C* tensiometer using a platinum ring for the stirrup method. The data was fitted with an exponential-decay function $\sigma = \sigma_0 + \sigma_1 \exp(-C_P/C_{P0})$, with $\sigma_0 = (11.81 \pm 0.21)$ mN m$^{-1}$, $\sigma_1 = 16.24 \pm 0.20$ mN m$^{-1}$, and $C_{P0} = 0.682 \pm 0.018$ mol. The Marangoni number was expressed as:

$$Ma = (d\sigma/dC_P)l^2 \nabla C_P / \eta(C_P)D \qquad (5.105)$$

with $\nabla C_P$ indirectly determined and $l$ the container width or twice its value depending on the experimental run with 3D surface waves.

The surface deformation caused by the waves was followed with a laser beam reflected at the liquid surface. The reflected part of the beam was detected on a recording plane. If the surface at the irradiation spot is deformed, the position of the reflected beam on the recording plane moves. The position of the spot on the plane was recorded by a CCD camera, stored on video tape, and evaluated, subsequently, by computer. The surface deformation was calculated from the shift of the laser spot on the recording plane and integrated with the phase velocity of the waves that travel underneath the laser beam.

For both interfacial and internal waves the shadowgraph technique was also used. The light coming from a monochromatic light source of 1 mm diameter was collimated to a parallel beam of 100 mm diameter. After traversing the liquid layer from underneath, the collimated beam was focused by a lens. Together with a zoom it adjusts the size of the beam arriving at the CCD chip of the (25 Hz) camera that records the image.

## ii. Experimental runs

Experiments started by covering the reservoir with a glass plate. Then the vapor concentration of pentane rose rapidly until its stationary value determined by the vapor pressure was achieved and a strong concentration difference between the vapor phase and the liquid was created. This difference decreased with time due to the absorption of pentane in the liquid, until the pentane surface concentration in the liquid was the same as in the vapor. At the beginning of this equilibration process a messy surface was observed. Subsequently, rather regular wave patterns formed in a clearly, spontaneous selforganizing manner. Since the pentane that is absorbed by the liquid toluene remains in the liquid, the experiment is nonstationary by its nature. This makes it difficult to determine the Marangoni number quantitatively. However, it is at least possible to estimate its order of magnitude and its time evolution.

During the experimental runs, quantities like the density, the viscosity, and the surface tension changed with time. They all depend on the pentane concentration within the liquid. The absolute amount of absorbed pentane was obtained by measuring the height of the surface level as a function of time. The evaporation of toluene was negligible due to its much lower vapor pressure compared to pentane. The temporal evolution of the pentane absorption rate was fitted to an exponential-association fit such that when the vapor had already reached its saturation value, the fit of the absorption rate $A_P$ through the surface of the annular ring was

$$A_P(t) = \frac{\rho_p}{u_p}\pi(r_o^2 - r_i^2)\frac{\Delta h}{\tau}\exp\left(-\frac{t}{\tau}\right), \qquad (5.106)$$

where $\rho_P$ and $u_P$ are the density of liquid pentane and its molecular mass, respectively, $r_o$ and $r_i$ denote the radii of the outer and inner container walls, and $t$ means time. The fit parmeters $\Delta h$ and $\tau$ represent the increment of the surface level and the time scale of the absorption of the pentane vapor by the liquid toluene, respectively. It was found that $h_1 = (3.03 \pm 0.02) \cdot 10^{-3}$m and $\tau = 895.0 \pm 7.3$s. Clearly, the Marangoni number (5.105) was a time-dependent quantity.

A typical example for the time evolution of a wave train and the number of its wave-crests is depicted in Fig. 5.37. At first, surface (interfacial) waves of rather long wavelength were generated indeed

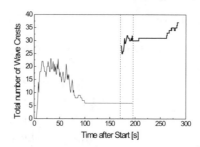

Figure 5.37:  Time evolution of the number of wave crests illustrating the passage from (mostly) surface waves to (mostly)
internal waves, all driven by the Marangoni effect.

by the Marangoni effect. From about 100 to 200 s a periodic wave
train was observed. Its contrast in the shadowgraph image decreased
until it faded away completely. At about 170 s very regular waves
with really shorter wavelength appeared, which can be considered as
internal rather than surface waves related indeed to the Marangoni
effect acting like shear in a stably stratified layer. From about 210
to 260 s a wave train of constant wave number remained. As time
proceeded further, the number of crests increased and the contrast
weakened until, finally, they were no longer observable.

## 5.7.2    Surface deformation, surface and internal waves

Depending on the experimental conditions one can observe one of
the three different types of waves earlier discussed, surface capillary-
gravity, surface dilational and internal (Table 5.2). With the set-
ups described above one can observe three-dimensional (3D) waves.
However, if the reservoir was left open to the ambient air only two-
dimensional (2D) surface waves were observed at high vapor pressures. The physical reason for this geometrical difference is that a
horizontal surfactant concentration gradient exists in the vapor phase
when the reservoir is covered, which seems to be less important when
the reservoir is left open. In both cases, internal waves become apparent after the genuine surface waves disappear. Characterization
of the different types of waves was done by measuring surface elevation, associated convective flow and dispersion relation. Even their
nonlinear behavior has been assessed (Wierschem *et al.*, 1999, 2000;
Linde *et al.*, 2001 a,b).

The dispersion relation of 3D surface waves was obtained by measuring the wavelength and frequency of the wave trains and the velocity of modulations in the wave trains. It is nonlinear as shown in Fig. 5.38. Noticeable is that 2D surface waves were, within experimental uncertainty, dispersion-free i.e. dilational waves, as can be seen in Fig. 5.39. Finally, the dispersion relation of internal waves was obtained from the phase velocity and the group velocity of the wave trains without knowing the Marangoni number. Fig. 5.40 shows that the ratio between phase and group velocities does not depend on wavelength, which is typical for internal gravity waves in two-layer systems.

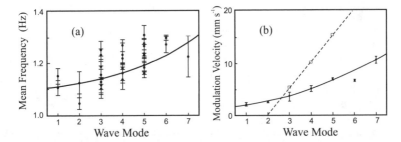

Figure 5.38: 3D surface waves. (a) Mean frequency of wave trains as a function of the wave mode. Each point corresponds to one experimental run. The error bars represent the highest and lowest measured frequency of each experimental run. The line is the integral of the fitted line from the phase velocity in (b) with fitted integral constant. (b) Velocity of the modulations for the trains with lowest frequency of the specific mode. Solid and open squares, respectivelly, indicate modulations traveling forward and backward relative to the wave crests. Mode six is not taken into account for the presented fit of quadratic order. The average values of the backward traveling modulations are fitted linearly.

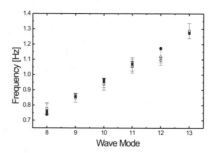

Figure 5.39: Dilational, dispersion–free waves. Mean frequency as a function of the wave mode. The theoretical values are indicated by solid squares and the experimental values by open squares. The error bars indicate the maximum and minimum values observed.

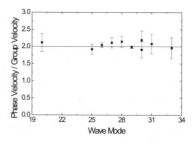

Figure 5.40: Ratio between phase velocity and group velocity of internal waves as a function of the wave mode. The ratio is determined for the modulations of the wave trains. The quotient is two, independent of the mode. This means for the dispersion relation that the angular frequency is proportional to the square-root of the wave number. Note that Table 5.2 refers to 2D waves.

**TABLE 5.2. TWO-DIMENSIONAL WAVE DISPERSION RELATIONS**
$[\omega = \omega(k)]$ **(LIQUID LAYER OPEN TO AMBIENT AIR)**
**SURFACE CAPILLARY-GRAVITY WAVES**
1) $\omega^2 = Gak(1 + k^2/Bo)\,\tanh k$
$Bo$ is the (static) Bond number
2) For infinitely deep liquid layers in dimensional form:
$\omega^2 = gk + \sigma k^3/\rho$
**SURFACE DILATIONAL WAVES (LUCASSEN-MARANGONI)**
$\omega^2 = |Ma|Prk^2/(1 + Pr^{1/2})$
$Ma$ and $Pr$ are the Marangoni and Prandtl numbers, respectively.
$Ma$ is taken positive if the heating comes from the liquid side
hence the bottom for standard liquids. For mass transfer problems
$Pr$ is to be replaced by the Schmidt number, $Sc$.
**INTERNAL WAVES (BRUNT-VÄISÄLÄ)**
$\omega^2 = |Ra|Prk^2/(k^2 + n^2\pi^2)$, $n = 1, 2, 3, \ldots$
$Ra$ is the Rayleigh number, taken positive if the heating comes
from below.


Fig. 5.41 presents a typical time series of the surface elevation
depicting both shadowgraph and laser-beam reflection measurements
of (3D) surface wave trains. The higher light-intensity peaks, shown
as open symbols, correspond to higher maxima of surface elevation
and the lower localized light-intensity peaks coincide with the smaller
crests of the surface elevation. The surface oscillates radially at the
same time that the crests travel in the azimuthal direction. The
displacement of the laser beam in the radial direction reaches its extreme positions when the surface elevation in the azimuthal direction
is lowest.

The peak-to-trough amplitude of the surface elevation in the azimuthal direction caused by a train of three surface waves is displayed
in Fig. 5.42. Originating from irregularly generated wave crests the
amplitude oscillates between the high and low crests. The amplitude
of the higher crests decreases while that of the lower crests slightly
increases until both reach equal size together decreasing further until,
finally, another set of waves is generated. The latter can be considered as the above mentioned *internal* waves in the already stably
stratified liquid layer presumably due to the shear induced by the
former surface waves and the Marangoni effect. Fig. 5.43 provides

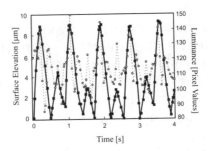

Figure 5.41: Surface elevation produced by a (3D) surface wave train. Solid (black) line comes from laser–beam reflectometry and dotted line (light intensity modulation) from the shadowgraph method.

the surface elevation of crests belonging to a wave train of (mostly) internal waves.

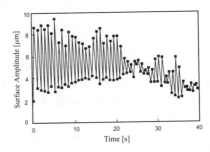

Figure 5.42: Surface elevation (peak-to-trough) amplitude caused by a wave train of (mostly) 3D surface waves. At the beginning the amplitude oscillates between the high and low wave crests, at the end the large amplitude waves become small and the overal amplitude decreases with time.

Experimental runs were also conducted by drastically altering the surface of the liquid in the following way. Rather than simply allowing the pentane vapor to be adsorbed, and subsequently absorbed, by the liquid toluene the ad/absorption processes were accelerated by suddenly placing on top of the liquid a glass cover box full of vapor. This procedure quite quickly forces the appearance of Marangoni stresses and a high value of the Marangoni number. Then the system is allowed to proceed on its own with the diffusion-adsorption processes but the time interval for observation was drastically shorter

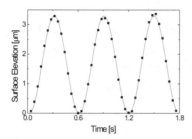

Figure 5.43: Surface elevation induced by a wave train of internal waves in a mass-transfer experiment with the Marangoni effect.

than with the other procedure. Such a drastic push on the Marangoni number yields a dramatic response of the surface. Hence a certain mess develops along it but, eventually, the system organizes itself and yields to a wavy pattern that according to the geometry and the amount of ad/absorbed vapor is in the form of a bunch of solitary waves or one or two countertraveling wave trains, and hence offering collisions and wall reflections as with the other procedure described below.

Experiments were also carried out with desorption rather than adsorption, with desorbed surfactants having higher surface tension than the liquid substrate, e.g. benzene vapor desorbed from liquid nonane.

### 5.7.3   Solitary waves and wave trains

Figures 5.44 – 5.47 illustrate the kind of solitary waves and wave trains observed for various gas-liquid pairs or heat transfer. In square or rectangular containers, solitary waves generally moved erratically thus leading to various collision and wall reflection events. An exceptional case is shown in Fig. 5.47 where a sequence of two wall reflections was suitably prepared on purpose. Further details about this sequence is given below. As earlier noted (Fig. 5.37) wave trains in cylindrical or annular cylindrical containers tended to start with a rather long wavelength to later yield to shorter wavelength trains most surely transforming genuine surface or interfacial waves into internal waves as also noted above. Either single wave trains or two counterrotating wave trains may appear. Fig. 5.46 illustrates the latter case in a heat transfer experiment to be discussed later on. It was remarked that in cylindrical containers either concentric waves (Fig. 5.45) or wave trains very much like those in the annular channel formed even when the inner ring was absent. Spontaneously, the system selforganizes pushing a high excess surfactant concentration to the center of the circular open surface thus creating by itself quite like a genuine annular wave channel.

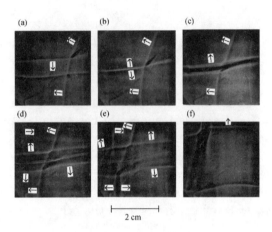

Figure 5.44: Dissipative solitons. Randomly formed (solitary) wave crests due to the Marangoni effect in a square container. Arrows indicate travel direction. Several head-on and oblique collisions can be observed (after Linde *et al.*, 2001).

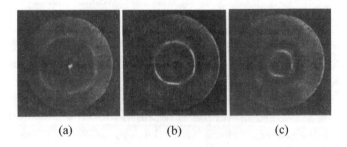

Figure 5.45:   Axisymmetric (solitary) waves due to the Marangoni effect in a cylindrical container. Waves appear near the outer periphery and dissappear at the center.

Finally, when the width of the container is large enough, two-dimensional traveling network patterns composed of synchronously

Figure 5.46: Dissipative solitons. Counterrotating and synchronously colliding wave trains in an annular container in a heat transfer experiment with the Marangoni effect (after Weh and Linde, 1997).

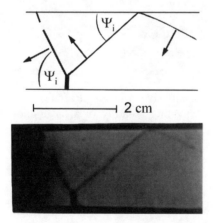

Figure 5.47: Surface tension gradient-driven (Marangoni) instability and dissipative solitons. Wave reflection pattern exhibiting a regular ($2\Psi_i < \pi/2$) and a Mach–Russell reflection ($2\Psi_i > \pi/2$) (after Linde $et$ $al$, 1993c).

colliding waves were observed (Fig. 5.48). Such traveling wave network patterns occurred when two periodic wave trains obliquely crossed. As in shocks and smoke track records of cellular detonation structures (Strehlo, 1970), they define patterned networks of either deformed or regular hexagons traveling with a corner ahead, or hexagons traveling with the sides having the strongest contrast ahead (oblique collisions with Mach-Russell third waves). Patterned patches depend on the crossing angle, as schematically illustrated with pairs of crossing waves and their directions shown with a simple arrow ($\rightarrow$). The traveling directions of the patches as a whole are also indicated by a double arrow ($\Rightarrow$).

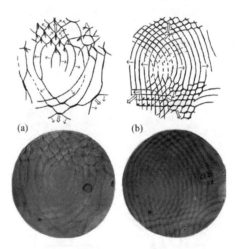

Figure 5.48: Dissipative solitons. Patches of wave collision network patterns traveling steadily in a large enough container exhibiting regular and Mach–Russell collisions (after Linde *et al.*, 2001a).

## 5.7.4   Collisions and reflections at walls

### i. General features

Let us now recall a few details of the experimentally observed collision events of two solitary waves and the reflection of a wave with the wall which, to a first approximation, can be considered the *virtual* collision of the wave with its mirror image. Experimentally, however, differences appear between collisions and wall reflections (Linde *et al.*, 2001a). The crests of synchronously colliding oppositely traveling wave trains show similar behavior.

Let us start with predictions and expected findings of general nature. Head-on collisions or frontal reflections at walls are best analyzed in a space-time diagram while collisions at an oblique angle can be easily analyzed in the real two-dimensional space. Fig. 5.49 illustrates the various events observed, together with the definition of some relevant quantities such as the so called phase shift. Fig. 5.49(a) depicts the time-dependent evolution of a head-on collision of two waves in a space-time plot. One can easily see how the waves approach each other traveling with equal velocity, along the same geometrical direction, though indeed opposite to each other. Upon experiencing the head-on collisions, they decelerate and eventually stop. After a time interval, denoted "residence" time (dotted line), they resume motion and accelerate until they reach a velocity about that before collision. The deceleration yields a change of trajectory in the space-time plot relative to their precollision paths. This is the so-called phase shift, which in this case, by convention, is considered negative. A frontal reflection at a wall appears very much like the head-on collision of the wave with its mirror image. Fig. 5.49(b) depicts, now in the real two-dimensional space, the oblique collision of two waves meeting with an incoming acute enough angle $\psi = 2\alpha < \pi/2$. The postcollision angle is $2\beta$. The picture illustrates a clear change in real trajectories hence leading to a phase shift also negative, by analogy with the convention used in the space-time plot for head-on collisions. In real space one sees the corresponding space lapse or "residence" length, $L_r$, in the oblique collision event (again depicted with a dotted line). As with the frontal reflection at a wall one can also, formally, consider the oblique reflection at a wall as the collision of the wave with its mirror image, and hence Fig. 5.49(b) illustrates both collision and wall reflection. Then the ratio $\sin\beta/\sin\alpha$ =velocity$_{after\,crossing}$ / velocity$_{before\,crossing}$ is a measure of the sudden but transient deceleration of the colliding waves immediately after the collision or reflection event. The collision-reflection analogy indicated above allows to define and equivalent residence length, $L_{r(eq)}$, for the reflection with transient deceleration of the waves. This can be done by extrapolation of the linear paths, which finally show each wave crest after the reflection. For both cases, collision and wall reflection, this residence length provides a measure of above defined *negative* phase-shift, $p_n$, by means of the following cascade of relationships: $p_n = L_r \sin \alpha = V_c t_r \sin \alpha = V_t t_r$ with $L_r$, $V_c$, $t_r$ and $V_t$ denoting, respectively, residence length, velocity of the crossing point, residence time and traveling velocity. For the reflection it suffices to replace $L_r$ by $L_{r(eq)}$.

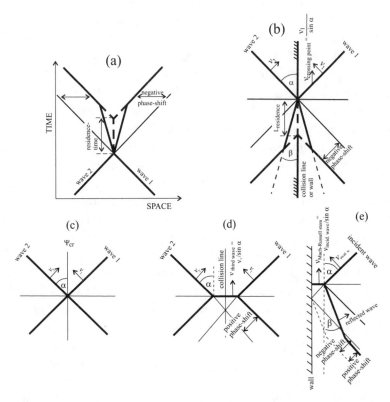

Figure 5.49: Dissipative solitons. Schematic classification of the varous possible collision and reflection events (a) space-time plot of a head-on collision, (b) real space 2D picture illustrating an oblique acute enough ($2\alpha < \pi/2$) collision, (c) neutral collision ($2\alpha \sim \pi/2$) , (d) Mach–Russell oblique collision ($2\alpha > \pi/2$), and (e) mostly Mach–Russell albeit complex wall reflection ($2\alpha > \pi/2$, ($2\beta < \pi/2$)) (after Linde *et al.*, 2001a).

For collisions and reflections one can use the already defined angles $\alpha$ and $\beta$ for the corresponding incoming and outgoing angles [*incident* and *reflected* angles correspond to the usual definition for standard waves], $\psi = 2\alpha$ (or $\psi_i$ for reflection). With increasing $\psi$ in collisions the *negative* phase shift decreases until it eventually vanishes at a critical value, $\psi_{cr}$ (earlier in subsection 5.2.2. it was denoted by $\psi_*$), where both waves experienced no change in their traveling direction and velocity (Fig. 5.49(c)), $\psi_{cr} = 2\alpha = \pi/2$. As earlier noted this may be considered a neutral or zero phase shift interaction. Experimentally, the collision-reflection analogy showed a difference. With increasing $\psi$ for reflections the negative phase shift decreases, and at the critical value $\psi_{cr}$ it reaches a minimum, not zero. Thus between $\psi_i = 0$ and $\psi_{cr}$ negative phase shift reflection was observed. Further increasing the collision or incident angle, when $\psi_i > \psi_{cr}$ (or $\psi_i > \psi_{cr}$), results in a change of trajectory opposite to that occuring in the acute or head-on case and hence can be considered as an event with *positive* phase shift. Such events appear with a phase-locked *third wave* or *stem*, sketched in Fig. 5.49(d) as a thick solid line. This stem discovered in shocks by Mach and collaborators (Mach and Wosyka, 1875) had been earlier described by Russell (1844) as a "lateral accumulation" in some of his laboratory experiments with solitary waves. It provokes a stronger contrast in the shadow image because of its higher intensity. Note the already introduced terminology, Mach-Russell third wave or stem. As earlier noted, shallow layer theory predicts a value $\psi_{cr}$ between $\pi/3$, the value for viscosity-free water waves, and $\pi/2$, the value for surface tension gradient-driven waves (see subsection 5.2.2).

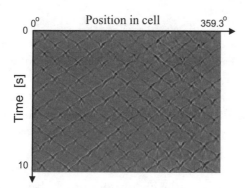

Figure 5.50: Space–time diagram of the modulation in a system of two counterrotating wave trains. The modulation travels from the upper right to the lower left (after Wierschem *et al.*, 2001).

The positive spatial phase shift, $p_p$, is related to the length of the stem or third wave, $L_{t-w}$, by the relationship $p_p = L_{t-w} \cos \alpha$ (is used $L_{stem}$ for reflections). The crossing pattern of the (incoming-[t-w]-outgoing) triad-wave travels along the crossing line with velocity $v_{t-w} = v_{wave\ before\ crossing}/\sin\alpha$ (Fig. 5.49(d)). A remarkable feature found, illustrated in Fig. 5.49(e), is that first the reflected wave transitorily changes trajectory with what can be considered a negative phase shift and subsequently takes an asymptotic trajectory with the expected positive phase shift. A Mach-Russell stem is depicted in Fig. 5.49(e).

Although the experiments are nonstationary, overtaking-collisions have not been reported by Linde *et al.* (1993, 1997, 2001a,b) or by Wierschem *et al.* (1999, 2000, 2001). Note that in the steady state all waves must have the same velocity if the supercritical value of the Marangoni number is not too high and if the wave velocity is not influenced by collisions and reflections. According to theory (subsection 5.2.2; Garazo and Velarde, 1991; Nepomnyashchy and Velarde, 1994) this is a general feature of dissipative waves when they survive thanks to a steady input-output energy balance like that exhibited by Eq. (5.30) at steady state. However, traveling modulation of solitonic wave train may be considered as a form of overtaking interaction (Fig. 5.50).

## ii.  Head-on and oblique collisions in a typical spontaneous sequence of collision events in a square container

Fig. 5.44 (see also Fig. 5.49) taken from the report by Linde *et al.* (2001) provides a wealth of information on several collision events occuring almost simultaneously in a square container with side wall length practically equal to the wavelength of the solitary-like wave. The particular gas-liquid pair used was hexane-benzene with liquid layer depth about that given in the above described cases. The container was, however, of square form. The pair of oppositely traveling wave crests, up and down in the image appears to be rather regularly synchronized by resonant-like frontal reflections between the two walls, $\lambda = L_{container}$, which is 49 mm. Indeed, one expects that these dissipative waves have a well-defined wavelength and are prone therefore to resonant behavior in the container. The two waves suffer head-on collisions with a negative phase shift nearly midway between the two walls as shown in Fig. 5.44(b). A sequence of 10 such sequentially repeated head-on collisions of this one wave-pair was observed. The other two oppositely traveling wave crests between the two other walls of the square container showed no apparent resonance with the walls and no mode coupling with the two resonant-like up-and-down reflected waves. They travel at a slightly oblique angle to the walls and obliquely cross the other regularly head-on reflected and head-on colliding waves. Another interesting event is the oblique crossing occurring in the lower right quarter (coming from below), which by chance is a collision with zero phase shift, a neutral one (Fig. 5.44(a)) at the experimental critical angle $\psi_{cr} = 107°$, which is a bit wider than the above mentioned, $\pi/2$, theoretical prediction (subsection 5.2.2). In the upper right quarter of the picture there is a crossing with the wave coming from above with an angle about 78° and hence there is a negative phase shift. After the three-wave interaction (including the overlapping of head-on collisions, acute-angle crossing, and neutral, no-phase-shift crossing (Fig. 5.44(b)), the angle of the event before the crossing with no phase shift is increased to the obtuse value of about 111°. This wider angle results from the transiently decelerated and, therefore, deflected upper part of the obliquely traveling wave after having experienced the earlier acute-angle crossing. This part of the wave forms another wave, which experiences further obtuse angle crossing after this three-wave collision. Thus as the critical angle is exceeded a positive phase shift appeared of about 1.2 mm (2.4% of the wavelength) with a 2.3 mm extended phase-locked third wave (t-w) (4.7% of the wavelength), visible in the upper part of Fig. 5.44(d). The event before the three-wave acute-angle crossing is unaltered because the crossing angle remains below critical (see the lower part of Fig. 5.44(d)).

At the end of this sequence (Figs. 5.44(e) and 5.44(f)), another wave crest belonging to the unsynchronized pair (in the figure it appears coming from the left) crosses the two patterns with corresponding acute and obtuse angle crossings, the latter occurring with a third wave. Both patterns are preserved after these triple collisions, but the third wave takes on a remarkable change of direction (about 20°). There is a slight increase in the length of the third wave (about 12%) because of an additional increase of the crossing angle (Fig. 5.44(f)). These stable (in the sense of laminar-like flows) and irregular (in the sense of presumably strong nonlinear interactions) triple collisions yield complex moving and alternating interaction patterns even of (periodic) wave trains. The dark traces after the acute angle crossing (Figs. 5.44(a) and 5.44(b)) and the head-on collision (Fig. 5.44(c)) were apparently, the result of strong deformations of the stably stratified concentration gradient in the liquid layer (hexane like pentane is lighter than toluene).

An alternative discussion of these events is the following. The sidewall length in the square container ($L_{container}$ = 49 mm) permitted excitation of what can be considered a bunch of independent solitary waves of wavelength $\lambda = L_{container}$. Indeed, strictly speaking a solitary wave is a traveling localized structure with practically flat wings, thus allowing for the possibility of other independent solitary waves being observable at the same in some other place in the container, and hence experiencing various interactions among themselves. Note that here we refer to solitary waves and not to wave trains. Solitary waves like the solutions of the BKdV equation are bumps, hence elevation waves, in the form of Sech$^2$ while wave trains are like cnoidal periodic waves. Leaving aside asymmetry in the profile (Fig. 5.5) and a wavy head/tail due to linear dissipative terms as earlier noted (subsection 5.2.2) the dissipation-modified BKdV equation (5.30) possesses solutions that capture the essence of such solitary waves and wave trains.

Isolated oblique collision events have also been observed by Linde et al. (1993). Fig. 5.51 illustrates three cases (a, b, c,) corresponding to other three sketched in Fig. 5.49 ((a), (c) and (d)) exhibiting negative, zero and positive phase shift, respectively. The latter clearly carries the phase-locked Mach-Russell third wave.

Fig. 5.52 illustrates the value attained by the phase shift, when two waves obliquely collide, as a function of the collision angle, $\psi$. It clearly appears that the phase shift goes from negative to positive when collision go from acute to obtuse. The critical value is $\psi_c \sim 105°$.

Similar collision events have also been observed by Santiago-Rosanne et al. (1997) in their experiment of drop dissolution of a partially miscible surfactant, nitroethane, on water. In Chapter 3 we have already mentioned this experiment when discussing spreading phenomena driven by the Marangoni effect. Incidentally in the related experiment of locally heating at the center of a circular container, Favre et al. (1997) reported a figure similar to Fig. 5.46. For yet another related experiment see Brimacombe and Weinberg (1972).

Fig. 5.53 shows the daisy-like pattern observed where the leaves are in fact wave fronts (Fig. 5.54). Noteworthy is that the crest along a given front may have an elevation profile fitting a Sech$^2$ smooth bump, near the drop bigger mass, or the form of a hydraulic jump or

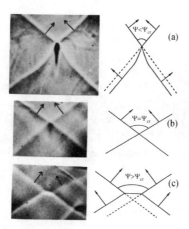

Figure 5.51: Dissipative solitons. Collision events due to waves excited by the Marangoni effect (a) Oblique acute enough collision $\Psi < \Psi_{cr} = \pi/2$, (b) neutral or critical collision with no change of trajectory $\Psi = \Psi_{cr} = \pi/2$, and (c) Mach-Russell oblique collision $\Psi > \Psi_{cr} = \pi/2$. Arrows indicate direction of traveling fronts (after Linde *et al.*, 1993a).

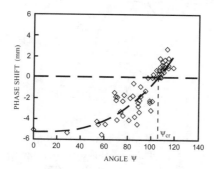

Figure 5.52: Dissipative solitons. Evolution of trajectory phase shift upon collision according to the collision angle $\Psi$ (after Linde *et al.*, 1993a).

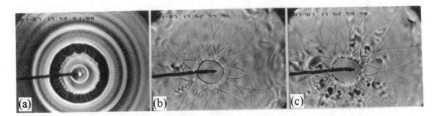

Figure 5.53:  Daisy-like pattern exhibited by colliding wave crests due to the Marangoni effect when a drop dissolves in another liquid (after Santiago–Rosanne *et al.*, 1997).

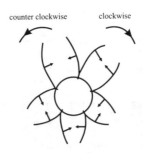

Figure 5.54:  Schematic diagram extracted from Fig.  5.53 to clearly show that daisy-like petals do form with colliding wave crests (after Santiago-Rosanne *et al.*, 1997).

a shock thus fitting a Tanh away enough from the center of the drop (Fig. 5.55). The phenomena described by Santiago-Rosanne *et al.* (1997) is very rich and dissipation seemed very strong. Accordingly, collisions exhibit variable degree of both inelasticity and sign in phase shifts (Fig. 5.56).

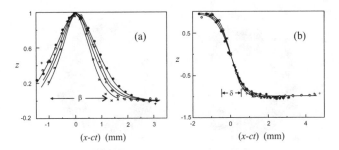

Figure 5.55: Sech$^2$ and Tanh fitting to wave crests occuring in the experiment of Fig. 5.53. Both may appear along one and the same wave front as we move away (b) from the center of the drop (a) (after Santiago-Rosanne *et al.*, 1997).

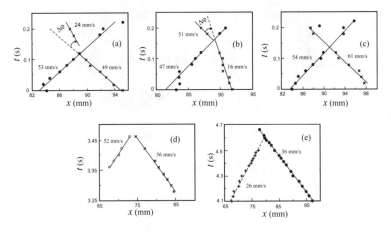

Figure 5.56: Dissipative solitons. Space–time plots corresponding to head-on collision of wave crests (Fig. 5.53) exhibiting variable degrees of inelasticity (after Santiago-Rosanne *et al.*, 1997).

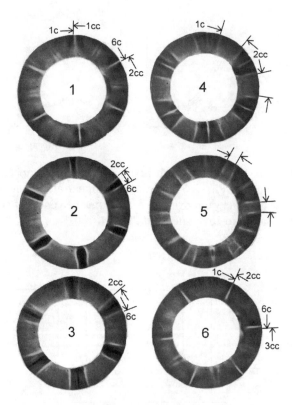

Figure 5.57: Dissipative solitons. Synchronous head-on collision of six wave crests in two counterrotating wave trains excited in an annular container by the Marangoni effect.

### iii. Collisions of counterrotating wave trains

Fig. 5.57 shows the synchronous head-on collision of six wave crests in an annular circular container where two wave trains counterrotate. Similar phenomena have been observed with many more wave crests or in plain cylindrical containers, hence in the absence of the inner ring, where the system itself spontaneously creates its own annular wave guide (Linde *et al.*, 1997). Fig. 5.58 depicts a typical space-time plot of head-on collisions of wave crests of such counterrotating wave trains illustrating the synchronous (negative) phase shifts in trajectories after collision. See also Figs. 5.49(a) and 5.60 but note the change in the space-time orientation.

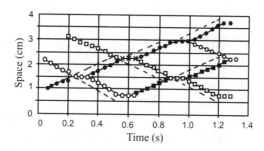

Figure 5.58: Dissipative solitons. Space-time plot illustrating phase shifts upon collision (Fig. 5.57).

### iv. Wall reflections

An experiment was conducted by Linde *et al.* (1993) to specifically assess the role played by the incident angle upon the reflected wave. The already shown Fig. 5.47 illustrates a suitably prepared short wave train displaying a regular reflection ($\psi_i < \pi/2$) with, however, wider than incident reflected angle and an anomalous reflection ($\psi_i > \pi/2$) exhibiting the *creation* of the Mach-Russell stem that, as with collisions, travels phase locked with the reflected wave. Note that the first reflected wave in one side of the container becomes the incident wave upon the opposite side. This Fig. 5.47 supports the view that the formation of the Mach-Russell stem is a purely kinematic property of solitary (and other nonlinear) waves and asymptotic trajectories, not related to the strength of the wave (or the hydraulic jump or shock).

## 5.7.5    Heat transfer results

Few experimental results exist about waves excited by the Marangoni effect due to the heat transfer across a Benard layer heated, from above, hence from the air side, or cooled from below, hence from the liquid side, thus opposite to the original Benard experiment, for standard liquids ($d\sigma/dT < 0$). Let us recall findings reported by Linde *et al.* (1997) and by Weh and Linde (1997). Square, rectangular and circular cylindrical, annular containers were used. Liquids used include octane with ambient air above and layer depths ranging from 3 to 8 mm. With the bottom layer cooled to 20°C and a quartz cover placed 3 to 15 mm above the liquid, heated to attain temperature gradients from 10 to 200 K cm$^{-1}$, solitary waves and wave trains were observed. The onset was found for Marangoni numbers about $10^5$ with, however, an estimated error of 50% (!). Experiments with liquid diphenyl in nitrogen or carbon dioxide atmosphere also permitted observation of waves for temperature gradients around 113 K cm$^{-1}$ in the liquid and 168 K cm$^{-1}$ in the gas layer, with corresponding layer depths of $d_{liquid} = 3$ mm and $d_{gas} = 15$ mm, respectively. The estimated Marangoni number for instability threshold was also in the range or slightly below that corresponding to mass transfer experiments in agreement with available theories (Chu and Velarde, 1988; Rednikov *et al.*, 1998; Velarde *et al.*, 2000; Reichenbach and Linde, 1981). Wave velocities measured were about 1–2 cm s$^{-1}$. Wave-collision and wave-reflection events, including the formation of Mach-Russell third waves or stems, were observed quite like those observed in mass-transfer experiments. For collisions, the experimentally observed neutral crossing angles (zero phase shift) were between $\pi/2$ and 100°, in good agreement with theory (Nepomnyashchy and Velarde, 1994). For the record, note that similar results were obtained by Schwarz as early as 1967 (Schwarz, unpublished) but no proper account of them was reported in the open literature.

   In experiments with liquid layers of n-octane and n-decane open to the ambient air in annular containers (6 and 5 cm, outer and inner diameters, respectively; channel width, 1 cm; liquid depth, 5 cm; air layer depth, 12 mm, and temperature gradients about 80 K cm$^{-1}$ with critical Marangoni numbers about $10^5$) head-on collisions of counterrotating wave trains were also reported by Weh and Linde (1997). Fig. 5.59 depicts details of the head-on collision of two wave crests of the wave trains shown in Fig. 5.46, and Fig. 5.60 is a space-time plot like those shown in Figs. 5.49 or 5.58. The trajectories in Fig. 5.60 have been superimposed to clearly illustrate the negative phase shift following the collision, from 18.5 to 18.9 mm s$^{-1}$.

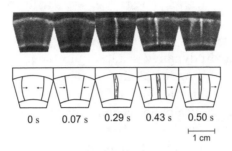

Figure 5.59: Dissipative solitons. Five snapshots of two head-on colliding wave crests before, at, and after interaction (after Weh and Linde, 1997).

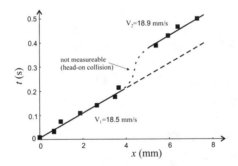

Figure 5.60: Dissipative solitons. Space-time plot showing the common trajectory of two counterrotating wave crests (Fig. 5.46 and 5.59) which upon collision exhibit negative phase shift (after Weh and Linde, 1997).

## 5.8 Hydrochemical surface waves due to the Marangoni effect

Let us conclude this chapter with a few words about the excitation of surface waves with apparent solitonic features due to Marangoni stresses induced by (surface) chemical reactions and hence the name of hydrochemical waves. They are, indeed, a form of dissipative surface tension gradient-driven (Marangoni) waves.

Surface reactions producing excess or depletion of surface surfactant concentration or releasing heat or cooling the surface are indeed known to lead to Marangoni stresses eventually creating instability as discussed in Chapter 3 when dealing with active drops. Instability and surface waves may be due to instability of a (surface) chemical reaction or to hydrodynamic instability in the presence of a chemical reaction not exhibiting instability by itself. The possibility also exists of an interfacial instability and waves occuring in a system where neither hydrodynamically nor chemically, taken separately, an instability is expected. The instability may occur when the hydrodynamics and adsorption-desorption, mass diffusion or heat are coupled to chemical reaction with processes having disparate and, eventually, largely different time scales. Note that those mass processes may bring surfactants that either lower or increase the interfacial tension. Building upon the earlier work by Sternling and Scriven (1959), theoretical predictions on hydrodynamic instability were provided by Sanfeld and collaborators (Steinchen-Sanfeld *et al.*, 1973; Hennenberg *et al.*, 1975; Sanfeld and Steinchen, 1975, 1984; Sanfeld *et al*, 1979; Dalle Vedove and Bisch, 1982) and by Ibanez and Velarde (1977). The latter considered a model-enzymatic, three-step chemical reaction obeying the Michaelis-Menten (Langmuir-Hinshelwood) kinetics with insoluble reactants with, however, inflow of matter from the bulk. Although a Rayleigh-Taylor hydrodynamic instability may appear, the striking prediction was that even if this is not the case and the chemical process is stabilizing, instability may appear due to the fact that the surface chemical reaction is so fast that it happens instantaneously relative to diffusional and convective relaxation times. Below we return to this question.

Experimental evidence and further theoretical studies illustrating phenomena occuring due to chemical reactions at interfaces (recall that we refrained from using the term chemocapillarity) in the form of hydrochemical waves has been reported by several groups of authors (Nakache *et al.*, 1983, 1984; Dupeyrat and Michel, 1971; Dupeyrat *et al.*, 1984; Pismen, 1984; Bekki *et al.*, 1990, 1992; Hu and Vignes-Adler, 1991; Muller *et al.*, 1984, 1985a-b; Warmuzinski and Tanczyk, 1991; Miike *et al.*, 1993; Kai *et al.*, 1985, 1991; Kai and Miike, 1994; Rabinovich *et al.*, 1995; Inomoto *et al.*, 1997; Mendes-Tatsis and Perez de Ortiz, 1992). In particular, Nakache *et al.*, (1983) discussed the case of a water-nitrobenzene interface where a reaction occurs with mass diffusion and adsorption bringing surfactants that lower the interfacial tension. Subsequently, due to the reaction and mass transfer across the interface, other surfactants appear that increase the tension. Then the periodic alternance of both processes leads to interfacial oscillations and surface waves. Incidentally, and just for the record, let us mention that Ertl and collaborators (Asakura *et al.*, 1995; Rotermund *et al.*, 1991; Bär *et al.*, 1992) have reported observation of apparent (dissipative) solitonic waves in solid surface chemical reactions like the catalytic oxidation of CO on Pt surfaces. Theory has been provided by Falcke *et al.*, (1992).

Müller *et al.* (1985a,b) have also observed steady and time-dependent convective patterns when dealing with oscillatory glycolitic reactions in thin layers of extract from yeast cells. They argued that the patterns were due to the coupling of the reaction with the Marangoni effect following the evaporative cooling of reactive products, and hence were influenced by variations in chemical composition and variable reaction rates.

Fig. 5.61 illustrates an apparent chemically-driven soliton observed by Kai *et al.* (1991) at the oil-water interface in a circular cylindrical annular container (outer diameter: 50-80 mm; width, 2.4-5.4 mm; two-layer liquid depth). The aqueous phase contained water with a surfactant (trimethyl stearyl ammonicum chloride: TSAC) and the oil was nitrobenzene containing iodine ($I_2$) saturated with potassium iodide (organic phase). Soon after the two immiscible solutions were placed in contact, one above the other, waves were observed. Kai *et al.* (1991) gave a plausible mechanism for the relaxation oscillation features of one of their hydrochemical waves. They assumed that the chemical reaction starts and stops at two different threshold values of a concentration product, called $K^+$ and $K^-$, respectively. When ions are transported to the interface by diffusion, the interfacial tension decreases, with excess surfactant concentration leading to formation of a monolayer. Reaction and hydrodynamics in the presence of gravity makes the interface prone to vibrate along the vertical. When the excess surface ion concentration reaches $K^+$, the reaction starts but ions are quickly consumed (fast reaction process) and it stops when $K$ reaches $K^-$. Then the slow diffusion process takes over and when, finally, $K$ reaches again $K^+$, once more the reaction starts, and so on.

Although no collisions or reflection events were studied and no

Figure 5.61: Solitonic hydrochemical wave train with four wave crests (amplitude 10 mm) (after Kai *et al.*, 1991).

quantitative data was provided by Kai *et al.* (1991) to allow clear-cut identifying solitonic signatures like a space-time diagram illustrating the sequence of events in a head-on collision of solitary waves (Fig. 5.49) the fact that their hydrochemical waves and wave trains have relatively big crests (10 mm relative peak amplitude) and travel mostly undeformed, point to solitonic behavior indeed. In view of the complex and, eventually, subtle interplay between chemistry, hydrodynamics and transport processes, theory is not yet available for a serious understanding of the observations reported about hydrochemical waves and their evolution and we shall not dwell further on this problem.

# Chapter 6

# Instabilities of parallel flows and film flows

In this chapter we shall consider the stability of parallel thermo-capillary flows described in Chapter 2. We shall also consider the influence of the Marangoni effect on the evolution and stability of (Kapitza-Shkadov) film flows.

## 6.1 Flows generated by a longitudinal surface tension gradient

In Chapter 2 we have described stationary parallel flows that appear in systems with an interface due to a longitudinal surface tension gradient. Let us now investigate the stability of these flows.

As in subsection 2.1, we consider a very long liquid layer (with the dimensionless length $L \gg 1$) on a horizontal rigid plane with a free surface subject to a horizontal temperature gradient $A$ in the direction of axis $x$. Except for a relatively small region near the lateral boundaries, the flow can be considered as parallel. Recall that one can distinguish between two typical situations:
(i) free *open flow* (2.27) with a zero pressure gradient and nonzero fluid flux ("unidirectional flow"); in the case of a purely thermocapillary convection ($R = 0$), this flow has a linear velocity profile, and it is called "linear flow"; (ii) flow in a very long but *closed* vessel with counteracting surface tangential stress and volume pressure gradient, which generate together a parabolic velocity profile with zero fluid flux ("return flow"; (2.23)). The heat advection by the parallel flows

produces some transverse temperature profiles, which essentially depend on the heat boundary conditions on the rigid boundary and the free boundary (subsection 2.1). In reality, the appearance of a horizontal pressure gradient on the free surface leads unavoidably to a non-uniformity of the layer thickness, but here this effect will be neglected (e.g., we can assume that the ratio of the Marangoni number, $M$, to the Galileo number, $G$, is much smaller than the inverse length of the layer $1/L$, see subsection 2.2).

First, we shall discuss the main types of instability in the case of purely thermocapillary flows which is appropriate mainly under low effective-gravity (microgravity) conditions. Then we shall consider the influence of buoyancy under the conditions of normal gravity. We shall show that even in the case of a very long horizontal layer the explanation of experiments is impossible if the finite length of the layer is not taken into account.

### 6.1.1   Purely thermocapillary flows

For purely thermocapillary flows ($R = 0$), the velocity profile is

$$u = -Mz \tag{6.1}$$

in the case of the unidirectional (linear) flow, and

$$u = M\left(-\frac{3z^2}{4} + \frac{z}{2}\right) \tag{6.2}$$

in the case of the return flow (2.27), (2.23). We shall discuss in detail the case where the rigid boundary is a really poor heat conductor. The temperature field generated by the parallel thermocapillary flow is described by Eq. (2.6),

$$T(x, z) = x + \tau(z), \tag{6.3}$$

where

$$\tau = M\left(-\frac{z^3}{6} + \frac{z}{2}\right) \tag{6.4}$$

in the case of the linear flow, and

$$\tau = -M\left(\frac{z^4}{16} - \frac{z^3}{12} + \frac{z}{48}\right) \tag{6.5}$$

in the case of the return flow (2.30), (2.25). According to the definition of the function $\tau(z)$, $\tau(1) = 0$ in both cases. It should be noted, that in the former case $\tau(z) > 0$, $\tau'(z) \le 0$ as $0 \le z < 1$, thus the layer is effectively "heated from below" by the parallel thermocapillary flow. Quite opposite, in the latter case $\tau(z) < 0$, $\tau'(z) \ge 0$ as $0 \le z < 1$, e.g. the layer is "cooled from below".

The investigation of the linear stability of thermocapillary flow uses the same technique as the investigation of the linear stability of the (mechanical) equilibrium state studied in Chapter 4. Because we have assumed already the non-deformability of the free surface, we shall formulate the eigenvalue problem in the case of a non-deformable free boundary. Let us emphasize that the condition $\tau(z) = 0$ on the free boundary is understood as the *choice of a reference value* for the temperature profile rather than the condition of perfectly conducting boundary. Thus, we can assume that heat exchange on the rigid boundary is characterized by an arbitrary heat exchange coefficient. Strictly speaking, in the case of the linear flow the heat exchange coefficient cannot be equal to zero on both horizontal boundaries, otherwise the stationary solution for the temperature field does not exist, but one can consider the limit $K \to 0$ rather than $B = 0$.

Linearizing the system of equations (1.28) - (1.30) around the solution corresponding to the parallel flow, we get the following problem for disturbances

$$\frac{1}{P}\left[\frac{\partial \tilde{\mathbf{v}}}{\partial t} + u(z)\frac{\partial \tilde{\mathbf{v}}}{\partial x} + \tilde{v}_z u'(z)\right] = -\nabla \tilde{p} + \nabla^2 \tilde{\mathbf{v}}, \qquad (6.6)$$

$$\frac{\partial \tilde{T}}{\partial t} + u(z)\frac{\partial \tilde{T}}{\partial x} + \tilde{v}_x + \tilde{v}_z \tau'(z) = \nabla^2 \tilde{T}, \qquad (6.7)$$

$$\nabla \cdot \tilde{\mathbf{v}} = 0; \qquad (6.8)$$

$$z = 0: \ \tilde{\mathbf{v}} = 0, \ \tilde{T} = 0; \qquad (6.9)$$

$$z = 1:$$

$$\frac{\partial \tilde{v}_x}{\partial z} + \frac{\partial \tilde{v}_z}{\partial x} - M\frac{\partial \tilde{T}}{\partial x} = 0; \qquad (6.10)$$

$$\frac{\partial \tilde{v}_y}{\partial z} + \frac{\partial \tilde{v}_z}{\partial y} - M\frac{\partial \tilde{T}}{\partial y} = 0; \qquad (6.11)$$

$$\tilde{v}_z = 0; \qquad (6.12)$$

$$\frac{\partial \tilde{T}}{\partial z} = -Bi\tilde{T}. \tag{6.13}$$

Assuming

$$(\tilde{\mathbf{v}}, \tilde{p}, \tilde{T}) = (\hat{\mathbf{v}}(\mathbf{z}), \hat{p}(z), \hat{T}(z)) \exp(i\mathbf{k} \cdot \mathbf{x}_\perp + \lambda t), \tag{6.14}$$

$$\mathbf{x}_\perp = (x, y),$$

and substituting (6.14) into (6.6) - (6.13), we obtain an eigenvalue problem which defines the growth rate, $\lambda$, as the function of the wavevector $\mathbf{k} = (k_x, k_y)$. An important difference between this eigenvalue problem and that of the stability of mechanical equilibrium state is the fact that in the case of the parallel flow the growth rate not only depends on the modulus of the wavevector $k = \sqrt{k_x^2 + k_y^2}$ but also on the direction of the wavevector, because the presence of the flow violates the isotropy of the problem. Let $k_x = k \sin \alpha$, $k_y = k \cos \alpha$. It is convenient to perform a transformation of horizontal coordinates $x = X \cos \alpha + Y \sin \alpha$, $y = -X \sin \alpha + Y \cos \alpha$. After this transformation, the wave vector $\mathbf{k}$ is directed along the $Y$ axis, and the disturbances do not depend on $X$. In the new system of coordinates, we obtain the following system of equations:

$$\frac{\lambda}{P}\tilde{v}_X + ik \sin \alpha \cdot u\tilde{v}_X + \cos \alpha \cdot u'\tilde{v}_z = \tilde{v}_X'' - k^2\tilde{v}_X; \tag{6.15}$$

$$\frac{\lambda}{P}\tilde{v}_Y + ik \sin \alpha \cdot u\tilde{v}_Y + \sin \alpha \cdot u'\tilde{v}_z = -ik\tilde{p} + \tilde{v}_Y'' - k^2\tilde{v}_Y; \tag{6.16}$$

$$\frac{\lambda}{P}\tilde{v}_z = -\tilde{p}_i' + \tilde{v}_z'' - k^2\tilde{v}_z; \tag{6.17}$$

$$\lambda\tilde{T}_i + ik \sin \alpha \cdot u\tilde{T} + \tau'\tilde{v}_z + \cos \alpha \cdot \tilde{v}_X + \sin \alpha \cdot \tilde{v}_Y = \tilde{T}'' - k^2\tilde{T}; \tag{6.18}$$

$$\tilde{v}_z' + ik\tilde{v}_Y = 0. \tag{6.19}$$

According to Smith and Davis (1983a,b; Davis 1987) the instabilities can be separated into several classes:

1. In the case of the linear flow, which produces some heating from below, the usual surface tension gradient-driven (Marangoni) instability may appear. It prevails for large Prandtl number liquids and generates stationary longitudinal rolls, e.g. the disturbances of all physical fields depend periodically on the transverse horizontal coordinate $y$ and the vertical coordinate $z$ but do not depend on the longitudinal direction.

2. The basic shear flows with a free boundary are subject to instabilities of *isothermal nature*. The prevailing type of the instability may be a short-wave instability (for the linear flow) or a long-wave instability (for the return flow). These instabilities appear if the capillary number is not too small.

3. There are also two types of instabilities which are essentially connected with both thermocapillarity and the existence of the flow. These instabilities are called *hydrothermal* instabilities (Smith, 1986; Ezersky *et al.*, 1993).

The first type of hydrothermal instabilities appears mainly for relatively low Prandtl number fluids and generates waves propagating *across* the basic flow. As earlier discussed, the origin of the oscillatory instability is the existence of compensating mechanisms tending to diminish the disturbances but acting with some time delay. Let us consider, for instance, the linear flow (hence leading to heating from below) and discuss the evolution of a surface temperature disturbance in the form of a hot line parallel to the flow direction. Accordingly, the surface flow directed away from this line generates *an upflow* supporting the temperature disturbance. However, it is necessary to take into account the fact that the fluid rises from the region with lower velocity of the base flow into a region with higher velocity of the base flow. That is why the upflow generates an *upstream* velocity disturbance on the surface which *cools* the surface. The time delay between two processes is the origin of oscillations.

The second type of hydrothermal instabilities is characteristic for high Prandtl number fluids. Let us consider the return flow (generating heating from above) and a surface hot line oriented across the stream. The surface temperature disturbance creates an upflow under the line itself and downflows on a certain distance from the line. The downflow arising upstream from the line moves upstream, generating an upstream traveling wave. Actually, because of the interaction of both types of instability, an oblique motion of rolls is observed (Smith and Davis, 1983a).

## 6.1.2 Combined action of thermocapillarity and buoyancy

In Chapter 2 we have obtained the profiles of the velocity and temperature for the parallel flow generated by a longitudinal temperature gradient under a simultaneous action of thermocapillarity and buoyancy. Now we shall present the main results of the instability analysis for such a flow. The case of a return flow between heat insulated boundaries is discussed (Parmentier et al., 1993). The deformation of the surface is disregarded.

### i. Purely buoyancy-driven instability

Figs. 6.1 – 6.3 depict the main results found. It turns out that there are two different types of hydrothermal waves, one for low values of the Prandtl number, $0 < P < P_1 \approx 0.41$, and another one for intermediate values of the Prandtl number, $P_1 < P > P_2 \approx 2.6$. The low Prandtl number instability is characterized by a nearly longitudinal orientation of rolls and relatively small values of the wavenumber. Quite opposite, the moderate Prandtl number waves rapidly tend to a nearly transverse orientation of rolls, as the Prandtl number grows, and their wavenumbers are never small.

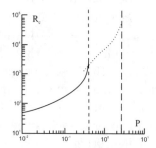

Figure 6.1: Critical Rayleigh number, $R_c$, vs $P$ in the purely thermogravitational case. (Redrawn after Parmentier et al., 1993).

### ii. Coupled buoyancy and thermocapillarity

Figs. 6.4 – 6.6 illustrate the main results found. The joint action of buoyancy and thermocapillarity leads to an instability whatever the value of $P$, including $P > P_2$. In the latter region, there is a remarkable transition from an obliquely inclined orientation of hydrothermal waves at low values of the dynamic Bond number $Bd = R/M$ to a transverse orientation at larger values of $Bd$.

Thus, one could expect that upon increasing $B_d$ one should observe a transition from inclined, three-dimensional, hydrothermal

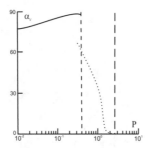

Figure 6.2: Angle of propagation, $\alpha_c$, of the disturbance vs $P$ in the purely thermogravitational case. (Redrawn after Parmentier *et al.*, 1993).

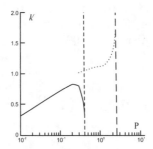

Figure 6.3: Critical number, $k_c$, vs $P$ in the purely thermogravitational case. (Redrawn after Parmentier *et al.*, 1993).

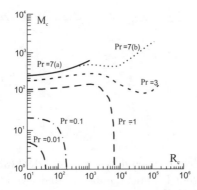

Figure 6.4: Critical Marangoni number, $M_c$, vs the critical Rayleigh number, $R_c$, at five different Prandtl numbers. The curve $P = 7$ is composed of two distinct solutions (a) and (b), which intersect at $R_c = 365$. (Redrawn after Parmentier *et al.*, 1993).

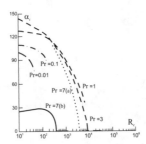

Figure 6.5: Angle of propagation, $\alpha_c$, of the disturbance vs critical Rayleigh number, $R_c$, at five different Prandtl numbers. (Redrawn after Parmentier *et al.*, 1993).

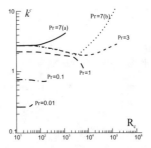

Figure 6.6: Critical wavenumber, $k_c$, vs critical Rayleigh number, $R_c$, at five different Prandtl numbers. (Redrawn after Parmentier *et al.*, 1993).

waves to transverse, two-dimensional, waves. Surprisingly, experiments (Schwabe *et al.*, 1992; De Saedeleer *et al.*, 1996; Riley and Neitzel, 1998) showed the appearance of *steady* transverse periodic patterns ("cat's-eyes flows"), and hence in apparent disagreement with the theory. The explanation of this paradox, given by Priede and Gerberth (1997), is discussed in the next subsection.

### 6.1.3  Influence of lateral boundaries

The instability criterion $\mathrm{Re}\lambda(k)=\mathrm{Im}\omega(k) > 0$, which we systematically use in this book, determines the boundary of a *convective* instability of flows (Lifshitz and Pitaevskii, 1981). After crossing this boundary, a perfect spatially periodic wave with wavenumber $k$, which fills the whole space, exponentially grows with time. However, a spatially localized *wave packet* with wavenumbers around $k$ moves with its group velocity $v_g = d\mathrm{Re}\omega(k)/dk$, and its amplitude grows with time only in the reference frame *moving* with the velocity $v_g$. The growth of the disturbance in a fixed spatial point is not granted by the criterion written above. In other words, the disturbance grows downstream, but it is washed out by the flow. The local growth of the disturbance starts after crossing another boundary, which is called the *absolute* instability boundary, which is located at higher values of the Marangoni number, or does not exist at all.

In the region of convective instability, the system is very sensitive to external perturbations. In their absence, the parallel flow is restored in each spatial point for any initial conditions, as if the system would be stable. But, e.g., a weak noise is amplified downstream, which may lead to the appearance of wave at a certain distance from the entrance.

In the case of convective instability of a parallel flow in a long but finite layer, the crucial point is the influence of rigid lateral walls. Near the lateral walls the flow is not parallel. The deviation of the flow from the parallel profile can be considered as a finite-amplitude *steady* disturbance. The response of the system to a boundary disturbance is described, in the linear approximation, by the same dispersion relation $\omega = i\lambda = \omega(k)$, but this time we are interested in *complex* roots $k = k(\omega)$ (in our case, $\omega = 0$). Note that a positive value of $\mathrm{Im}k$ for a certain real $\omega$ may correspond either to a wave which moves to the right and decays in space ("non-transparency") or to a wave which moves to the left and grows in space ("amplification of waves"). The criterion for distinguishing between these two cases is given by Lifshitz and Pitaevskii (1981), and we shall not reproduce it here. What matters is that in the region of parameter values where the flow is stable or convective unstable with respect to *waves*, one can observed a boundary-generated *steady* structure on the distances of order $1/|\mathrm{Im}k(0)|$ from the boundary. This structure may compete with the wavy flows predicted by the "standard" linear stability theory. If there is an oscillatory instability of the flow, but the frequency $\omega(k) = -\mathrm{Im}\lambda(k)$ vanishes at a certain point $k = k_*$ on the neutral stability curve $M = M(k)$, that means that when the Marangoni number reaches the value $M(k_*)$, the distance $1/|\mathrm{Im}k(0)|$ diverges to infinity, i.e. the boundary-generated steady pattern fills the whole layer forming a *steady multicellular convection flow* with the wavenumber $k_*$, rather than a wavy flow.

The quantitative theory of the phenomenon described above was given by Priede and Gerbeth (1997). As it is seen in Fig. 6.7), for relatively small values of the dynamic Bond number, $Bd < 0.2$, the boundary of absolute instability is rather close to that of convective instability, so that the standard linear stability theory gives a rather good prediction. In the region $0.2 < Bd < 0.3$ the boundary of absolute instability is still lower than that of the development of the steady multicellular flow, but it is much higher than the boundary of the convective instability, thus the standard linear stability theory is not sufficient for the calculation of the actual instability threshold. For $Bd > 0.3$ the parallel flow is first replaced by the lateral-boundary-induced steady flow. The theoretical predictions agree well with observations made by Riley and Neitzel (1998).

## 6.2   Film flows

If the deformation of the free surface (or interface) is allowed, a specific longwave deformational instability may arise corresponding to surface waves supported by Reynolds stresses. A paradigmatic example of such an instability is the instability of the parallel flow in a *film flowing down an inclined plane* thouroughly studied by Kapitza (1948; Kapitza and Kapitza, 1949) and later on by Shkadov (1967, 1968, 1977. For more recent theoretical studies and experiments see Benjamin, 1957; Fulford, 1964; Stainthrop and Allen, 1965; Benney, 1966; Bankoff, 1971; Petviashvili and Tsvelodub, 1978; Trifonov,

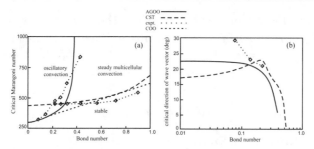

Figure 6.7: (a) Experimental and theoretical critical Marangoni numbers for the onset of both steady multicellular (CST) and oscillatory convection (AGOO), and (b) the corresponding angle between the critical wave vector and the $x$ axis depending on the dynamic Bond number for 1-cS silicone oil ($P = 13.9$). The convective oscillating oblique wave instability (COO) results from the standard theory. (Redrawn after Priede and Gerbeth, 1997).

1989; Tsvelodub, 1980a,b; Demekhin et al., 1983, 1987; Pumir et al., 1983; Lin and Wang, 1985; Joo et al., 1991; Trifonov and Tsvelodub, 1991; Frenkel, 1992; Joo and Davis, 1992; Tsvelodub and Trifonov, 1992; Liu et al., 1993, 1995; Chang, 1994; Liu and Gollub, 1994; Salamon et al., 1994; Liu et al., 1995; Zeytounian, 1995; Chang and Demekhin, 1996; Ruyer-Quil and Manneville, 2000; Shkadov and Sisoev, 2000). The instability arises for the longwave disturbances with the wavevector parallel to the direction of the flow, as the Galileo number exceeds the critical value

$$Ga_c = \frac{5\cos\beta}{2\sin^2\beta},\tag{6.20}$$

where $\beta$ is the angle of layer inclination (Yih, 1963). The primary instability generates two-dimensional wavy motions (Benjamin, 1957; Alekseenko et al., 1994; Chang, 1994), which might be unstable with respect to three-dimensional disturbances (Nepomnyashchy, 1974a,b; Trifonov, 1991; Joo and Davis, 1992; Chang et al., 1993) and are replaced downstream by three-dimensional waves (Nepomnyashchy, 1974b; Lin and Krishna, 1977; Trifonov and Tsvelodub, 1986; Trifonov, 1990; Chang et al., 1994) and solitons (Petviashvili and Tsvelodub, 1978).

Let us consider the influence of the Marangoni effect upon the stability of a falling liquid film. The thermocapillary effect generates new instability modes which can interact with Kapitza's instability mode. For illustration we shall consider the case of a surfactant aqueous solution with surfactant or solutocapillarity but most arguments and salient results should carry over to the case of heat transfer. We shall follow a recent study by Shkadov *et al.* (2001).

### 6.2.1   Formulation of the problem

#### i. Governing equations

Let $x, y, z$ be the orthogonal coordinate system with origin located on the rigid solid wall. The axis $x$ is directed along the wall, positive down in the direction of gravity. The film flow of the aqueous solution is described by the Navier-Stokes equations

$$\frac{du}{dt} = -\frac{1}{\rho}\frac{\partial p}{\partial x} + \nu \triangle u + g$$

$$\frac{dv}{dt} = -\frac{1}{\rho}\frac{\partial p}{\partial y} + \nu \triangle v \qquad (6.21)$$

$$\frac{\partial u}{\partial x} + \frac{\partial v}{\partial y} = 0$$

together with the following boundary conditions on the rigid wall

$$y = 0, \quad u = 0, \quad v = 0 \qquad (6.22)$$

and on the film surface

$$y = h\left(x, t\right), \quad \frac{\partial h}{\partial t} + u\frac{\partial h}{\partial x} = v$$

$$p = p_a + 2\mu\frac{1}{b^2}\left[\left(1 - b^2\right)\frac{\partial v}{\partial y} - \frac{\partial h}{\partial x}\left(\frac{\partial v}{\partial x} + \frac{\partial u}{\partial y}\right)\right] - \sigma\frac{1}{b^3}\frac{\partial^2 h}{\partial x^2} \quad (6.23)$$

$$\mu\left(1 - b^2\right)\left(\frac{\partial u}{\partial y} + \frac{\partial v}{\partial x}\right) + 2\mu\left(\frac{\partial v}{\partial y} - \frac{\partial u}{\partial x}\right)\frac{\partial h}{\partial x} - b\frac{\partial \sigma}{\partial x} = 0$$

$$b = \left[1 + \left(\frac{\partial h}{\partial x}\right)^2\right]^{\frac{1}{2}}$$

where $p_a$ denotes the pressure at the gas phase.

The bulk concentration of surfactant, $C$, obeys Fick's diffusion equation

$$\frac{dC}{dt} = D\triangle C \qquad (6.24)$$

with boundary condition on the rigid wall

$$y = 0, \quad \frac{\partial c}{\partial y} = 0 \tag{6.25}$$

Note that in (6.21) and (6.24) $d/dt$ is the material derivative. On the surface of the liquid film as in earlier similar cases treated here, the equations of mass conservation of surfactant and Fick's law (subsection 1.2.2) yield

$$\frac{\partial \Gamma}{\partial t} + \frac{1}{b}\frac{\partial}{\partial x}\left[\frac{\Gamma}{b}\left(u + v\frac{\partial h}{\partial x}\right)\right] - D_s\frac{1}{b}\frac{\partial}{\partial x}\left(\frac{1}{b}\frac{\partial \Gamma}{\partial x}\right) + \frac{k_g}{m}\overline{C} = j \tag{6.26}$$

$$j = -D\frac{1}{b}\left(\frac{\partial C}{\partial y} - \frac{\partial h}{\partial x}\frac{\partial C}{\partial x}\right), \quad y = h(x,t) \tag{6.27}$$

The bar over $C$ denotes the bulk value at the surface. $\Gamma$ accounts for surface excess surface concentration. A term due to surface deformation has been ignored in (6.26).

In the initial section of the film a uniform distribution of surfactant concentration is assumed

$$x = x_0, \quad C = C_* = const \tag{6.28}$$

The last term on the left hand side of (6.26) accounts for the desorption of the surfactant from the liquid phase to the gas phase, $k_g$ is the gas phase mass transfer coefficient of the surfactant, $m$ is the ratio of the concentration in the liquid phase to the concentration in the gas phase at equilibrium. Two assumptions when deriving this boundary equation are that the concentration of the surfactant in the gas phase far from interface is zero and that there is equilibrium at the interface between the concentration in the gas phase and the concentration in the liquid film. To obtain $(\partial \sigma/\partial x)$ an equation of state, $\sigma = \sigma(x,t) = \sigma(\bar{C})$, must be specified thus leading to a closed mathematical formulation.

## ii. Dimensionless formulation

To state the problem in dimensionless form we introduce suitable scales and hence

$$x, y, h, t \rightarrow \frac{1}{n_*}lx, \quad ly, \quad lh, \quad \frac{l}{n_*U_*}t; \tag{6.29}$$

$$u, v, C, \Gamma \to U_* u, \quad n_* U_* v, \quad C_* \left(1 + C\right), \quad \Gamma_* \left(1 + \Gamma\right)$$

Dimensional quantities stand in the left sides of (6.29) and their corresponding dimensionless values are contained in the right sides. The scales $l, U_*, C_*$, and $\Gamma_*$ together with a stretching parameter $n_*$ need also are to be prescribed. From (6.29) it follows

$$b = \left(1 + n_*^2 (\partial h / \partial x)^2\right)^{\frac{1}{2}}, \tag{6.30}$$

where $\partial h / \partial x = o\left(1\right)$ in accordance with the choice of the stretching coefficient $n_*$. This is taken as $n_*^2 \ll 1$ in what follows. For wavy motions the latter condition means that the wavelength of disturbances is much larger than the liquid layer thickness. Thus for a wave number $\alpha$ this implies that $n_* \alpha \leq 1$. Then the cumbersome boundary conditions (6.23), (6.26), (6.27) could be simplified by taking $b = 1$ to order $o\left(n_*^2\right)$. After introducing (6.29) in (6.21)-(6.22), the dimensionless formulation of the problem yields the equations:

$$\frac{\partial u}{\partial t} + u \frac{\partial u}{\partial x} + v \frac{\partial u}{\partial y} = -\frac{\partial p}{\partial x} + \frac{1}{n_* \mathrm{Re}} \left(\frac{\partial^2 u}{\partial y^2} + n_*^2 \frac{\partial^2 u}{\partial x^2}\right) + \frac{1}{n_* \mathrm{Fr}}$$

$$n_*^2 \left(\frac{\partial v}{\partial t} + u \frac{\partial v}{\partial x} + v \frac{\partial v}{\partial y}\right) = -\frac{\partial p}{\partial y} + n_*^2 \frac{1}{n_* \mathrm{Re}} \left(\frac{\partial^2 v}{\partial y^2} + n_*^2 \frac{\partial^2 v}{\partial x^2}\right) \tag{6.31}$$

$$\frac{\partial u}{\partial x} + \frac{\partial v}{\partial y} = 0 \qquad \frac{\partial C}{\partial t} + u \frac{\partial C}{\partial x} + v \frac{\partial C}{\partial y} = \frac{1}{n_* \mathrm{Pe}} \left(\frac{\partial^2 C}{\partial y^2} + n_*^2 \frac{\partial^2 C}{\partial x^2}\right) \tag{6.32}$$

As usual, the surface tension, $\sigma$, is taken as a linear function of the bulk surfactant concentration on the free surface, $\overline{C}\left(x, t\right)$,

$$\sigma = \sigma_* + \frac{d\sigma}{d\overline{C}} \left(\overline{C} - C_*\right) \tag{6.33}$$

where the starred quantities are related reference values. Then due to (6.29) from (6.25) - (6.28) the dimensionless formulation of the boundary conditions become

$$y = 0, \quad u = 0, \quad v = 0, \quad \frac{\partial C}{\partial y} = 0; \quad x = 0, \quad C = 0$$

$$y = h\left(x, t\right), \quad \frac{\partial h}{\partial t} + u \frac{\partial h}{\partial x} = v$$

$$p = -\frac{n_*^2}{\text{We}} \left(1 - \frac{\text{MaWe}}{\text{Re}}\overline{C}\right) \frac{\partial^2 h}{\partial x^2} - \frac{2n_*^2}{n_*\text{Re}} \left(\frac{\partial v}{\partial y} - \frac{\partial h}{\partial x}\frac{\partial u}{\partial y} - n_*^2\frac{\partial h}{\partial x}\frac{\partial v}{\partial x}\right) \tag{6.34}$$

$$\frac{\partial u}{\partial y} + n_*^2 \left(\frac{\partial v}{\partial x} + 4\frac{\partial h}{\partial x}\frac{\partial v}{\partial y}\right) = -n_*\text{Ma}\frac{\partial \overline{C}}{\partial x}$$

$$\frac{\partial C}{\partial y} + \text{Bi}\left(1 + \overline{C}\right) + n_*\text{G}\left[\frac{\partial \Gamma}{\partial t} + \frac{\partial}{\partial x}\left(u + n_*^2 v\frac{\partial h}{\partial x}\right)(1 + \Gamma)\right] - n_*^2\text{Di}\frac{\partial^2 \Gamma}{\partial x^2} = 0$$

$$-\frac{\partial c}{\partial y} + n_*^2\frac{\partial h}{\partial x}\frac{\partial C}{\partial x} = \pi_1\left(1 + \overline{C}\right) - \pi_2\left(1 + \Gamma\right)$$

In (2.18) and (2.19) the following ten dimensionless parameters have been introduced:

$$\text{Re} = \frac{U_*l}{\nu}, \quad \text{Pe} = \frac{U_*l}{D}, \quad \text{We} = \frac{\rho l U_*^2}{\sigma}, \quad \text{Fr} = \frac{U_*^2}{gl}$$

$$\text{Ma} = -\frac{d\sigma}{d\overline{C}}\frac{C_*}{\eta U_*}, \quad \text{G} = \frac{\Gamma_* U_*}{c_* D}, \quad \text{Bi} = \frac{k_g l}{mD} \tag{6.35}$$

$$\text{Di} = \frac{D_s\Gamma_*}{Dl C_*}, \quad \pi_1 = \frac{k_a l}{D}, \quad \pi_2 = \frac{k_d\Gamma_* l}{c_* D}$$

where Re, Pe (for simplicity we have omitted the subscript $D$), We, Fr, Ma, and Bi stand for the appropriate definitions here of the Reynolds, Peclet, Weber, Froude, Marangoni and Biot numbers; Di refers to diffusion.

To investigate the multiparameter problem (6.31)-(6.35) a simplifying method introduced long ago by Shkadov (1967, 1977) can be used.

Terms in (6.31) - (6.34), which have an order $o\left(n_*^2\right)$ can be omitted. However the products $n_*\text{Re}$, $n_*\text{Ma}$, $n_*\text{Fr}$, $n_*^2\text{We}^{-1}$ and $n_*^2\text{Di}$, which could have order unity or even higher, i.e. of the order of $n_*\text{Pe}$ and $n_*\text{G}$ must be retained.

After omitting all terms in (6.31)-(6.34) of order $n_*^2$, the boundary layer approximation, with self-induced pressure, is obtained. That aproximation includes hydrodynamic and diffusion parts which are connected by the Marangoni stress due to the boundary condition for tangential forces in (6.34). The hydrodynamic part is

$$\frac{\partial u}{\partial t} + u\frac{\partial u}{\partial x} + v\frac{\partial u}{\partial y} = -\frac{\partial p}{\partial x} + \frac{1}{n_*\text{Re}}\frac{\partial^2 u}{\partial y^2} + \frac{1}{n_*\text{Fr}} \tag{6.36}$$

$$\frac{\partial p}{\partial y} = 0 \tag{6.37}$$

$$\frac{\partial u}{\partial x} + \frac{\partial v}{\partial y} = 0 \tag{6.38}$$

$$y = 0, \quad u = 0, \quad v = 0, \tag{6.39}$$

$$y = h\,(x,t)\,, \quad \frac{\partial h}{\partial t} + u\frac{\partial h}{\partial x} = v \tag{6.40}$$

$$p = -n_*^2/\mathrm{We}\left(1 - \mathrm{MaWe}/\mathrm{Re}\overline{C}\right)\frac{\partial^2 h}{\partial x^2} \tag{6.41}$$

$$\frac{\partial u}{\partial y} = -n_*\mathrm{Ma}\frac{\partial \overline{C}}{\partial x} \tag{6.42}$$

while the diffusion part is

$$\frac{\partial C}{\partial t} + u\frac{\partial C}{\partial x} + v\frac{\partial C}{\partial y} = \frac{1}{n_*\mathrm{Pe}}\frac{\partial^2 C}{\partial y^2} \tag{6.43}$$

$$y = 0, \quad \frac{\partial C}{\partial y} = 0; \quad x = 0, \quad c = 0 \tag{6.44}$$

$$y = h\,(x,t)\,, \tag{6.45}$$

$$\frac{\partial C}{\partial y} + \mathrm{Bi}\left(1 + \overline{C}\right) + n_*\mathrm{G}\left[\frac{\partial \Gamma}{\partial t} + \frac{\partial u\,(1 + \Gamma)}{\partial x}\right] - n_*^2\mathrm{Di}\frac{\partial^2 \Gamma}{\partial x^2} = 0 \tag{6.46}$$

$$-\frac{\partial C}{\partial y} = \pi_1\left(1 + \overline{C}\right) - \pi_2\left(1 + \Gamma\right) \tag{6.47}$$

The base solution of (6.36) corresponds to a plane parallel film flow of constant thickness. To further simplify the problem, one can take this solution in the form

$$u = \frac{3}{2}\left(2y - y^2\right), \quad h = 1 \tag{6.48}$$

Then using (2.25) it follows from (2.21)

$$\frac{1}{n_*\mathrm{Fr}} = \frac{3}{n_*\mathrm{Re}} \tag{6.49}$$

and hence viscous and gravitational forces are of the same order. Then with the help of (6.35) using Re, Fr and (6.49) one obtains

$$l = \left(\frac{3\nu^2}{g}\right)^{\frac{1}{3}}\mathrm{Re}^{\frac{1}{3}}, \quad U_* = \left(\frac{g\nu}{3}\right)^{\frac{1}{3}}\mathrm{Re}^{\frac{2}{3}} \tag{6.50}$$

Considering that capillary forces are of the same order as the viscous and the gravitational ones, then the following relations hold:

$$\frac{n_*^2}{\text{We}} = \frac{3}{n_*\text{Re}} = \frac{1}{n_*\text{Fr}} = \frac{1}{5\delta} \qquad (6.51)$$

as a generalization of (6.49). From (6.51), (6.50) the two quantities $n_*$ and $\delta$ are found

$$n_* = \gamma^{-\frac{1}{3}}(3\text{Re})^{\frac{2}{9}}, \quad \delta = \frac{1}{45}\gamma^{-\frac{1}{3}}(3\text{Re})^{\frac{11}{9}}, \quad \gamma = \frac{\sigma}{\rho}\left(\nu^4 g\right)^{-\frac{1}{3}} \qquad (6.52)$$

The parameter $n_*$ which plays a crucial role in the procedure of simplification of the Navier–Stokes equations, (6.31)-(6.34), can be expressed in terms of the capillary number, that to avoid confusion is here denoted by Ca

$$\text{Ca} = \frac{\eta U_*}{\sigma} = \frac{\text{We}}{\text{Re}} \qquad (6.53)$$

and thus

$$n_* = (3\text{Ca})^{\frac{1}{3}}. \qquad (6.54)$$

The relevance of the above introduced quantities and approximations can be seen by considering a water film, $\gamma = 2850$, for a sequence of Re values. Table 6.1 provides numerical estimates of the main parameters. It appears that the conditions $n_*^2 \ll 1$ and $n_*^2\text{We}^{-1} \sim 1$ are fulfilled for Re values, $30 > \text{Re} > 5$, or for $\delta$ values, $0.4 > \delta > 0.043$. Noteworthy is that the Kapitzas in their (1949) pioneering experiments on wavy film flows worked precisely in such interval of Re numbers. From (6.54) follows that

$$\text{Ma}\frac{\text{We}}{\text{Re}} = \text{MaCa} = \frac{1}{3}n_*^3\text{Ma} \qquad (6.55)$$

For the $n_*$ values in Table 6.1 this product is small, of order $o\left(n_*^2\right)$ for Ma $\leq 10$. Thus MaWe/Re in (6.41) could be omitted for all $\delta$ values under consideration. Due to (6.51) only the parameter $\delta$ enters the formulation (6.36)-(6.46) instead of Re, Fr and We. Then the formulation given below applies for any liquid provided $\gamma \gg 1$.

**Table 6.1. Parameter (Re, $\delta$) values.** Range of values of Re and $\delta$ where the conditions $n_*^2 \ll 1$, $n_*\mathrm{Re} \sim 1$, $n_*^2\mathrm{We}^{-1} \sim 1$ are fulfilled for falling water films.

| $l_{mm}$ | Re | $\delta$ | $n_*^2$ | $n_*\mathrm{Re}$ | $n_*^2/\mathrm{We}$ |
|---|---|---|---|---|---|
| 0.115 | 5 | 0.043 | 0.0165 | 0.642 | 4.673 |
| 0.145 | 10 | 0.101 | 0.023 | 1.511 | 1.985 |
| 0.183 | 20 | 0.235 | 0.031 | 3.526 | 0.851 |
| 0.209 | 30 | 0.386 | 0.037 | 5.787 | 0.518 |
| 0.230 | 40 | 0.545 | 0.042 | 8.170 | 0.367 |
| 0.248 | 50 | 0.720 | 0.046 | 10.8 | 0.278 |

## 6.2.2   Galerkin approach

To investigate the solutions of the differential equations with boundary and initial conditions, (6.36)-(6.46), Shkadov *et al.* (2001) applied the Galerkin method in coordinate $y$. For the spectral representation of the unknown functions a system of polynomials was used as a basis set. For the coefficients of spectral representations which are functions of $x$ and $t$, a system of one-dimensional time-dependent differential equations follow from the Galerkin method. Direct numerical solutions of the initial problem (6.36)-(6.46) for Ma $= 0$ by Demekhin *et al.*, (1987) and experimental measurements by Alekseenko *et al.*, (1994) have shown that the velocity profile in coordinate $y$ for the wavy flow could be approximated by the simplest polynomial satisfying the boundary conditions. Thus for the spectral representation of $u$ it is enough to take into account only the first term.

### i. Hydrodynamic modes (Kapitza)

For the velocity field $u(x, y, t)$ obeying the boundary condition (6.41) one can take

$$u = \overline{u}\left(2\eta - \eta^2\right) + \mathrm{M}h\left(\eta - \eta^2\right)\overline{C}_x \qquad (6.56)$$

where the Marangoni number, $\mathrm{M} = n_*\mathrm{Ma}$, has been rescaled. No confusion is expected due to the use of $\eta$ here.

By inserting (6.56) in equations (6.36) and by integrating the resulting expressions from $y = 0$ to $y = h$, the equations for $h(x,t)$, $\overline{u}(x,t)$ or $q(x,t)$ are obtained with the help of (6.51). One has

$$\frac{\partial h}{\partial t} + \frac{\partial q}{\partial x} = 0 \qquad (6.57)$$

$$\frac{\partial q}{\partial t} + \frac{\partial Q}{\partial x} = \frac{1}{5\delta}\left(h\frac{\partial^3 h}{\partial x^3} + h - \frac{2}{3h}\left(\overline{u} + \mathrm{M}h\overline{C}_x\right)\right)$$

with

$$q = \int_0^h u\,dy = \frac{2}{3}\overline{u}h + \frac{1}{6}\mathrm{M}h^2\overline{C}_x$$

$$Q = \int_0^h u^2\,dy = \frac{8}{15}\overline{u}^2 h + \frac{7}{30}\mathrm{M}h^2\overline{u}\overline{C}_x + \frac{1}{30}h\left(\mathrm{M}h\overline{C}_x\right)^2 \qquad (6.58)$$

$$y = \eta h, \quad \overline{C}_x = \frac{\partial \overline{C}}{\partial x}$$

As expected, for $\mathrm{Ma} = 0$, equations (6.57) reduce to the original results obtained by Shkadov (1967,1977) for wavy film flows with no surfactants. For $\mathrm{M}h\overline{C}_x = \tau$, where $\tau$ is prescribed, the equations derived by Esmail and Shkadov (1971) are also recovered.

Most of the Kapitzas (1948, 1949) experiments and more recent experimental observations on regular wave film flows have been explained theoretically on the basis of periodic solutions of the nonlinear system (6.57). A detailed comparison of theory and experiments concerning wavy films has been recently provided by Shkadov and Sisoev (2000). The base state of the flow with surfactant mass transfer is expressed by the stationary solution of (6.57)

$$\overline{u} = \frac{3}{2}, \quad h_0 = 1, \quad \overline{c}_0, \quad \Gamma_0 \qquad (6.59)$$

As $\overline{C}_0$ and $\Gamma_0$ are slowly varying functions, $\overline{C}_0 = \overline{C}_0\left(\varepsilon_1^2 x\right)$, and $\Gamma_0 = \Gamma_0\left(\varepsilon_1^2 x\right)$, $\varepsilon_1 \ll 1$, one can take $\overline{C}_0 = const = \overline{C}_0\left(\varepsilon_1^2 x_0\right)$, $\Gamma_0 = const$. Then the stability of the base state (6.59) to infinitesimal disturbances can be carried out. The stability analysis must be repeated for various sections $x = x_0$ as $x$ grows from the initial section $x = 0$. Introducing disturbances in the form of primed quantities,

$$\overline{u} = \overline{u}_0 + u', \ h = 1 + h', \ \overline{C} = \overline{C}_0 + C', \ q = 1 + q', \ \Gamma = \Gamma_0 + \Gamma' \quad (6.60)$$

their linearized evolution is given by the following set of equations

$$\frac{\partial h'}{\partial t} + \frac{\partial q'}{\partial x} = 0$$

$$5\delta\left(\frac{\partial q'}{\partial t} + \frac{\partial Q'}{\partial x}\right) = \frac{\partial^3 h'}{\partial x^3} + 2h' - \frac{2}{3}u' - \frac{2}{3}\mathrm{M}C_x' \qquad (6.61)$$

$$q' = h' + \frac{2}{3}u' + \frac{1}{6}MC'_x$$

$$Q' = \frac{6}{5}h' + \frac{8}{5}\overline{u}' + \frac{7}{20}MC'_x$$

To solve Eqs. (6.61) normal modes are introduced

$$(u', h', C', \Gamma') = \left(\tilde{u}, \tilde{h}, \tilde{C}, \tilde{\Gamma}\right) \exp i\alpha \left(x - \omega t\right) \qquad (6.62)$$

with $\omega$ complex. Using (6.61) the following equations for the unknown amplitudes, $\tilde{u}, \tilde{h}$ and $\tilde{C}$, are obtained

$$\frac{2}{3}\tilde{u} + (1 - \omega)\tilde{h} + \frac{1}{6}z\tilde{C} = 0$$

$$\left(\frac{2}{3}\theta - \frac{2}{3}\omega\beta + \frac{8}{5}\beta\right)\tilde{u} + \left[-2\theta + \beta\alpha^2\theta + \beta\left(\frac{6}{5} - \omega\right)\right]\tilde{h}+ \qquad (6.63)$$

$$z\left(\frac{2}{3}\theta - \frac{1}{6}\omega\beta + \frac{7}{20}\beta\right)\tilde{C} = 0$$

with $\theta = 1/(5\delta)$, $\beta = i\alpha$ and $z = M\beta$.

Demekhin *et al.* (1989) have shown that equations (6.63) give quite a good approximation to the Orr-Sommerfeld formulation of the instability problem for prescribed $\tau$ and for low enough values of $\alpha$.

Let us now assume that $\tilde{C}$ and $\tilde{u}$, obey the following relationship:

$$\tilde{C} = \Re\tilde{u} \qquad (6.64)$$

where the factor $\Re$ is to be obtained from the diffusion part of Eqs. (6.43) and (6.46).

As the problem is homogeneous, the existence of nontrivial solution to (6.63) demands that the determinant of this system vanishes. From this condition the dispersion equation is obtained for $\xi = 1.5-\omega$

$$-\frac{2}{3}\beta\xi^2 + \left(\frac{2}{5}\beta - \frac{2}{3}\theta\right)\xi + \frac{1}{10}\beta - \theta + \frac{2}{3}\beta\alpha^2\theta+$$

$$z\Re\left[-\frac{1}{6}\beta\xi^2 + \left(\frac{3}{20}\beta - \frac{2}{3}\theta\right)\xi + \frac{1}{6}\beta\alpha^2\theta\right] = 0 \qquad (6.65)$$

When Ma $= 0$, equation (6.65) reduces to the corresponding equation for a falling film with no Marangoni effect and no surfactant, as expected. From (6.65) the instability interval is determined, $0 < \alpha < \alpha_0$, with neutral curve

$$\alpha_0 = \sqrt{15\delta} \qquad (6.66)$$

and with $w_r = 3$ as wave velocity of the neutral disturbance. Shkadov et al. (2001) have checked the relations (6.66) for neutral disturbances and the growth rate $(\alpha w_i)_{max}$ of the fastest growing disturbances with the results from the direct numerical solution of the full Navier–Stokes formulation (6.31)-(6.34) and with the results obtained by Demekhin et al. (1987) for the boundary-layer with self-induced pressure approximation (6.36)-(6.41). These three approaches give practically identical results for $0 < \delta < 0.9$. Such an agreement shows the utility of the approximations used to obtain the dispersion equation (6.65).

## ii. Surface tension gradient-driven (Marangoni) modes

By omitting all terms of order $n_*^2$ in (6.31)-(6.34) the boundary value problem, (6.43) and (6.46), for the mass transfer of surfactant in the aqueous solution film is greatly simplified. The small parameter $\varepsilon = (n_* \text{Pe})^{-1/2}$ in equation (6.43) estimates the diffusion boundary layer near the free film surface. It is useful to introduce the stretched coordinate near the surface $y = h(x, t)$,

$$y = h - \varepsilon\zeta \qquad (6.67)$$

For the concentration field, $C(x, y, t)$, inside the diffusion boundary layer the following boundary conditions can be used

$$\zeta = 0, \ C = \overline{C}(x, t)\zeta = \Delta(x, t), \ C = 0, \ \frac{\partial C}{\partial \zeta} = 0 \qquad (6.68)$$

In (6.68), $\Delta(x, t)$ is the standard thickness of the diffusion boundary layer, hence $C = 0$ for $y > 0$. The simplest polynomial representation of $C(x, y, t)$ which satisfies boundary conditions (6.68) is

$$C = \overline{C}\left(1 - \frac{\zeta}{\Delta}\right)^2 \qquad (6.69)$$

Then the corresponding representation for the velocity, $u(x, y, t)$, from (3.1) is

$$u = \overline{u} + \varepsilon \text{M}\overline{C}_x\zeta - \left(\overline{u} + \text{M}h\overline{C}_x\right)\left(\frac{\varepsilon}{h}\right)^2\zeta^2 \qquad (6.70)$$

Integrating equation (2.23) from $\zeta = 0$ to $\zeta = \Delta$ yields

$$\frac{\partial}{\partial t} \int_0^\Delta C d\zeta + \frac{\partial}{\partial x} \int_0^\Delta u C d\zeta = \left.\frac{\partial C}{\partial \zeta}\right|_0^\Delta \qquad (6.71)$$

Using (6.69) and (6.70) with (6.46) and (6.71), then the system of equations for $\overline{C}, \overline{\Gamma}, \varphi$ and $\Delta$, as functions of $x$ and $t$, is reduced to

$$\frac{\partial \varphi}{\partial t} + \frac{\partial}{\partial x}\left[\left(A\overline{u} + BMh\overline{C}_x\right)\varphi\right] = 2\frac{\overline{C}}{\Delta}$$

$$n_*G\left(\frac{\partial \overline{\Gamma}}{\partial t} + \frac{\partial}{\partial x}\left(\overline{u}\overline{\Gamma}\right) + \frac{\partial \overline{u}}{\partial x} - n_*^2 \mathrm{Di}\frac{\partial^2 \overline{\Gamma}}{\partial x^2}\right) + \mathrm{Bi}\left(1 + \overline{C}\right) = -2\frac{\overline{C}}{\varepsilon\Delta}$$

$$\qquad (6.72)$$

$$\pi_1\left(1 + \overline{C}\right) - \pi_2\left(1 + \Gamma\right) = -2\frac{\overline{C}}{\varepsilon\Delta}, \quad \varphi = \frac{1}{3}\overline{C}\Delta$$

with

$$A = 1 - \frac{1}{10}\left(\frac{h_1}{h}\right)^2, \quad B = \frac{1}{4}\frac{h_1}{h} - \frac{1}{10}\left(\frac{h_1}{h}\right)^2 \quad h_1 = \varepsilon\Delta, \quad \varphi = \int_0^A C d\zeta$$

$$\qquad (6.73)$$

In view of the thickness of the diffusion boundary layer $(h_1/h) \ll 1$ one can take

$$A = 1, \; B = \frac{1}{4}\frac{h_1}{h} \qquad (6.74)$$

The system of equations (6.72) must be solved with the original boundary conditions

$$x = 0, \; \overline{C} = 0, \; \Gamma = 0, \; \varphi = 0, \; \Delta = 0 \qquad (6.75)$$

The first two conditions (6.75) together with (6.28) imply that the dimensional quantities $\overline{C}$ and $\Gamma$ at the point $x = 0$ are taken as reference values $C_* = \overline{C}(0, t)$ and $\Gamma_* = \Gamma(0, t)$, respectively.

### iii. Stationary diffusion

Using stretched variables, $x_1 = \varepsilon^2 x$ and $\varphi_1 = \varepsilon\varphi$, the system (6.72) becomes

$$\overline{u}\frac{\partial \varphi_1}{\partial x_1} - \frac{1}{2}\varepsilon^2 \mathrm{M}\frac{\partial}{\partial x_1}\left(\frac{\overline{C}\varphi_1}{s}\overline{C}_x\right) = -s$$

$$s = \mathrm{Bi}\left(1 + \overline{C}\right) + \varepsilon^2 n_* \mathrm{G}\overline{u}\frac{\partial \Gamma}{\partial x_1} - \varepsilon^4 n_*^2 \mathrm{Di}\frac{\partial^2 \Gamma}{\partial x_1^2} \qquad (6.76)$$

$$s = \pi_1 \left(1 + \overline{C}\right) - \pi_2 \left(1 + \Gamma\right)$$

$$\varphi = -\frac{2}{3}\frac{\overline{C}^2}{s}$$

For stationary film flow one has

$$\overline{u} = \frac{3}{2} \tag{6.77}$$

Neglecting the terms of order $\varepsilon^2$, from (6.76) and (6.77) yields that

$$\frac{d}{dx_1}\left(\frac{\overline{C}^2}{1 + \overline{C}}\right) = \mathrm{Bi}^2 \left(1 + \overline{C}\right)\pi_1\left(1 + \overline{C}\right) - \pi_2\left(1 + \Gamma\right) = \tag{6.78}$$

$$\mathrm{Bi}\left(1 + \overline{C}\right)s = \mathrm{Bi}\left(1 + \overline{C}\right)$$

The differential equation (6.2.2) with the initial condition (6.75) has a solution in closed form

$$\ln\left(1 + \overline{C}\right) + \frac{1}{2}\left(\frac{1}{1 + \overline{C}}\right)^2 = \frac{1}{2} + \mathrm{Bi}x_1 \tag{6.79}$$

From the second equation, (6.2.2), one gets

$$\pi_1 - \pi_2 = \mathrm{Bi}, \ \Gamma = \overline{C} \tag{6.80}$$

The concentrations, $\overline{C}$ and $\Gamma$, for the stationary diffusion conditions can be calculated from (6.79) and (6.80) for every $x_1$ value, if the value of the parameter Bi is specified. From (6.2.2) follows that the desorption of surfactant to air is the only reason for the adsorption-desorption to deviate from the equilibrium state corresponding to Bi=0. Thus we deal with the evolution and subsequent effects of the surfactant mass transfer from the equilibrium conditions at the initial section x=0. To estimate the accuracy of the approximate solution (6.79), comparison of (6.79) with the exact solution of the stationary diffusion problem given by Ji and Setterwall (1994) can be done. For $\mathrm{Pe} = 10^6$, $\mathrm{Bi} = 10$ and $\overline{C}_0 = -0.25$ the exact solution gives $x = 1000l$ and $h_1 = 0.09$ ($h_1$ is obtained only approximately from the graphical dependence $c(y)$). The appropriate values from (6.79) are $x = 1012l$ and $h_1 = 0,0666$.

## iv. Diffusion instability

The base state corresponds to the diffusion boundary layer near the surface of the falling film. The linear or nonlinear behavior of the aqueous solution film system with surfactant desorption to air is described by the equations (6.72). For the base state one takes (6.59), (6.79), (6.80)

$$\overline{u}_0 = \frac{3}{2}, \quad h_0 = 1, \quad \overline{C}_0 = \overline{C}\left(\varepsilon^2 x_0\right),$$

$$\Gamma_0 = \overline{C}\left(\varepsilon^2 x_0\right), s_0 = \mathrm{Bi}\left(1 + \overline{C}_0\right), \quad \varphi_0 = \frac{2}{3}\frac{\overline{C}_0^{\,2}}{s_0} \qquad (6.81)$$

For the hydrodynamic stability analysis the base state (6.81) is assumed to have no $x$-dependence as $\overline{C}\left(\varepsilon^2 x_0\right)$ is a slowly varying function. Because $x$ enters in (6.81) as a parameter, the stability analysis must be carried out for various sections $x = x_0$.

In addition to (6.60) small disturbances of $s$ and $\varphi$ as functions of $x$ and $t$ need to be introduced

$$s = s_0 + s', \quad \varphi = \varphi_0 + \varphi' \qquad (6.82)$$

After linearizing (6.72) the equations for disturbances are

$$\frac{\partial \varphi'}{\partial t} + \overline{u}_0 \frac{\partial \varphi'}{\partial x} + \varphi_0 \frac{\partial u'}{\partial x} - \frac{1}{2}\mathrm{M}\frac{\overline{C}_0 \varphi_0}{s_0}\frac{\partial^2 c'}{\partial x^2} = -s'$$

$$\frac{1}{\varepsilon}s' = \mathrm{Bi}C' + n_* \mathrm{G}\left[\frac{\partial \Gamma'}{\partial t} + (1 + \Gamma_0)\frac{\partial u'}{\partial x} + \overline{u}_0 \frac{\partial \Gamma'}{\partial x}\right] - n_*^2 \mathrm{Di}\frac{\partial^2 \Gamma'}{\partial x^2} \qquad (6.83)$$

$$\frac{1}{\varepsilon}s' = \pi_1 C' - \pi_2 \Gamma'$$

Using normal modes

$$(\varphi', \Gamma', C', u') = \left(\tilde{\varphi}, \tilde{\Gamma}, \tilde{C}, \tilde{u}\right)\exp i\alpha\left(x - \omega t\right) \qquad (6.84)$$

the amplitudes obey the following algebraic system

$$\beta \xi \tilde{\varphi}_1 - \left[\mathrm{B}_1^{\,2} + \mathrm{M}_1 \alpha^2 \left(1 + \overline{C}_0\right)\right]\tilde{C}-$$

$$\mathrm{B}_1 \mathrm{G}_1\left(\beta \xi + \frac{d}{\mathrm{T}}\right)\tilde{\Gamma} + \beta\left(1 - \mathrm{B}_1 \mathrm{G}_1\right)\left(1 + \overline{C}_0\right)\tilde{u} = 0$$

$$\tilde{\varphi}_1 - \frac{2 + \overline{C}_0}{\overline{C}_0}\tilde{C} - \frac{G_1}{B_1}\left(\beta\xi + \frac{d}{T}\right)\tilde{\Gamma} + \frac{G_1}{B_1}\left(1 + \overline{C}_0\right)\beta\tilde{u} = 0 \quad (6.85)$$

$$\tilde{C} - (1 + T\beta\xi + d)\tilde{\Gamma} - T\beta\left(1 + \overline{C}_0\right)\tilde{u} = 0$$

with

$$B_1 = \varepsilon VBi, \quad G_1 = \varepsilon n_* VG, \quad D_1 = \varepsilon n_*{}^2 Di,$$

$$M_1 = -\frac{n_* \overline{C}_0 Ma}{2Bi\left(1 + \overline{C}_0\right)}, \quad \tilde{\varphi} = \gamma_2 \tilde{\varphi}_1,$$

$$V = \sqrt{\frac{3}{2}\frac{1 + \overline{C}_0}{|\overline{C}_0|}}, \quad \gamma_1 = \frac{2 + \overline{C}_0}{\overline{C}_0}\gamma_2, \quad \gamma_2 = -\frac{2}{3}\frac{\overline{C}_0{}^2}{Bi\left(1 + \overline{C}_0\right)^2}, \quad (6.86)$$

$$T = \frac{n_* G}{\pi_2}, \quad d = TD_1\alpha^2$$

From (6.85) the relation (6.64) connecting $\tilde{C}$ and $\tilde{u}$ follows. Then using the expression for $\Re$ and from (6.63) follows the dispersion equation albeit a cumbersome one. It provides four eigenvalues, $\xi_k$, that depend on the wave number $\alpha$, and on the dimensionless parameters Bi, G, T, Di, $\delta, n_*, \overline{C}_0$, Ma and $\varepsilon$.

### 6.2.3 Numerical results

As mentioned earlier, too many particular cases can be considered due to the nine independent dimensionless groups entering the problem. One can take the following quantities as free parameters

$$Re, Ma, Pe, \gamma, Bi, G, \overline{C}_0, T, D_1 \quad (6.87)$$

For every specified set of values (6.87), both the growth ratefactor, $\omega_i(\alpha)$, and the wave phase velocity $\omega_r(\alpha)/\alpha$, are obtained. If the parameter Ca is specified instead of $\gamma$, then the value of $\gamma$ could be obtained using the equations

$$n_* = (3Ca)^{\frac{1}{3}}, \quad \delta = \frac{Re}{15}n_*, \quad \gamma^{\frac{1}{3}} = (3Re)^{\frac{1}{9}}n_* \quad (6.88)$$

Relations exist to assign numerical values to free parameters, like

$$Pe = \left(\frac{\nu}{D}\right)Re, \quad G = \frac{\Gamma_*}{c_* l}Pe \quad (6.89)$$

From the analysis of the growing adsorption sublayer follows

$$\Gamma_* = c_* \frac{k_a - k_g}{k_a}. \tag{6.90}$$

As the key parameter connecting the hydrodynamic and diffusion parts is the Marangoni number, Ma, both cases, positive (Ma > 0) and negative (Ma < 0) Marangoni numbers have been considered.

The main hydrodynamic parameters are Re and $\gamma$, or equivalently $\delta$ and $\gamma$. These two quantities determine the mean film thickness, $l$, mean velocity, $U_*$ and flow rate, $lU_*$, as well as parameter $n_*$. The parameters $\overline{C}_0$ and Pe determine the local thickness of the diffusion boundary layer, $h_1$, and the smallness parameter $\varepsilon$. Two quantities, T and $D_1$, characterize, respectively, the mass transfer of surfactant by the adsorption-desorption and the intensity of dissipation by the surface diffusion. The intensity of the surfactant desorption to the gas phase is determined by the parameter Bi. The remaining parameter G gives an indication of the typical value of surface excess concentration $\Gamma_*$ relative to $c_*$.

For every prescribed set of parameter values (6.87) the eigenvalues $\omega(\alpha)$ could be computed for arbitrary high $\alpha > 0$ values. But the assumptions on the long wave approximation introduce some limitations on $\alpha$. Note that for the problem under consideration there exist two length scales $h$ and $h_1$. Thus, due to the inequality $h_1 << h$, short waves in the $h$ scale could be considered as long waves in the $h_1$ scale. In view of this a cut-off $\alpha < 10$ was introduced in the numerical computations.

A few remarks about the parameter T are of interest. Actually, the third equation (6.85) is the disturbed equation of the adsorption-desorption kinetics. From (6.35) and (6.85) follows that

$$T = n_* \frac{U_*}{k_d l}, \quad TD_1 = n_*^2 \frac{D_i}{k_d l^2} \tag{6.91}$$

Accordingly, the parameter T characterizes the ratio of rates of the surface excess concentration, $\Gamma$, transfer by two transfer processes: one is convective flow along the film surface, and the other is desorption inside the liquid bulk. For $T \to 0$ the case of diffusion controlled adsorption-desorption kinetics is obtained

$$\Gamma = \overline{C} \tag{6.92}$$

Equation (6.92) corresponds to a fast desorption process leading to local kinetic equilibrium. In the opposite limiting situation, $T \to \infty$, it follows from (6.85)

$$\tilde{\Gamma} = \frac{1 + \overline{C}_0}{\overline{u}_0 - \omega - \beta D_1} \tilde{u} \qquad (6.93)$$

Equation (6.93) corresponds to the kinetically frozen desorption. Only for $1 + \overline{C}_0 = 0$ it is possible to consider the surface excess concentration of surfactant, $\Gamma$, as constant and hence $\tilde{\Gamma} = 0$ for $j \neq 0$. Besides the limiting cases of fast desorption ($T = 0$) and slow desorption ($T \to \infty$) the more general case $T \sim 1$ can also be considered.

It is worth noting that the method used (Shkadov, 1967) leading to equations (6.57) and (6.72) and, subsequently to the dispersion equation, is highly efficient for large $\gamma$ values and for small Ca values. According to (6.54) the crucial parameter $n_*$ is small for small Ca values. For water $\gamma = 2850$ instead of $\gamma = 29.241$ for a liquid metal, and Ca $= 0.0024$ instead of Ca $= 0.2$. Shkadov *et al.* used the following parameter values

$$\gamma = 2904, \quad \delta = 0.412, \quad n_* = 0.193$$

These values apply well to a water film flow with a soluble volatile surfactant.

By setting $\sigma = \sigma(\bar{C})$, the dispersion relation can be solved explicitly, and we shall describe the salient results found by Shkadov *et al.* (2001). The parameter T estimates the role of adsorption-desorption kinetics on the hydrodynamic instability. For finite of values T the mass rate transfer of the adsorbed surfactant concentration $\Gamma$ by convective flow is of the same order as that by desorption inside the liquid bulk. From the dispersion equation four eigenvalues, $\xi_k = \overline{u}_0 - \omega_k$, have been obtained with corresponding growth or decay rates. The salient features obtained in numerical experiments with aqueous solutions for various Ma, T and $\alpha$ values are depicted in the figures (6.8)–(6.10). For sake of completeness results were obtained for positive (Ma $> 0$) as well as for negative (Ma $< 0$) Marangoni numbers.

For every value of the Marangoni number, one hydrodynamic instability mode exists together with one to three growing, unstable diffusion modes. Fig. 6.8 shows for the small negative Marangoni numbers (Ma $= -1$) three unstable modes, namely, the hydrodynamic mode and two diffusion modes. The influence of the Marangoni effect on hydrodynamic waves is rather weak for this Marangoni number. The curves $\omega_r(\alpha)$ and $\alpha\omega_i(\alpha)$ are practically the same for Ma = 0 and Ma = ±1. We can call them $Cr$ and $f$, respectively. A pair of diffusion modes exists on the interval $\alpha > \alpha_{**}$ ($\alpha_{**} \approx 4$). For one of them the wave phase velocity is slightly above the value $\omega_r = 1.5$ while for the other it is slightly below this value and both are growing. For Ma = 1 in addition to the hydrodynamic mode, only one diffusion mode with $\alpha_{**} \approx 1$ on Fig. 6.8 could be seen. Note that on the wave number interval $\alpha_{**} < \alpha < 10$ the diffusion modes have a growth rate about that of the hydrodynamic mode.

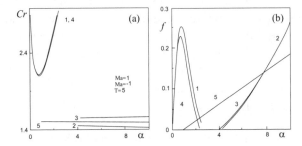

Figure 6.8: Falling film and the Marangoni effect. (a) Phase velocity, $Cr$, and (b) growth rate, $f = m\alpha\omega_i$, versus wave number, $\alpha$, for $\sigma = \sigma(\bar{C})$; $Bi = 10$, $\gamma = 2904$, $G = 2000$, $Pe = 10^6$, $Re = 32$, $D_1 = 10^{-2}$, $\bar{C}_0 = -0.25$, $T = 5$, $Ma = 1$ (1,5), $Ma = -1$ (2-4), $m = 1$ (1-5) (after Shkadov et al., 2001).

A drastic modification of the instability curves $\omega_r(\alpha)$ and $\alpha\omega_i(\alpha)$ with increasing Marangoni number is shown in Fig. 6.9. New instability modes for Ma = $-10$ and T = 1.0 are observed. There exists a slow diffusion mode at $\alpha > \alpha_{**}$ ($\alpha_{**} \approx 1$) which moves with the liquid on the film surface ($\omega_r = 1.5$). Its growth rate, $\alpha\omega_i$, is of the same order as that for the similar mode in Fig. 6.8. Another diffusion mode appears at $\alpha > \alpha_n$ ($\alpha_n \approx 1.5$) which has very different nature. The corresponding disturbances represent fast waves with phase velocity $\omega_r(\alpha)$, linearly growing, when $\alpha$ increases. For example, the phase velocity $\omega_r$ varies from $\omega_r \approx 2$ to $\omega_r \approx 8$ as $\alpha$ increases from $\alpha = 2$ to $\alpha = 10$. The growth rates of the fast and slow diffusion modes are about the same.

The salient effect of the Marangoni number on the instability of film flow in is the appearance of a strong combined mode with growth

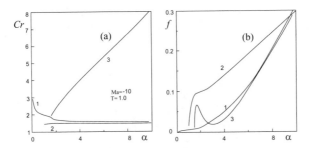

Figure 6.9: Falling film and the Marangoni effect. (a) Phase velocity, $Cr$, and (b) growth rate, $f$, versus waves number, $\alpha$. $Ma = -10$, $T = 1.0$ (1-3), $m = 10^{-2}$ (1), $m = 1$ (2,3). Other parameter values as in Fig. 6.8 (after Shkadov et al., 2001).

rate two orders higher relative to the case of low Ma values. This strong combined mode begins at $\alpha = 0$ as an ordinary hydrodynamic mode with phase velocity $\omega_r = 3$ and then diminishes to $\omega_r \approx 2$ as $\alpha$ grows to $\alpha \approx 1.5$. But then instead of growing, $\omega_r$ falls down to $\omega_r \approx 1.5$ with subsequent asymptotic behavior $\omega_r \to 1.5$ for $\alpha$ growing as it occurs for the diffusive mode. Such transverse hydrodynamic wave gets over to a longitudinal wave due to $|\omega| >> 1$. In the vicinity of $\alpha \approx 1.5$ this instability mode exhibits features of both transverse and longitudinal waves. A jump from $T = 1$ to $T = 0.5$ does not appreciably change the picture of instability modes.

In Fig. 6.10 the curves $\omega_r(\alpha)$ and $\alpha\omega_i(\alpha)$ obtained from the eigensolutions of (3.31) for $Ma = 10$ and $T = 0.5$ are shown. Three instability modes can be distinguished noting their phase velocity. The first two modes were discussed for negative Ma numbers, namely, a hydrodynamic mode with phase velocity $3 > \omega_r > 2$ in the interval $0 < \alpha < \alpha_*$ and a diffusive mode with phase velocity $\omega_r \approx 1.5$ in the interval $\alpha > \alpha_{**}$. There is strong damping effect of the Marangoni number, $Ma > 0$, on the hydrodynamic waves. The maximum growth rate is equal to one half and the boundary value of the wave number $\alpha_*$ is one third of the corresponding values for $Ma = 0$. Accordingly, the Marangoni effect generates the diffusion mode which exists for $\alpha > \alpha_{**}$ ($\alpha_{**} \approx 0.7$) and for which at $\alpha > 4$ the growth rate $\alpha\omega_i$ is higher than that of the hydrodynamic mode. It should be noted that for $Ma>0$ the hydrodynamic and the diffusive modes do not interact to form a combined mode as it was the case for $Ma < 0$.

The third unstable mode at $\alpha > \alpha_n$ in 6.10 is unusual because

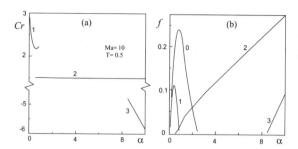

Figure 6.10: Falling film and the Marangoni effect.(a) Phase velocity, $Cr$, and (b) growth rate, $f$, versus waves number, $\alpha$. $Ma = 0$ (0), $Ma = 10$, $T = 0.5$ (1-3), $m = 1$ (0-3). Other parameter values as in Fig. 6.8 (after Shkadov et al., 2001).

its phase velocity is negative. Accordingly, the corresponding wave moves upstream. These waves have short lengths ($\alpha_n \approx 8.5$), are fast moving ($|\omega_r| > 5$), and grow fast ($\alpha \omega_i$ is of the same order as for hydrodynamic mode). As $|\omega| \gg 1$ these waves are longitudinal ones.

### 6.2.4   Flows with a transverse thermal gradient

Let us discuss now the stability of a liquid layer flowing down an inclined plane and heated from below. The flow is subject to three instability modes: the short-wave (longitudinal) and long-wave (transverse) surface tension gradient–driven–Marangoni instabilities discussed in Chapter 4. The base parallel flow violates the isotropy characteristic for the Benard–Marangoni problem. One can show that the critical Marangoni number does not depend on the intensity of the motion in the case where the wavevector $k$ of the disturbance is parallel to the case where the wavevector $k$ of the disturbance is parallel to the direction of the basic flow. At the same time, for the disturbance with the wavevector $k$ orthogonal to the direction of flow, the critical Marangoni number may increase or decrease. One can expect helical flows (stationary longitudinal rolls) in the former case and traveling waves in the latter case.

It turns out that the short-wave surface tension gradient–driven (Marangoni) instability appears in the form of longitudital rolls and is not influenced by the motion (Sreenivasan and Lin, 1978). The long-wave deformational Marangoni instability, which was stationary in absence of flow, becomes oscillatory because of the drift of the surface deformation by the base flow. This instability generates traveling surface waves, as well as the Kapitza instability. It turns out that both long-wave instabilities support one another, and the instability starts on the critical curve (in the limit $Bi \to 0$:

$$\frac{2\sin^2 \beta Ga}{5\cos\beta} + \frac{3M}{2GaP\cos\beta} = 1. \tag{6.94}$$

(Goussis and Kelly, 1991). A condition like (6.20) is a particular case of the more general condition (6.94). Let us note that the increase of the Galileo number suppresses the surface tension gradient–driven (Marangoni) instability but generates Kapitza's instability. For low enough values of $M$ there are two separate instability regions: for relatively low values of $Ga$ (mostly Marangoni instability) and for relatively high values of $Ga$ (mostly Kapitza instability). For larger values of $M$, the flow is unstable for any $G$.

In the nonlinear region, one can expect a competition between the Marangoni instability tending to film breakdown and hence producing a "dry spot", and the mean flow preventing the dryout. Joo et al (1991) derived a one-dimensional equation for long-wave surface deformations which is a hybrid between Davis equation describing the Marangoni instability and Benney equation for isothermal falling films (Benney, 1966) (evaporation effects were also taken into account). The numerical simulation showed wave breaking, creation of secondary structures, and suppression of dryout by the mean flow. However, the original problem is essentially non-one-dimensional, and it is still unsolved. The linear stability analysis was performed also for a a system of two horizontal layers of immiscible fluids heated across the layers (Gumerman and Homsy, 1974).

## 6.3   Flows in two-layer systems

Let us investigate now the stability of the parallel thermocapillary flow (Sect. 2.1.3). We shall follow a recent study by Nepomnyashchy et al. (2001).

We use the assumption $z = 0$, and hence we disregard the deformational instabilities studied by Smith and Davis (1983b). One can expect these instabilities to be unimportant if the capillary and Galileo numbers are low and high enough, respectively. Here they are $C_j = \eta_j \kappa_j / a_j \sigma_0 \ll 1$, and $G_j = g a_j^3 / \nu_j \kappa_j \gg 1$, $j = 1, 2$.

## 6.3.1 Linear stability theory

We linearize the fields of all the variables that appear in the boundary value problem (2.31) - (2.39) around the stationary solution (2.46) - (2.51):

$$\mathbf{v}_i(\mathbf{x}, z, t) = U_i^{(0)}(z)\mathbf{e}_x + \tilde{\mathbf{v}}_i(z) \exp(i\mathbf{k} \cdot \mathbf{x} + \lambda t),$$

$$T_i(\mathbf{x}, z, t) = \Theta_i^{(0)}(z) + \tilde{T}_i(z) \exp(i\mathbf{k} \cdot \mathbf{x} + \lambda t),$$

$$p_i(\mathbf{x}, z, t) = B_i^{(0)}x + \tilde{p}_i(z) \exp(i\mathbf{k} \cdot \mathbf{x} + \lambda t),$$

where $\mathbf{x} = (x, y)$, $\mathbf{k} = (k_x, k_y)$ are horizontal two-dimensional vectors.

Let $k_x = k \sin \alpha$, $k_y = k \cos \alpha$. It is convenient to perform a transformation of horizontal coordinates $x = X \cos \alpha + Y \sin \alpha$, $y = -X \sin \alpha + Y \cos \alpha$. After this transformation, the wave vector $\mathbf{k}$ is directed along axis $Y$, and the disturbances do not depend on $X$.

We obtain the following equations for disturbances of the parallel flow:

$$\lambda \tilde{v}_{Xi} + ik \sin \alpha \cdot U_i^{(0)} \tilde{v}_{Xi} + \cos \alpha \cdot U_i^{(0)\prime} \tilde{v}_{zi} = c_i(\tilde{v}_{Xi}'' - k^2 \tilde{v}_{Xi}); \quad (6.95)$$

$$\lambda \tilde{v}_{Yi} + ik \sin \alpha \cdot U_i^{(0)} \tilde{v}_{Yi} + \sin \alpha \cdot U_i^{(0)\prime} \tilde{v}_{zi} = -ike_i\tilde{p}_i + c_i(\tilde{v}_{Yi}'' - k^2 \tilde{v}_{Yi}); \quad (6.96)$$

$$\lambda \tilde{v}_{zi} + ik \sin \alpha \cdot U_i^{(0)} \tilde{v}_{zi} = -e_i\tilde{p}_i' + c_i(\tilde{v}_{zi}'' - k^2 \tilde{v}_{zi}); \quad (6.97)$$

$$\lambda \tilde{T}_i + ik \sin \alpha \cdot U_i^{(0)} \tilde{T} + \Theta_i^{(0)\prime} \tilde{v}_{zi} + \cos \alpha \cdot \tilde{v}_{Xi} + \sin \alpha \cdot \tilde{v}_{Yi} = \frac{d_i}{P}(\tilde{T}'' - k^2 \tilde{T}); \quad (6.98)$$

$$\tilde{v}_{zi}' + ik\tilde{v}_{Yi} = 0; \quad i = 1, 2; \quad (6.99)$$

where $c_1 = d_1 = e_1 = 1$, $c_2 = 1/\nu$, $d_2 = 1/\kappa$, $e_2 = \rho$, $'$ denotes differentiation with respect to $z$.

Introducing the stream function disturbance $\tilde{\psi}_i$, $\tilde{v}_{zi} = -ik\tilde{\psi}_i$, $\tilde{v}_{Yi} = \tilde{\psi}'$ and eliminating the pressure disturbance, we obtain the following eigenvalue problem describing the stability of the parallel flow:

$$c_i(\tilde{\psi}_i'''' - 2k^2 \tilde{\psi}_i'' + k^4 \tilde{\psi}_i'') - ik \sin \alpha[U_i^{(0)}(\tilde{\psi}_i'' - k^2 \tilde{\psi}_i) - U_i^{(0)\prime\prime}\tilde{\psi}_i] \quad (6.100)$$

$$-\lambda(\tilde{\psi}_i'' - k^2 \tilde{\psi}_i) = 0;$$

$$\lambda \tilde{v}_{Xi} + ik \sin \alpha U_i^{(0)} \tilde{v}_{Xi} - ik \cos \alpha U_i^{(0)\prime} \tilde{\psi}_i = c_i(\tilde{v}_{Xi}'' - k^2 \tilde{v}_{Xi}); \quad (6.101)$$

$$\lambda \tilde{T}_i + ik\sin\alpha U_i^{(0)}\tilde{T}_i - ik\Theta_i^{(0)'}\tilde{\psi}_i + \cos\alpha\tilde{v}_{Xi} + \sin\alpha\tilde{\psi}_i' = (d_i/P)(\tilde{T}_i'' - k^2\tilde{T}_i);$$

$$\tag{6.102}$$

$$z = -1: \ \tilde{\psi}_1 = 0; \ \tilde{\psi}_1' = 0; \ \tilde{v}_{X1} = 0; \ \tilde{T}_i = 0; \tag{6.103}$$

$$z = a: \ \tilde{\psi}_2 = 0; \ \tilde{\psi}_2' = 0; \ \tilde{v}_{X2} = 0; \ \tilde{T}_i = 0; \tag{6.104}$$

$$z = 0: \ \tilde{v}_{X1}' - \eta^{-1}\tilde{v}_{X2}' = 0; \ -\tilde{\psi}_1'' + \eta^{-1}\tilde{\psi}_2'' = (ikM/P)\tilde{T}_1; \tag{6.105}$$

$$\tilde{v}_{X1} = \tilde{v}_{X2}; \ \tilde{\psi}_1 = \tilde{\psi}_2 = 0; \ \tilde{\psi}_1' = \tilde{\psi}_2'; \tag{6.106}$$

$$\tilde{T}_1 = \tilde{T}_2; \ \kappa\tilde{T}_1' = \tilde{T}_2'. \tag{6.107}$$

For illustration let us recall the results obtained for an air-water system (at $20°$ and $1\,bar$) with the following parameter values: $\eta = 55.3$, $\nu = 0.0659$, $\lambda = 23.3$, $\kappa = 0.00667$, $P = 6.96$, $a = 1$ (Perry and Chelton, 1993).

### i. Horizontal temperature gradient

In the case $b = 0$, which corresponds to the thermocapillary flow generated by a horizontal temperature gradient, the instability appears with respect to *oblique* hydrothermal waves, i.e. the direction of the critical wavevector is characterized by a certain value of $\alpha$, $0 < |\alpha < 90|°$. For any orientation of the disturbance wave vector characterized by the angle $\alpha$, the minimum of the neutral stability curve $M = M_o(k, \alpha)$ was found, and the corresponding values $M_c$, $\omega_c = -\text{Im}\lambda_c$ and $k_c$ are presented as functions of $\alpha$ in Fig. 6.11. Only one of two curves which correspond to opposite values of $\omega_c$ at opposite values of $\alpha$ are shown.

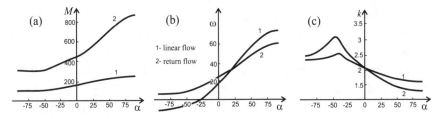

Figure 6.11: Dependences of (a) critical Marangoni number, (b) critical frequency and (c) critical wavenumber on the direction of the wavevector in the case of a horizontal temperature gradient.

In the case of the return flow, the sign of $\omega_c$ does not depend on $\alpha$. The minimum of the curve $M_c(\alpha)$ occurs at $\alpha = -57.6°$. The positive value of $\omega_c$ for negative $\alpha$ (as well as the negative value of $\omega_c$ for positive $\alpha$, which characterizes the branch which is not shown in Fig. 6.11 corresponds to the propagation of the oblique hydrothermal wave towards the hot end, i.e. opposite to the direction of the flow at the interface (Davis, 1987).

In the case of the linear flow, the sign of $\omega_c$ depends on $\alpha$. Near the minimum of the curve $M_c(\alpha)$, $\omega_c$ is negative. That means that the hydrothermal wave propagates towards the cold end. Such behavior may be caused by the difference in the transverse temperature profiles of the linear flow and the return flow. Note that the dependence of $M_c$ on $\alpha$ is very weak in the region $-90° < \alpha < -40°$.

## ii. Obliquely inclined temperature gradient

Let us investigate now the case $b \neq 0$ (obliquely inclined temperature gradient). We shall consider only the case of the return flow. Note that for the chosen set of parameter values $1 - \kappa a^2 > 0$, and hence in the base flow the vertical component of the temperature gradient in the upper layer is always negative, while in the lower layer there exists a region with a positive value of the temperature gradient, if $b/M < 0.864$. The temperature, $\Theta_s$, on the interface is positive if $b/M < 0.477$ and negative in the opposite case.

The instability with respect to oblique hydrothermal waves is important in the interval $0 < b < b_F$, $b_F \approx 162.4$. A typical neutral curve calculated for the critical inclination angle is shown in Fig. 6.12(a) (line 2). The dependence of the frequency $\omega = |\text{Im}\lambda|$ on the wavenumber $k$ for the critical inclination angle is shown in Fig. 6.12(b) (line 2). The critical value of the Marangoni number $M_c$, which corresponds to the minimum of the neutral curve, grows from $M = M_H \approx 263.9$ as $b = 0$ to $M = M_F \approx 314.9$ as $b = b_F$ (see line 3 in Fig. 6.13), while the angle $|\alpha|$ decreases from $\alpha_H \approx 57.6°$ down to $\alpha_F \approx 30°$. The critical wavenumber $k_c$ and the critical frequency $\omega_c$ both decrease slowly when $b$ grows ($k_H \approx 2.67$, $\omega_H \approx 5.91$; $k_F \approx 2.58$, $\omega_F \approx 4.95$). Note that the phase velocity of the hydrothermal wave $v_{ph} = \omega_c/k_c$ is much smaller than the fluid velocity at the the interface $v_s$. For instance, in the point H $v_{ph} \approx 2.21$, $v_s \approx 9.31$ while at F $v_{ph} \approx 1.92$, $v_s \approx 11.1$. The group velocity $v_{gr} = d\omega/dk$ calculated in the point $k = k_c$, $M = M_c$ is rather low for hydrothermal waves (at H, $v_{gr} \approx 0.507$ while at F $v_{gr} \approx 0.0885$).

The case $b \to \infty$ corresponds to the problem solved by Pearson

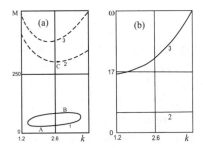

Figure 6.12: (a) Neutral curves for $b = 100$: the closed neutral curve corresponding to stationary longitudinal rolls, $\alpha = 0$, (line 1); neutral curve for hydrothermal waves at the critical angle $\alpha = \alpha_c$ (line 2); neutral curve for hydrothermal waves, $\alpha = 0$, (line 3). Points A and B determine the existence interval of stationary longitudinal rolls and point C determines the onset threshold of oblique hydrothermal waves. (b) Dependence $\omega(k)$ for hydrothermal waves at $\alpha = \alpha_c$ (line 2) and at $\alpha = 0$ (line 3); $b = 100$. (Redrawn after Nepomnyashchy *et al.*, 2001).

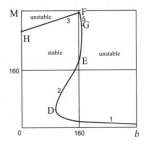

Figure 6.13: Boundaries of instability with respect to stationary longitudinal rolls($\alpha = 0$; lines 1 and 2), oblique hydrothermal waves ($\alpha = \alpha_c$; line 3), and transverse traveling rolls $\alpha = 90°$; lines 4 and 5). Point D corresponds to the disappearance of longitudinal rolls; point E determines the transition between longitudinal rolls and transverse traveling rolls; point F determines the transition between oblique hydrothermal waves and transverse traveling rolls; and point G corresponds to the disappearance of transverse traveling rolls. (Redrawn after Nepomnyashchy *et al.*, 2001).

(1958) when studying Benard-Marangoni convection which we recall corresponds to a layer heated from below in the absence of a horizontal temperature gradient. In this limit, the critical Marangoni number $M_c \sim M_c'/b$, where $M'$ is the Marangoni number defined using the transverse temperature difference, which remains finite and does not depend on the direction of the wavevector. The instability is monotonic, and the stability boundary can be calculated analytically (Smith 1966). For the air-water system with $a = 1$ $M_c' \approx 21$, $k_c \approx 2$ (Gilev, Nepomnyashchy and Simanovskii, 1987a,b). At arbitrary large but finite $b$, the degeneracy of $\lambda(\alpha)$ is broken. Only longitudinal rolls with the axis parallel to the horizontal component of the temperature gradient ($\alpha = 0$) grow monotonically, while any other disturbances grow in an oscillatory way because of the drift by the main flow. We have found that at large values of $b$ the disturbances with $\alpha = 0$ are the most unstable ones. The corresponding stability boundary is shown in Fig. 6.13 (line 1).

Surprisingly, the instability predicted by Pearson disappears at high enough values of parameter $b$, $b = b_D \approx 88.5$ (Fig. 6.13, point D). Note, that for $b = b_D$ the horizontal temperature gradient is two orders of magnitude smaller than the *mean* vertical one. Nevertheless, such a small horizontal temperature gradient turns out to be sufficient for a complete suppression of Pearson's instability. The latter kind of instability appears only at $b > b_D$ in a small closed region in the space of parameters $(\alpha, k, M)$ around the point $\alpha = 0$, $k = k_D \approx 2.31$, $M = M_D \approx 47.9$ (Fig. 6.12(a), line 1). The explanation of this paradox is as follows. Because the heat conductivity of the water is much larger than that of the air ($\kappa = 23.3$), the vertical gradient in the water is actually much less than $b$. Using formula (2.55), which determines the temperature on the interface $\Theta_s$ (note that $\Theta_s \approx -b/\kappa + M/48$ since $\kappa \gg 1 \gg \chi, \eta \gg 1$) and evaluating expression (2.55) in the point $(b = b_D, M = M_D)$, we find that Pearson's instability is damped when the mean vertical temperature gradient *in the water layer* is only 2.7 times stronger than the horizontal temperature gradient.

For the fixed value of $\alpha = 0$, the stability boundary is a closed curve in the plane $(k, M)$ (Fig. 6.12(a), line 1), which is located much lower than the stability boundary for the hydrothermal waves (Fig. 6.12(a), line 2). The minimum and maximum points of the closed neutral curve (points A and B in Fig. 6.12(a)) determine the interval of stationary instability $M_A < M < M_B$. If the heating is enhanced while maintaining constant the ratio of characteristic vertical and horizontal temperature differences, $b$, the parallel flow first becomes unstable with respect to longitudinal rolls at $M = M_A$, but then it is *restabilized* as $M > M_B$. The parallel flow becomes unstable again (this time with respect to inclined hydrothermal waves) only for much higher values of the Marangoni number $M > M_C$, where point C corresponds to the minimum of the neutral curve for hydrothermal waves at the critical angle $|\alpha| = \alpha_c$ (Fig. 6.12(a), line 2). How $M_A$, $M_B$ and $M_C$ depend on $b$ is shown in Fig. 6.13 (lines 1, 2 and 3, respectively). Lines 1 and 2 merge at $(b_D, M_D)$ (point D in Fig. 6.13).

When $b$ increases, the interval of angles $|\alpha| < \alpha_m$ where Pearson's instability is expected, grows. At $b = b_1$, $b_1 \approx 118$, $\alpha_m$ reaches $90°$, i.e. at $b > b_1$ for any direction of the wavevector there exists an interval of instability. However, the critical Marangoni number, $M_A$, corresponds to $\alpha = 0$ in the whole region $b > b_D$. The maximum of the neutral surface $M = M(k, \alpha)$, $M_B$, corresponds to the value $\alpha = 0$ only in the interval $b_D < b < b_E$, $b_E \approx 154.6$.

For values of $b$ slightly smaller than $b_E$, a new maximum appears at $|\alpha| = 90°$. It is caused by a new instability mode which corresponds to two-dimensional rolls oriented perpendicularly to the direction of the base flow. For $b = b_E$, both maxima have the same height $M = M_E \approx 181.5$ (point E in Fig. 6.13). As $b > b_E$, the maximum at $|\alpha| = 90°$ becomes higher than the maximum at $\alpha = 0$. Thus, the lower boundary of the stability gap between the regions of Pearson's instability and the instability with respect to hydrothermal waves is connected with two-dimensional rolls. The dependence of the corresponding critical Marangoni number on $b$ is shown in Fig. 6.13 as line 4. Similarly, when $b$ increases, an additional minimum appears at $|\alpha| = 90°$ on the neutral surface $M = M(k, \alpha)$ for hydrothermal waves which competes with the minimum at $|\alpha| = \alpha_c$, $\alpha_c \neq 90°$. The former minimum corresponds to waves moving in the direction of the flow at the interface, i.e. to drifted rolls. Both minima provide the same critical Marangoni number $M = M_F$ as $b = b_F$ (it is a codimension-2 point; point F in Fig. 6.13). As $b > b_F$, the upper boundary of the stability gap is connected with two-dimensional disturbances (Fig. 6.13, line 5) (as well as its lower boundary). At $b = b_G$, $b_G \approx 164.3$, lines 4 and 5 merge with $M = M_5 \approx 276.1$ (point G in Fig. 6.13), and the stability gap disappears.

Note that for the two-dimensional instability mode described above, $\omega \neq 0$, because the rolls are driven by the base flow. Unlike hydrothermal waves, the drifted convective rolls move in the *same* direction as the flow at the interface. Moreover, the phase velocity of disturbances can be even *higher* than the fluid velocity on the interface $v_s$ (2.52). At the same time, the group velocity of waves is always smaller than $v_s$. For instance, in the point F $v_s \approx 11.1$, $v_{ph} \approx 13.4$, and $v_{gr} \approx 10.1$.

Similarly, when $b$ increases, an additional minimum appears at $|\alpha| = 90°$ on the neutral surface $M = M(k, \alpha)$ for hydrothermal waves which competes with the minimum at $|\alpha| = \alpha_c$, $\alpha_c \neq 90°$. The former minimum corresponds to waves moving in the direction of the flow at the interface, i.e. to drifted rolls. Both minima provide the same critical Marangoni number $M = M_F$ as $b = b_F$, $b_F \approx 162.4$ (it is a codimension-2 point; point F in Fig. 6.13). As $b > b_F$, the upper boundary of the stability gap is connected with two-dimensional disturbances (Fig. 6.13, line 5) (as well as its lower boundary).

At $b = b_G$, $b_G \approx 164.3$, lines 4 and 5 merge with $M = M_5 \approx 276.1$ (point G in Fig. 6.13), and the stability gap disappears.

The transition between the inclined hydrothermal waves moving upstream and the two-dimensional waves moving downstream takes place at the value $b_F / M_F \approx 0.515$. Note that this value is close to the value $b/M \approx 0.477$ where the quantity $\Theta_s$, which characterizes the mean vertical temperature gradient in water, changes its sign. Indeed, the explanation of the upstream motion of hydrothermal waves in a one-layer system is based on the fact that a downward flow generates a subsurface heating, because of the positive temperature gradient (Davis, 1987). In our case, as the heat diffusivity of air is much larger than that of water, the heat advection by a flow disturbance in air can be ignored, thus the direction of the temperature gradient in water is crucial. For relatively small $b/M$, the mean vertical temperature gradient in water is mostly positive, and the direction of the wave propagation can be explained as in the case of a one-layer system. For larger values of $b/M$, the downward flow causes a subsurface cooling, which influences the propagation direction of a temperature disturbance in an opposite way.

Thus for relatively small values of $b$ ($0 < b < b_F$) and large values of $M$, linear theory predicts the excitation of inclined hydrothermal waves (Fig. 6.13, line 3). These waves move in the opposite direction to that of the flow at the interface. For relatively large values of $b$ ($b > b_D$) and low enough values of $M$, the theory predicts the appearance of stationary convective rolls due to the Pearson's instability (Fig. 6.13, lines 1, 2). The axes of the rolls are ordered by the thermocapillary flow along the direction of the imposed horizontal temperature gradient. For intermediate values of $M$, the convective rolls are ordered *across* the direction of the horizontal temperature gradient, and they are drifted by the thermocapillary flow. Unlike the hydrothermal waves, the drifted rolls move in the same direction as the flow at the interface.

## 6.3.2   Numerical exploration of nonlinear patterns

Let us see now the flow regimes predicted by the linear stability theory.

### i. Description of the method

Nepomnyashchy *et al.* (2001) have performed numerical simulations of the flow regimes predicted by the above given linear stability theory. We shall describe now the numerical approach for nonlinear simulations of the longitudinal rolls (spiral flows) which appear due to the instability of the base thermocapillary flow with respect to monotonically growing disturbances with $\alpha = 0$. One can expect that the corresponding solutions of the nonlinear boundary value problem (2.31) - (2.39) have the following structure:

$$\mathbf{v}_i = \mathbf{v}_i(y, z),\ p_i = p_i(y, z) + B_i x,\ T_i = x + \Theta_i(y, z),\ i = 1, 2. \quad (6.108)$$

We shall assume that the motion is spatially periodic in $y$ with period $L = l/a_1$. The constants $B_i$ are unknown and are determined from the conditions of vanishing mean horizontal fluxes of fluids:

$$\int_{-L/2}^{L/2} dy \int_{-1}^{0} dz\, v_{x1}(y, z) = 0,\ \int_{-L/2}^{L/2} dy \int_{0}^{a} dz\, v_{x2}(y, z) = 0. \quad (6.109)$$

For spiral flows (6.108) the continuity equations

$$\frac{\partial v_{yi}}{\partial y} + \frac{\partial v_{zi}}{\partial z} = 0,\ i = 1, 2$$

do not include $v_{xi}$. Accordingly, the stream functions $\psi_i$ of the transverse flow can be defined in the following way:

$$v_{yi} = \frac{\partial \psi_i}{\partial z},\ v_{zi} = -\frac{\partial \psi_i}{\partial y}. \quad (6.110)$$

After appropriate elimination of the pressure fields $p_i(y, z)$ we obtain the following nonlinear boundary value problem:

$$\frac{\partial}{\partial t}\Delta_\perp \psi_i + \frac{\partial \psi_i}{\partial z}\frac{\partial}{\partial y}\Delta_\perp \psi_i - \frac{\partial \psi_i}{\partial y}\frac{\partial}{\partial z}\Delta_\perp \psi_i = c_i \Delta_\perp^2 \psi_i; \qquad (6.111)$$

$$\frac{\partial}{\partial t}U_i + \frac{\partial \psi_i}{\partial z}\frac{\partial}{\partial y}U_i - \frac{\partial \psi_i}{\partial y}\frac{\partial}{\partial z}U_i = c_i \Delta_\perp^2 U_i - e_i B_i; \qquad (6.112)$$

$$\frac{\partial}{\partial t}\Theta_i + \frac{\partial \psi_i}{\partial z}\frac{\partial}{\partial y}\Theta_i - \frac{\partial \psi_i}{\partial y}\frac{\partial}{\partial z}\Theta_i + U_i = \frac{d_i}{P}\Delta_\perp \Theta_i; \qquad (6.113)$$

$$z = -1 : \psi_1 = 0, \frac{\partial \psi_1}{\partial z} = 0, U_1 = 0, \Theta_1 = 0; \qquad (6.114)$$

$$z = a : \psi_2 = 0, \frac{\partial \psi_2}{\partial z} = 0, U_2 = 0, \Theta_2 = -b; \qquad (6.115)$$

$$z = 0 : \psi_1 = \psi_2 = 0, \frac{\partial \psi_1}{\partial z} = \frac{\partial \psi_2}{\partial z}, U_1 = U_2; \qquad (6.116)$$

$$\eta \frac{\partial^2 \psi_1}{\partial z^2} = \frac{\partial^2 \psi_2}{\partial z^2} - \frac{M\eta}{P}\frac{\partial \Theta_1}{\partial y}; \qquad (6.117)$$

$$\eta \frac{\partial U_1}{\partial z} = \frac{\partial U_2}{\partial z} - \frac{M\eta}{P}; \qquad (6.118)$$

$$\Theta_1 = \Theta_2; \qquad (6.119)$$

$$\kappa \frac{\partial \Theta_1}{\partial z} = \frac{\partial \Theta_2}{\partial z}; \qquad (6.120)$$

$$\psi_i(y + L, z) = \psi_i(y, z); \; U_i(y + L, z) = U_i(y, z); \qquad (6.121)$$
$$\Theta_i(y + L, z) = \Theta_i(y, z), \, i = 1, 2,$$

with

$$\Delta_\perp = \frac{\partial^2}{\partial y^2} + \frac{\partial^2}{\partial z^2}, \; U_i = v_{xi}, \, c_1 = d_1 = e_1 = 1,$$

$$c_2 = 1/\nu, \, d_2 = 1/\chi, \, e_2 = \rho.$$

The constants $B_i$ $(i = 1, 2)$ are obtained from the conditions

$$\int_{-L/2}^{L/2} dy \int_{-1}^{0} dz \, U_1 = 0, \; \int_{-L/2}^{L/2} dy \int_{0}^{a} dz \, U_2 = 0. \qquad (6.122)$$

For the calculation of $B_i$ and the fields $U_i$ that satisfy conditions (6.122), the following procedure is applied. Let $U_i$ be of the form

$$U_i = \tilde{U}_i + \sum_{j=1}^{2} e_j B_j V_{ij}, \, i = 1, 2, \, j = 1, 2, \qquad (6.123)$$

where $\tilde{U}_i$ and $V_{ij}$ satisfy the following equations and boundary conditions:

$$\frac{\partial}{\partial t}\tilde{U}_i + \frac{\partial \psi_i}{\partial z}\frac{\partial}{\partial y}\tilde{U}_i - \frac{\partial \psi_i}{\partial y}\frac{\partial}{\partial z}\tilde{U}_i = c_i \Delta_\perp \tilde{U}_i; \qquad (6.124)$$

$$z = -1: \tilde{U}_1 = 0: z = a: \tilde{U}_2 = 0; \qquad (6.125)$$

$$z = 0: \tilde{U}_1 = \tilde{U}_2; \eta\frac{\partial \tilde{U}_1}{\partial z} = \frac{\partial \tilde{U}_2}{\partial z} - \frac{M\eta}{P}; \qquad (6.126)$$

$$\tilde{U}_i(y + L, z) = \tilde{U}_i(y, z); \qquad (6.127)$$

$$\frac{\partial}{\partial t}V_{ij} + \frac{\partial \psi_i}{\partial z}\frac{\partial}{\partial y}V_{ij} - \frac{\partial \psi_i}{\partial y}\frac{\partial}{\partial z}V_{ij} = c_i \Delta_\perp V_{ij} - \delta_{ij} - \frac{\partial \ln B_j}{\partial t}V_{ij}; \quad (6.128)$$

$$z = -1: V_{1j} = 0: z = a: V_{2j} = 0; \qquad (6.129)$$

$$z = 0: V_{1j} = V_{2j}; \eta\frac{\partial V_{1j}}{\partial z} = \frac{\partial V_{2j}}{\partial z}; \qquad (6.130)$$

$$V_{ij}(y + L, z) = V_{ij}(y, z); \qquad (6.131)$$

$i = 1, 2; j = 1, 2; \delta_{ij}$ is the Kronecker symbol. After the calculation of $\tilde{U}_i$ and $V_{ij}$ (the last term in (6.128) is evaluated from the previous time steps), expression (6.123) is substituted into conditions (6.122), and the system of two linear algebraic equations for $B_j$ is solved.

Solutions of the boundary value problem (6.111) - (6.122) are obtained by the finite difference method. The variables "stream function - vorticity" are used to solve the equation (6.111) with corresponding boundary conditions (Simanovskii and Nepomnyashchy, 1993). Equations and boundary conditions (6.111) - (6.121) are approximated on a uniform mesh using a second order approximation for the spatial coordinates. The integration of evolution equations is performed by means of an explicit scheme. A rectangular mesh $28 \times 56$ was used. The time step was chosen from the stability conditions.

For simulation of inclined hydrothermal waves, it is best to make a transformation of variables

$$x = X\cos\alpha + Y\sin\alpha, \ y = -X\sin\alpha + Y\cos\alpha \qquad (6.132)$$

and search the solution in the form:

$$\mathbf{v}_i = \mathbf{v}_i(Y, z), \ p_i = p_i(Y, z) + B_i X, \qquad (6.133)$$

$$T_i = X \cos \alpha + Y \sin \alpha + \Theta_i(Y, z), \ i = 1, 2.$$

In this case, equation (6.113) is replaced by the following equation:

$$\frac{\partial}{\partial t} \Theta_i + \frac{\partial \psi_i}{\partial z} \left( \frac{\partial}{\partial y} \Theta_i + \sin \alpha \right) - \frac{\partial \psi_i}{\partial y} \frac{\partial}{\partial z} \Theta_i + U_i \cos \alpha = \frac{d_i}{P} \Delta_\perp \Theta_i,$$

$$(6.134)$$

and then the boundary conditions (6.117), (6.118) take the form:

$$\eta \frac{\partial^2 \psi_1}{\partial z^2} = \frac{\partial^2 \psi_2}{\partial z^2} - \frac{M\eta}{P} \left( \frac{\partial}{\partial y} \Theta_1 + \sin \alpha \right); \qquad (6.135)$$

$$\eta \frac{\partial U_1}{\partial z} = \frac{\partial U_2}{\partial z} - \frac{M\eta}{P} \cos \alpha. \qquad (6.136)$$

This boundary value problem was solved in the same way as in the case of longitudinal rolls.

In the simulation of drifted transverse rolls $\alpha$ was taken equal to $90°$ in equations (6.132) - (6.136).

### ii. Nonlinear flow regimes

Nepomnyashchy *et al.* (2001) have investigated the nonlinear flow regimes for the same air-water system as in subsection 6.3.1.

First, let us consider hydrothermal waves with an oblique wave vector ($0 < \alpha < 90°$). This type of waves appears for relatively small values of $b/M$, when there exists a region with a positive vertical component of the temperature gradient in the lower layer, and the interfacial temperature $\Theta_s$ is positive (2.55) Fig. 6.14 provides a snapshot of the fields of (a) the stream functions $\psi_i(Y, z)$, (b) *disturbances* of the temperature $\Theta_i(Y, z) - \Theta_i^{(0)}(z)$, and (c) the longitudinal velocity $U_i(Y, z)$ ($i = 1, 2$). The motion is a traveling wave, $f_i(Y, z, t) = f_i(Y - ct, z)$, $i = 1, 2$, where $f_i = (\psi_i, U_i, \Theta_i - \Theta_i^{(0)})$, with *positive* phase velocity $c$, i.e. the waves move in the direction *opposite* to the $Y$-component of the flow velocity at the interface. The explanation of this phenomenon given by Davis (1987) is based on the fact that the vertical component of the temperature gradient in water is positive in a certain region. Let us consider the fields of variables shown in Figs. 6.14(a) and 6.14(b). If a negative disturbance of the *interfacial* temperature appears to the left of the computation region centre (Fig. 6.14(b)) it generates a corresponding descending motion in the water. Because of the incompressibility of the fluid,

an ascending motion appears in the region to the right of the centre
(Fig. 6.14(a)). The latter flow leads to a *cooling* in the region *under the interface*, where the vertical component of the temperature
gradient is positive. This effect tends to shift the minimum of the interfacial temperature to the right. Thus the wave moves to the right,
despite the advection of the interfacial temperature field by the $Y$-
component of the interfacial velocity. Because of the relatively high
thermal diffusivity of air, the influence of the motion in the upper
fluid can be neglected.

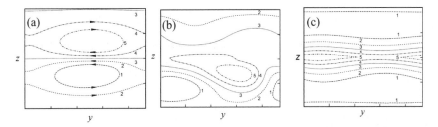

Figure 6.14: (a) Isolines of the stream function, (b) the temperature disturbance, and (c) the longitudinal velocity for oblique
hydrothermal waves ($M = 328$, $b = 100$, $\alpha = 41°$), $L = 2.35$.
(Redrawn after Nepomnyashchy *et al.*, 2001).

Let us discuss now the results of nonlinear simulations of two-
dimensional (transverse) rolls moving in the same direction as the
interfacial flow. This kind of waves appears at larger values of $b/M$,
where the region of a positive vertical temperature gradient is relatively small, and the sign of the interfacial temperature (2.55) in
water is negative (i.e. the external heating from below prevails the
"heating from above" caused by the base flow). A snapshot of the
fields of variables, (a) stream function and (b) temperature disturbance, is shown in Fig. 6.15. Let us emphasize that the temperature
in the region of the ascending flow is now higher than that in the region of the descending flow. Thus, the mechanism driving the wave
in the direction opposite to the interfacial flow is switched off. The
minimum of the interfacial temperature moves now in the same direction as the interfacial flow, i.e. the wave moves to the left ($c < 0$).

Finally, let us consider longitudinal rolls (spiral flows). Simula-
tions heve been carried out for $b = 100$ and $L = 2.75$ (i.e. $k = 2\pi/L \approx 2.285$). Linear theory predicts an instability in the inter-

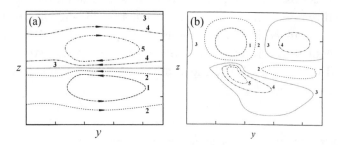

Figure 6.15: (a) Isolines of the stream function and (b) the temperature disturbance for transverse traveling rolls ($M = 200$, $b = 200$, $\alpha = 90°$), $L = 2.35$. (Redrawn after Nepomnyashchy *et al.*, 2001).

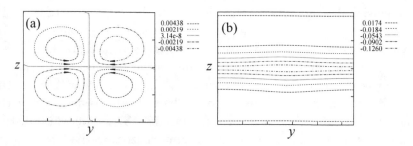

Figure 6.16: (a) Isolines of the stream function and (b) longitudinal velocity field for stationary longitudinal rolls ($M = 61.2$, $b = 100$, $\alpha = 0$), $L = 2.75$. (Redrawn after Nepomnyashchy *et al.*, 2001).

val between $M = M_A \approx 28.47$ (with critical wavenumber $k = 2.11$) and $M = M_B \approx 82.59$ (with critical wavenumber $k = 2.825$). The numerical simulations show the existence of stable spiral flows with the chosen period $L$ in the interval $M_1 < M < M_2$, $M_1 \approx 30$, $M_2 \approx 80$. No subcritical flows have been found with either $M > M_A$ or $M < M_B$. Thus, the prediction of the linear theory concerning the return to stability of the parallel flow above the line 2 of Fig. 6.13 is supported by the numerical simulations. Fig. 6.16 depicts (a) the isolines of the stream function fields, $\psi_i$, and (b) the longitudinal velocity fields $U_i$ calculated for $M = 61.2$. The secondary motion $\psi_i(y, z)$ has a four-vortex structure. The fields of variables satisfy the following symmetry conditions:

$$\psi_i(y, z) = -\psi_i(-y, z), \; U_i(y, z) = U_i(-y, z), \; \Theta_i(y, z) = \Theta_i(-y, z).$$
$$(6.137)$$

# Chapter 7

# Outlook

In the present, final chapter let us conclude our book with a brief discussion, with no pretension of completeness, of some new, rapidly developing topics.

## 7.1 Interfacial turbulence and dissipative waves

Intensive chaotic convective motions caused by interfacial effects ("interfacial turbulence") are observed in a variety of experiments (e.g. Linde *et al.*, 1979; Orell and Westwater, 1962) and engineering processes like metallurgy (Ready, 1965; Brimacombe and Weinberg, 1972; Szekely, 1979; van Gurp *et al.*, 1979; Heiple *et al.*, 1983; Hammerschmid, 1987; Choo and Szekely, 1991; Hondros *et al.*, 1998; Mills, 2001 and references therein), evaporation and drying processes (Berg *et al*, 1965; Palmer and Maheshri, 1981; Marra and Huethorst, 1991), electrochemistry (Mityushev and Krylov, 1986) and crystal growth (Ostrach, 1982, 1983; Walter, 1987; Regel, 1987; Ratke *et al*, 1989). Like it happens at gas-melt or melt-slag interface in various separation processes like distillation, absorption, chemisorption and extraction (Sherwood and Wei, 1957; Sawistowski, 1971, 1981; Thornton, 1987; Perez de Ortiz, 1992; Golovin, 1992) spontaneous interfacial turbulence leads to a sharp increase of mass transfer rates between the two phases.

Still, understanding interfacial turbulence remains a challenge for theoreticians. In contradistinction to the Kolmogorov-Reynolds *inertial turbulence* which is characterized by an energy transfer from long to short scales mostly dissipation-free (within the inertial interval; dissipation occurs only at the end of the cascade where viscosity kills the small eddies), the interfacial turbulence is *intrinsically strongly dissipative*, i.e. it is from the very beginning characterized by the generation/dissipation of energy at any relevant scales. Thus, interfacial turbulence should be similar to the dissipative *spatio-temporal chaos*, which is observed, i.e., in experiments with reaction-diffusion systems and in numerical simulations of the Kuramoto-Sivashinsky and related equations (e.g.  Hyman and Nicolaenko, 1986, 1987). Nevertheless, one can expect that such a strongly dissipative turbulence may have some universal properties as with the "usual" (inertial) turbulence.  In the case of Kuramoto-Sivashinsky equation, it has been established that there exists a certain kind of energy cascade from *short to long scales* which determines the long-scale properties of the spatio-temporal chaos (Bohr *et al.*, 1998; Boos and Thess, 1999).  More precisely, the nonlinear cascade is governed by the *strongly-dissipative* Burgers equation with a certain effective positive *turbulent* viscosity, while the instability manifests itself as a uncorrelated short-scale ("white") Gaussian noise.  In other words, the long-scale behavior of KS chaos can be modeled by the universal Kardar-Parisi-Zhang equation (1986), and it is characterized by corresponding scaling properties.

It is a challenge to find the universal scaling properties of (strongly dissipative) interfacial turbulence.  In subsection 4.5 we discussed some attempts to treat this problem by constructing approximate models and direct numerical computations. The latter approach has been recently carried out in the three-dimensional case (Boeck and Thess, 2001).  However, all these attempts are related to the case of time-dependent patterns in a layer with a non-deformable open surface, whereas interfacial turbulence in surface tension gradient–driven systems manifests itself as a disordered array of propagating and interacting nonlinear waves, ripples and solitons. Among factors affecting such dissipative interfacial turbulence, surface deformation may play an important role. Accordingly, one can expect that instead of non-dissipative vortices which compose the "usual" isothermal turbulence, the building elements leading to interfacial turbulence are *dissipative waves* (which includes soliton-like and shock-like waves). In Chapter 5 we discussed such waves and described them by means of *weakly-nonlinear* dissipative wave equations. However, in the case of fully developed, interfacial turbulence, waves should be probably described by some *strongly-nonlinear* equations not yet known. Incidentally, the field of not so highly dissipative waves and solitons demands serious efforts in the mathematical study. Text books on fluid mechanics or on waves (see, e.g., Whitham, 1974; Lamb, 1980; Drazin and Jonhson, 1989; Shen, 1993) consider viscosity as a mere damping agent while we have shown how viscosity together with e.g. the Marangoni effect can play a dynamic role past an instability threshold in indeed damping wave disturbances albeit in balance with appropriate and sustained energy supply. Also, the future theory of interfacial turbulence should take into account the diversity of instability mechanisms characteristic for fluid systems with an interface as amply illustrated along the chapters of this book.

## 7.2 Control of instabilities

In this book, we have considered interfacial instabilities that appear and develop in "a natural way". As already noted in the Preface, for the engineering applications there is the need, however, of *controlling* or delaying instabilities, in some cases to be suppressed and in others to be enhanced. The above mentioned problems have been studied in the context of the influence of interfacial convection on crystal growth, specifically under very reduced effective-gravity or microgravity conditions. In order to control interfacial convection, authors have investigated the influence of magnetic fields (Series and Hurle, 1991; Wilson, 1993a, 1993b; Priede and Gerbeth, 2000), vibrations (Zharikov *et al.*, 1990; Lyubimov *et al.*, 1997) and other factors. The control was understood just as a suppression of instability, without any feedback. Recently, another approach based on *feedback (online) control* has been suggested, first for buoyancy-driven (Rayleigh-Benard) convection (Tang and Bau, 1994), and later for surface tension gradient-driven (Benard–Marangoni) convection (Bau, 1999; Or *et al.*, 1999). The nonlinear feedback control strategy enables to shift the onset of the instability and to eliminate the subcritical nature of a nonlinear instability. It seems that controlling in order to sustain a definite type of motion, unstable in the absence of the control is of great engineering interest. On the other hand, in general, even in the absence of instabilities understanding how to handle large scale flows in the drastically reduced effective-gravity of e.g. the International Space Station, as emphasized by S. Ostrach, is a worthwhile direc-

tion of investigation for a new engineering of fluid flow with heat and
solutal transport, in disperse or multicomponent media is yet to be
developed.

## 7.3 Interfacial phenomena in the presence of phase transitions and chemical reactions, multiphase flows, etc.

In engineering applications, interfacial convection appears often in
the course of some other physico-chemical processes (Hondros *et
al.*, 1998; Mills, 2001). We have mentioned already the process
of *crystallization* which is accompanied by convection in the melt
(Kuhlmann, 1999). Another important kind of phase transition
which produces interfacial convection is *evaporation* or boiling. Because the evaporation leads to having very *thin liquid films*, the classical *macroscopic* models of fluid mechanics should be revised and
*microscopic* processes and forces should be appropriately taken into
account (Shikhmurzaev, 1994, 1997; Seppecher, 1996; Oron *et al.*,
1997; Pismen and Pomeau, 2000; Thiele *et al.*, 2001a-c). We have
mentioned some extensions of classical models in Chapter 3 while
discussing spreading processes. Understanding physical phenomena
and flows on very small scales, including the phenomena of drying
and contact line motion, is still far from completion.

As noted in Chapters 3 and 5 the interface is often subject to
*chemical reactions* which produce interfacial convection due to the
dependence of surface tension on the concentrations of reactants with
heat release or absorption (Ruckenstein and Berbente, 1964; Dalle
Vedove and Sanfeld, 1981; Rabinovich *et al.*, 1995; Pismen, 1997). In
turn, convection influences the rate of chemical reactions (Sherwood
and Wei, 1957). Interfacial phenomena in the presence of chemical reaction exhibit the appearance of patterns (Avnir and Kagan,
1995), fronts (Pismen, 1997), plumes and fingers (Eckert and Grahn,
1999), etc. Some are probably produced by double-diffusion instabilities which appear in *multicomponent fluids* with cross-transport
processes acting together with the Marangoni effect (Chen and Su,
1992; Tanny *et al.*, 1995). For drop and bubble migration, with
or without surface chemical reaction and the Marangoni effect, the
theory presented, although providing striking and appealing results,
suffers from oversimplification in the mathematical formulation of

the really complex phenomena appearing in engineering processes.

## 7.4   "Exotic" patterns and defects

Some new directions in the field of interfacial pattern formation are worth mentioning.

First, a clear understanding of the role of defects in pattern formation and the transitions between patterns is needed. In Chapter 4, the transition between hexagonal and square patterns was briefly discussed. It turns out that in reality this transition is deeply rooted in the dynamics and evolution of defects. Specifically, before this transition the penta-hepta defects, which are bound states of dislocations on different sublattices, are "dissolved" into separate dislocations connected by a new kind of *linear* defects, pentalines (Eckert and Thess, 1999). This effect has no theoretical explanation yet. The pentalines act as nuclei of the square pattern. The next stage in the description of the transition would be the investigation of the evolution of a front between two patterns, the new and the old. In Chapter 4, such a problem was studied using amplitude equations derived from appropriate perturbation theory. However, non-perturbative effects, like pinning of the front by the short-scale structure, are crucial for the development of finite-amplitude patterns (Pomeau, 1986) and should be studied.

It was also mentioned in Chapter 4 that there are arguments in favor of the appearance of quasiperiodic patterns with incommensurable basic wavevectors (Golovin et al., 1995), and periodic "superlattices" with the set of basic wavevectors which have different lengths but commensurable components (Dionne et al., 1997). Recently, Echebarria and Riecke (2000) investigated the sideband instabilities and defects of quasipatterns. This is a new promising field for a theoretical analysis and computer simulations.

While the approach for the investigation of the evolution of defects in steady patterns has been well developed (Pismen, 1999), much less is known on the *defects in wavy patterns*, except the case of longwave instabilities governed by the isotropic Ginzburg-Landau equation, where spirals appear. The effect of anisotropy was studied by Faller and Kramer (1998, 1999). Only few works are still devoted to defects in short-wavelength wavy patterns (Coullet et al., 1992; Malomed, 1997).

An interesting open question is that of pattern or wave selection under spatially *inhomogeneous* conditions. In this context, let us mention results obtained about thermocapillary flows driven by a local heat source (Favre *et al.*, 1997; Bratukhin *et al.*, 2000). Patterns composed of stationary and rotating petals (also appearing in mass-transfer processes as discussed in Chapter 5; Santiago-Rosanne *et al.*, 1997; Brimacombe and Weinberg, 1972), circular and spiral waves, and daisy- or star–shape standing waves have been observed. The appearance of such patterns seems caused by the instability of an axisymmetric thermocapillary flow whose nonlinear description and that of the observed patterns demands further work.

# Bibliography

[1] Acheson, D. J. (1990). *Elementary Fluid Dynamics* (Clarendon, Oxford).

[2] Acrivos, A. and Taylor, T. D. (1962). Heat and mass transfer from single spheres in Stokes flow, *Phys. Fluids* **5**, 387-394.

[3] Adamson, A. W. (1982). *Physical Chemistry of Surfaces* (Wiley, N. Y.).

[4] Afenchenko, O. V., Ezersky, A. B., Nazarovsky, A. V. and Velarde, M. G. (2001). Experimental evidence on the structure and evolution of penta-hepta defect in hexagonal lattices due to Marangoni-Benard convection, *Int. J. Bifurcation Chaos* **11**, 1261-1273.

[5] Agble, D. and Mendes-Tatsis, M. A. (2000). The effect of surfactants on interfacial mass transfer in binary liquid-liquid systems, *Int. J. Heat Mass Transfer* **43**, 1025-1034.

[6] Ahlers, G. and Behringer, R. P. (1978). Evolution of turbulence from the Rayleigh-Benard instability, *Phys. Rev. Lett.* **40**, 712-716.

[7] Alekseenko, S. V., Nakoryakov, V. E. and Pokusaev, B. G. (1994). *Wave Flow of Liquid Films* (Begell House, Wallingford, U. K.).

[8] Alexeev, A. A. and Vinogradova, O. I. (1996). Flow of a liquid in a nonuniformly hydrophobized capillary, *Colloids Surfaces A* **108**, 173-179.

[9] Anderson, D. M. and Davis, S. H. (1995). The spreading of volatile liquid droplets on heated surfaces, *Phys. Fluids* **7**, 248-265.

[10] Aoosey, D. W., Skinner, S. R., Cooney, J. L., Williams, J. E., Gavin Andersen, D. R. and Lonngren, K. E. (1992). Properties of soliton-soliton collisions, *Phys. Rev. A* **45**, 2606-2610.

[11] Argentina, M., Coullet, P. and Krinsky, V. (2000). Head-on collisions of waves in an excitable FitzHugh-Nagumo system: a transition from wave annihilation to classical wave behavior, *J. Theor. Biol.* **205**, 47-52.

[12] Aris, R. (1962). *Vectors, Tensors, and the Basic Equations of Fluid Mechanics* (Prentice-Hall, Englewood Cliffs, N. J.).

[13] Aronson, D. G. and Weinberger, H. F. (1978). Multidimensional nonlinear diffusion arising in population genetics, *Adv. Math.* **30**, 33-76.

[14] Asakura, K., Lautenbach, J., Rotermund, H. H. and Ertl, G. (1995). Spatiotemporal concentration patterns associated with the catalytic oxidation of CO and Au covered Pt (110) surfaces, *J. Chem. Phys.* **102**, 8157-8184.

[15] Avnir, D., Kagan, M. L. (1995). The evolution of chemical patterns in reactive liquids, driven by hydrodynamic instabilities, *Chaos* **5**, 589-601.

[16] Azouni, M. A. and Petre, G. (1998). Experimental evidence of the effect of evaporation-condensation on thermal Marangoni flows in aqueous fatty alcohol, *J. Colloid Interface Sci.* **206**, 332-338.

[17] Bacri, L., Debregeas, G. and Brochard-Wyart F. (1996). Experimental study of the spreading of a viscous droplet on a nonviscous liquid, *Langmuir* **12**, 6708-6711.

[18] Bakker, C. A. P., van Buytenen, P. M. and Beek, W. J. (1966). Interfacial phenomena and mass transfer, *Chem. Engng. Sci.* **21**, 1039-1046.

[19] Balasubramaniam, R., Lacy, C. E., Wozniak, G. and Subramanian, R. S. (1996). Thermocapillary migration of bubbles and drops at moderate values of the Marangoni number in reduced gravity, *Phys. Fluids* **8**, 872-880.

[20] Balasubramaniam, R. and Subramanian, R. S. (1996). Thermo-capillary bubble migration - Thermal boundary layers for large Marangoni numbers, *Int. J. Multiphase Flow* **22**, 593-612.

[21] Balasubramaniam, R. and Subramanian, R. S. (2000). The migration of a drop in a uniform temperature gradient at large Marangoni numbers, *Phys. Fluids* **12**, 733-740.

[22] Bankoff, S. G. (1971). Stability of a liquid flow down a heated inclined plate, *Int. J. Heat Mass Transfer* **14**, 377-385.

[23] Bar, D. E. and Nepomnyashchy, A. A. (1995). Stability of periodic waves governed by the modified Kawahara equation, *Physica D* **86**, 586-602.

[24] Bar, D. E. and Nepomnyashchy, A. A. (1999). Stability of periodic waves generated by long-wavelength instabilities in isotropic and anisotropic systems, *Physica D* **132**, 411-427.

[25] Bar, M., Eiswirth, M. Rotermund, H. H. and Ertl, G. (1992). Solitary-wave phenomena in an excitable surface reaction, *Phys Rev. Lett.* **69**, 945-948.

[26] Barenblatt, G. I. (1996). *Scaling, self-similarity, and intermediate asymptotics* (University Press, Cambridge).

[27] Barton, K. D. and Subramanian, R. S. (1989). The migration of liquid drops in a vertical temperature gradient, *J. Colloid Interface Sci.* **133**, 211-222.

[28] Batchelor, G. K. (1967). *An Introduction to Fluid Dynamics* (University Press, Cambridge).

[29] Bau, H. H. (1999). Control of Marangoni-Benard convection, *Int. J. Heat Mass Transfer* **42**, 1327-1341.

[30] Bazin, H. (1865). Recherches experimentales sur la propagation des ondes, *Mem. presentes par divers savants a l'Acad. Sci. Inst.* France (Paris) **19**, 495-644.

[31] Becher, P. (1990). *Dictionary of Colloid and Surface Science* (Marcel Dekker, N. Y.) p. 148.

[32] Bekki, S., Vignes-Adler, M., Nakache, E. and Adler, P. M. (1990). Solutal Marangoni effect: I. Pure interfacial transfer, *J. Colloid Interface Sci.* **140**, 492-506.

[33] Bekki, S., Vignes-Adler, M. and Nakache, E. (1992). Solutal Marangoni effect: II. Dissolution, *J. Colloid Interface Sci.* **152**, 314-324.

[34] Ben Hadid, H. and Roux, B. (1990). Thermocapillary convection in long horizontal layers of low-Prandtl number melts subject to a horizontal temperature gradient, *J. Fluid Mech.* **221**, 77-85.

[35] Ben Hadid, H. and Roux, B. (1992). Buoyancy- and thermocapillary-driven flows in differentially heated cavities for low-Prandtl number fluids, *J. Fluid Mech.* **235**, 1-15.

[36] Benard, H. (1900). Les tourbillons cellulaires dans une nappe liquide. Premiere partie: Description generale des phenomenes, *Rev. Gen. Sci. Pures Appl.* (Paris) **11**, 1261-1271; Deuxieme partie: Procedes mecaniques et optiques d'examen. Lois numeriques des phenomenes, *ibidem*, **11**, 1309-1328.

[37] Benard, H. (1901). Les tourbillons cellulaires dans une nappe liquide transportant de la chaleur par convection en regime permanent, *Ann. Chim. Phys.* (Paris) **23**, 62-143.

[38] Benard, H. (1928). Sur les tourbillons cellulaires, les tourbillons en bandes et la theorie de Rayleigh, *Bull. Soc. Franc. Phys.* (Paris) No. 266, 112S-115S.

[39] Benard, H. (1930). Sur la limite theorique de stabilite de l'equilibre preconvectif de Rayleigh, comparee a celle que donne l'etude experimentale des tourbillons cellulaires de Benard, *Proc. 3rd Int. Congr. Appl. Mech.*, vol. **1**, 120-121.

[40] Benjamin, T. B. (1957). Wave formation in laminar flow down an inclined plane, *J. Fluid Mech.* **2**, 554-574.

[41] Benjamin, T. B. (1972). The stability of solitary waves, *Proc. Royal Soc. London A* **328**, 153-183.

[42] Benjamin, T. B. (1982). The solitary wave with surface tension, *Quart. Appl. Math.* **40**, 231-234.

[43] Benjamin, T. B. (1993). Note on formulas for the drag of a sphere, *J. Fluid Mech.* **246**, 335-342.

[44] Benney, D.J. (1966). Long non-linear waves in fluid flows, *J. Math. & Phys.* **45**, 52-63.

[45] Berg, J. C. and Acrivos, A. (1965). The effect of surface active agents on convection cells induced by surface tension, *Chem. Engng. Sci.* **20**, 737-745.

[46] Berg, J. C., Acrivos, A. and Boudart, M. (1965). Evaporative convection, *Adv. Chem. Engng.* **6**, 61-123.

[47] Bergman, T. L. and Keller, J. R. (1988). Combined buoyancy, surface tension flow in liquid metals, *Numer. Heat Transfer* **13**, 49-63.

[48] Bergman, T. L. and Ramadhyani, S. (1986). Combined buoyancy- and thermocapillary driven convection in open square cavities, *Numer. Heat Transfer* **9**, 441-451.

[49] Bestehorn, M. (1993). Phase and amplitude instabilities for Benard-Marangoni convection in fluid layers with large aspect ratio, *Phys. Rev. E* **48**, 3622-3634.

[50] Bestehorn, M. (1996). Square patterns in Benard-Marangoni convection, *Phys. Rev. Lett.* **76**, 46-49.

[51] Bestehorn, M., Fantz, M., Friedrich, R. and Haken, H. (1993). Hexagonal and spiral patterns of thermal convection, *Phys. Lett. A* **174**, 48-52.

[52] Bestehorn, M. and Friedrich, R. (1998). Natural patterns in nonequilibrium systems, in *Perspective Look at Nonlinear Media. From Physics to Biology and Social Sciences* (Springer-Verlag, Berlin) 31-44.

[53] Bestehorn, M. and Friedrich, R. (1999). Rotationally invariant order parameter equations for natural patterns in nonequilibrium systems, *Phys. Rev. E* **59**, 2642-2652.

[54] Birikh, R. V. (1966). Thermocapillary convection in horizontal layer of liquid, *J. Appl. Mech. Tech. Phys.* **7**, 43-49.

[55] Blake, T. D. (1984). Wetting, in *Surfactants*, edited by (Academic, London) 221-275.

[56] Block, M. J. (1956). Surface tension as the cause of Benard cells and surface deformation in a liquid film, *Nature* (London) **176**, 650-651.

[57] Bodenschatz, E., de Bruyn, J. R., Ahlers, G. and Cannell, D. S. (1991). Transitions between patterns in thermal convection, *Phys. Rev. Lett.* **67**, 3078-3081.

[58] Bodenschatz, E., Cannell, D. S., de Bruyn, J. R., Ecke, R., Hu, Y.-C., Lerman, K. and Ahlers, G. (1992). Experiments on three systems with non-variational aspects, *Physica D* **61**, 77-93.

[59] Boeck, Th., Nepomnyashchy, A. A., Simanovskii, I., Golovin, A., Braverman, L. and Thess, A. (2001). Three-dimensional convection in a two-layer system with anomalous thermocapillary effect (preprint).

[60] Boeck, Th. and Thess, A. (1997). Inertial Benard-Marangoni convection, *J. Fluid Mech.* **350**, 149-175.

[61] Boeck, Th. and Thess, A. (2001). Power-law scaling in Benard-Marangoni convection at large Prandtl numbers (preprint).

[62] Bohr, T., Jensen, M. H., Paladin, G. and Vulpiani, A. (1998). *Dynamical Systems Approach to Turbulence* (University Press, Cambridge).

[63] Bojadjiev, Ch. and Beshkov, V. (1984). *Mass Transfer in Liquid Film Flows* (Publ. House Bulg. Acad. Sci., Sofia).

[64] Bond, W. N. (1927). Bubbles and drops and Stokes' law, *Phil. Mag.* **7**, 889-898.

[65] Boos, W. and Thess, A. (1999). Cascade of structures in long-wavelength Marangoni instability, *Phys. Fluids* **11**, 1484-1494.

[66] Borgas, M. S. and Grotberg, J. B. (1988). Monolayer flow on a thin film (lung application), *J. Fluid Mech.* **193**, 151-170.

[67] Bouasse, H. (1924). *Houle, Rides, Seiches et Marees* (Delagrave, Paris) 291-292, where reference to experiments by Bazin is provided.

[68] Boussinesq, J. V. (1871). Theorie de l'intumescence liquide appelee onde solitaire ou de translation se propageant dans un canal rectangulaire, *C. R. Hebd. Seances Acad. Sci.* (Paris) **72**, 755-759.

[69] Boussinesq, J. V. (1872). Theorie des ondes et des remous qui se propagent le long d'un canal rectangulaire horisontal en communiquant au liquid contenu dans ce canal des vitesses sensiblement pareilles de la surface au fond, *J. Math. Pures Appl.* (Paris) **17**, 55-108.

[70] Boussinesq, J. V. (1877). Essai sur la theorie des eaux courantes, *Mem. presentes par divers savants a l'Acad. Sci. Inst. France* (Paris) **23**, 1-680. (1878) Additions et eclaircissements au memoire intitule (supra), *ibidem* **24**, No. 2, 1-64.

[71] Boussinesq, J. V. (1901). Mise en equation des phenomenes de convection calorifique et apercu sur le povoir refroidissant des fluides, *C. R. Hebd. Seances Acad. Sci.* (Paris) **132**, 1382-1385.

[72] Boussinesq, J. V. (1903). *Theorie Analytique de la Chaleur* (Gauthier-Villars, Paris), vol. **2**, 172.

[73] Boussinesq, J. V. (1913a). Sur l'existence d'une viscosite superficielle, dans la mince couche de transition separant un liquide d'un autre fluide contigu, *Ann. Chim. Phys.* (Paris) **29**, 349-357.

[74] Boussinesq, J. V. (1913b). Application des formules de viscosite superficielle a la surface d'une goutte liquide spherique, tombant lentement, d'un mouvement devenu uniforme, au sein d'une masse fluide indefinie en repos, d'un poids specifique moindre, *Ann. Chim. Phys.* (Paris) **29**, 357-364.

[75] Boussinesq, J. V. (1913c). Vitesse de la chute lente, devenue uniforme, d'une goutte liquide spherique, dans un fluide visqueux de poids specifique moindre, *Ann. Chim. Phys.* (Paris) **29**, 364-372.

[76] Bragard, J. and Lebon, G. (1993). Non-linear Marangoni convection in a layer of finite depth, *Europhys. Lett.* **21**, 831-836.

[77] Bragard, J. and Velarde, M. G. (1997). Benard convection flows, *J. Non-Equilib. Thermodyn.* **22**, 1-19.

[78] Bragard, J. and Velarde, M. G. (1998). Benard-Marangoni convection: Theoretical predictions about planforms and their relative stability, *J. Fluid Mech.* **368**, 165-194.

[79] Brand, H. R. (1989). Envelope equations near the onset of a hexagonal pattern, *Progr. Theor. Phys. Suppl.* **99**, 442-449.

[80] Bratukhin, Yu. K. (1975). Thermocapillary drift of a viscous fluid droplet, *Izv. Akad. Nauk SSSR, Mekh. Zhidk. Gaza* **5**, 156-161 (in Russian).

[81] Bratukhin, Yu. K., Briskman, V. A., Zuev, A. L., Pshenichnikov, A. F., and Rivkind, V. Ya. (1982). Experimental studies of the thermocapillary migration of gas bubbles in fluid, in *Hydrodynamics, Heat and Mass Transfer in Microgravity* (Nauka, Moscow) 98-109.

[82] Bratukhin, Yu. K., Makarov, S. O. and Mizyov, A. I. (2000). Oscillating thermocapillary convection regimes driven by a point heat source, *Fluid Dynamics* **35**, 232-241.

[83] Braun, B., Ikier, Ch., Klein, H. and Woermann, D. (1993). Thermocapillary migration of droplets in a binary mixture with miscibility gap during liquid/liquid phase separation under reduced gravity, Rocket Experiments in Fluid Science and Material Sciences, TEXUS 28 to 30, MASER 5 & MAXUS 1, *J. Colloid Interface Sci.* **159**, 515-516.

[84] Braverman, L. M., Eckert, K., Nepomnyashchy, A. A., Simanovskii, I. B. and Thess, A. (2000). Convection in two-layer systems with anomalous thermocapillary effect, *Phys. Rev. E.* **62**, 3619-3631.

[85] Brimacombe, J. K. and Weinberg, F. (1972). Observations of surface movements of liquid copper and tin, *Metall. Trans.* **3**, 2298-2299.

[86] Briskman, V. A., Zuev, A. L., Lyubimova, T. P. and Nepomnyashchy, A. A. (1991). Thermocapillary flows and deformations of the surface in the systems of fluid layers with the longitudinal temperature gradient in microgravity, *Int. J. Microgravity Res. Appl.* **IV/2**, 98-99.

[87] Brochard-Wyart, F., Hervet, H., Redon, C. and Rondelez, F. (1990). Spreading of Heavy Droplets. I. Theory, *J. Colloid Interface Sci.* **142**, 518-527.

[88] Bush, J. W. M. (1997). The anomalous wake accompanying bubbles rising in a thin gap: a mechanically forced Marangoni flow, *J. Fluid Mech.* **352**, 283-303.

[89] Busse, F. H. (1967). The stability of finite amplitude cellular convection and its relation to an extremum principle, *J. Fluid Mech.* **30**, 625-649.

[90] Busse, F. H. (1978). Non-linear properties of thermal convection, *Rep. Prog. Phys.* **41**, 1929-1967 and references therein.

[91] Campbell, T. A. and Koster, J. N. (1995). Modelling of liquid encapsulated gallium melts, *Acta Astronaut.* **35**, 805-812.

[92] Camel, D., Tison, P. and Favier, J. J. (1986). Marangoni Flow regimes in liquid metals, *Acta Astronaut.* **13**, 723-726.

[93] Canosa, J. and Gazdag, J. (1977). The Korteweg-de Vries-Burgers Equation, *J. Comput. Phys.* **23**, 393-403.

[94] Canright, D. (1994). Thermocapillary flow near a cold wall, *Phys. Fluids* **6**, 1415-1424.

[95] Caroli, B., Caroli, C. and Roulet, B. (1984). On the stability of hexagonal interfacial patterns in directional solidification of binary mixtures, *J. Crystal Growth* **68**, 677-690.

[96] Carpenter, B. M. and Homsy, G. M. (1989). Combined buoyant-thermocapillary flow in a cavity, *J. Fluid Mech.* **207**, 121-132.

[97] Castans, M. (1991). Obtencion de ecuaciones diferenciales mediante el Analisis Dimensional, *Annu. Cient. Grupo Interuniv. Anal. Dim.* (Escuela Tecnica Superior de Arquitectura, Madrid) Part 1, 77-80; Part 2, 81-86.

[98] Castillo, J. L. and Velarde, M. G. (1982). Buoyancy-thermocapillary instability: the role of interfacial deformation in one- and two-component fluid layers heated from below or above, *J. Fluid Mech.* **125**, 463-474.

[99] Castillo, J. L., Garcia-Ybarra, P. L. and Velarde, M. G. (1988). Thermohydrodynamic instabilities, in *Synergetics and Dynamic Instabilities*, edited by G. Caglioti, H. Haken, and L. A. Lugiato (North Holland, Amsterdam) 219-243.

[100] Cazabat, A. M. and Cohen Stuart, M. A. (1986). Dynamics of wetting: effects of surface roughness, *J. Phys. Chem.* **90**, 5845-5849.

[101] Cazabat, A. M. (1987). How does a droplet spread?, *Contemp. Phys.* **28** (4), 347-364.

[102] Cazabat, A. M., Heslot, F., Troian, S. M. and Carles, P. (1990). Fingering instability of thin spreading films driven by temperature gradients, *Nature* (London) **346**, 824-826.

[103] Cerisier, P., Occelli, R., Perez-Garcia, C. and Jamond, C. (1987a). Structural disorder in Benard-Marangoni convection, *J. Phys.* (Paris) **48**, 569-576.

[104] Cerisier, P., Perez-Garcia, C., Jamond, C. and Pantaloni, J. (1987b). Wavelength selection in Benard-Marangoni convection, *Phys. Rev. A* **35**, 1949-1952.

[105] Chandrasekhar, S. (1961). *Hydrodynamic and Hydromagnetic Stability* (Clarendon, Oxford).

[106] Chang, C. E. and Wilcox, W. R. (1976). Analysis of surface tension driven flow in floating zone melting, *Int. J. Heat Mass Transfer* **19**, 355-365.

[107] Chang, H.-C. (1994). Wave evolution on a falling film, *Annu. Rev. Fluid Mech.* **26**, 103-136.

[108] Chang, H.-C. (1986). Traveling waves on fluid interfaces: Normal form analysis of the Kuramoto-Sivashinsky equation, *Phys. Fluids* **29**, 3142-3147.

[109] Chang, H.-C., Demekhin, E. A. and Kopelevich, D. I. (1993). Laminarizing effects of dispersion in an active-dissipative nonlinear medium, *Physica D* **63**, 299-320.

[110] Chang, H.-C., Cheng, M., Demekhin, E. A. and Kopelevich, D. I. (1994). Secondary and tertiary excitation of three-dimensional patterns on a falling film, *J. Fluid Mech.* **270**, 251-275.

[111] Chang, H.-C. and Demekhin, E. A. (1996). Solitary wave formation and dynamics of falling films, *Adv. Appl. Mech.* **32**, 1-58.

[112] Chao, B. T. (1962). Motion of spherical gas bubbles in a viscous liquid at large Reynolds numbers, *Phys. Fluids* **5**, 69-79.

[113] Chate, H. and Manneville, P. (1987). Transition to turbulence via spatiotemporal intermittency, *Phys. Rev. Lett.* **58**, 112-115.

[114] Chen, C. F. and Su, T. F. (1992). Effect of surface tension on the onset of convection in a double-diffusive layer, *Phys. Fluids A* **4**, 2360-2367.

[115] Chen, J. C. and Lee, Y. T. (1992). Effect of surface deformaton on thermocapillary bubble migration, *AIAA J.* **30**, 993-998.

[116] Choo, R. T. C. and Szekely, J. (1991). The effect of gas shear-stress on Marangoni flows in arc-welding, *Welding J.* **70**, S223-S233.

[117] Christov, C. I. and Velarde M. G. (1992). On localized solutions of an equation governing Benard-Marangoni convection, *Appl. Math. Modelling* **17**, 311-318.

[118] Christov, C. I. and Velarde, M. G. (1995). Dissipative solitons, *Physica D* **86**, 323-347.

[119] Christov, C. I., Maugin G. A. and Velarde, M.G. (1996). Well-posed Boussinesq paradigm with purely spatial higher-order derivatives, *Phys. Rev. E* **54**, 3621-3638.

[120] Chu, X.-L. and Velarde, M. G. (1988). Sustained transverse and longitudinal waves at the open surface of a liquid, *Physicochem. Hydrodyn.* **10**, 727-737.

[121] Chu, X.-L. and Velarde, M. G. (1989). Transverse and longitudinal waves induced and sustained by surfactant gradients at liquid-liquid interfaces, *J. Colloid Interface Sci.* **131**, 471-484.

[122] Chu, X.-L. and Velarde, M. G. (1991). Korteweg-de Vries soliton excitation in Benard-Marangoni convection, *Phys. Rev. A* **43**, 1094-1096.

[123] Churaev, N. V., Sobolev, V. D. and Zorin, Z. M. (1971). Measurement, of viscosity of liquids in quartz capillaries, *Trans. Faraday Soc.* **52**, 213-220.

[124] Churaev, N. V., Sobolev, V. D. and Somov, A. N. (1984). Slippage of liquids over lyophobic solid surfaces, *J. Colloid Interface Sci.* **97**, 574-581.

[125] Cini, R., Lombardini, P. P., Manfredi, C. and Cini, E. (1987). Ripples damping due to monomolecular films, *J. Colloid Interface Sci.* **119**, 74-80.

[126] Cloot, A. and Lebon, G. (1984). A nonlinear stability analysis of the Benard-Marangoni problem, *J. Fluid Mech.* **145**, 447-469.

[127] Cohen, B. I., Krommes, J. A., Tang, W. M. and Rosenbluth, M. N. (1976). Non-linear saturation of the dissipative trapped-ion mode by mode coupling, *Nucl. Fusion* **16**, 971-972.

[128] Colinet, P., Georis, Ph., Legros, J. C. and Lebon, G. (1996), Spatially quasiperiodic convection and temporal chaos in two-layer thermocapillary instabilities, *Phys. Rev. E* **54**, 514-524.

[129] Colinet, P., Nepomnyashchy, A. A. and Legros, J. C. (1997). Multiplication of defects in hexagonal patterns (preprint).

[130] Cooker, M. J., Weidman, P.D. and Bale, D. S. (1997). Reflection of a high-amplitude solitary wave at a vertical wall, *J. Fluids Mech.* **342**, 141-158.

[131] Coullet, P., Emilsson, K. and Walgraef, D. (1992). Phase instability and defect behavior in modulated wave patterns, *Physica D* **61**, 132-139.

[132] Courant, R. and Friedrichs, K. O. (1948). *Supersonic Flow and Shock Waves* (Wiley-Interscience, N.Y.).

[133] Cowley, S. J., and Davis, S. H. (1983). Viscous thermocapillary convection at high Marangoni numbers, *J. Fluid Mech.* **135**, 175-188.

[134] Cox, R. G. (1983). The spreading of a liquid on a rough solid surface, *J. Fluid Mech.* **131**, 1-26.

[135] Cox, R. G. (1986a). The dynamics of the spreading of liquids on a solid surface. Part. 1. Viscous flow, *J. Fluid Mech.* **168**, 169-194.

[136] Cox, R. G. (1986b). The dynamics of the spreading of liquids on a solid surface. Part. 2. Surfactants, *J. Fluid Mech.* **168**, 195-220.

[137] Cox, R. G. (1998). Inertial and viscous effects on dynamic contact angles, *J. Fluid Mech.* **357**, 249-278.

[138] Crighton, D. G. (1995). Applications of KdV, *Acta Appl. Math.* **39**, 39-67.

[139] Cross, M. C., Daniels, P. G., Hohenberg, P. C. and Siggia, E. D. (1980). Effect of distant sidewalls on wave-number selection in Rayleigh-Benard convection, *Phys. Rev. Lett.* **45**, 898-902.

[140] Cross, M. C. and Hohenberg, P. C. (1993). Pattern formation outside of equilibrium, *Rev. Mod. Phys.* **65**, 851-1112.

[141] Crossley, H. E. (1949). Analogy between surface shock waves in a liquid and shocks in compressible gases, *CALTECH Rep. N-54.1* (Pasadena, California).

[142] Cuvelier, C. and Driessen, J. M. (1986). Thermocapillary free boundaries in crystal growth, *J. Fluid Mech.* **169**, 1-26.

[143] Dalle Vedove, W. and Sanfeld, A. (1981). Hydrodynamic and chemical stability of fluid-fluid reacting interfaces. I. General theory for a periodic regimes, *J. Colloid Interface Sci.* **84**, 318-327.

[144] Daniels, P. G. (1977). The effect of distant sidewalls on the transition to finite amplitude Benard convection, *Proc. Royal Soc. London A* **358**, 173-197.

[145] Dauby, P. C., Lebon, G., Colinet P. and Legros, J. C. (1993). Hexagonal Marangoni convection in a rectangular box with slippery walls, *Q. J. Mech. Appl. Math.* **46**, 683-707.

[146] Dauzere, C. (1908). Recherches sur la solidification cellulaire, *J. Phys.* (Paris) **7**, 930-934.

[147] Dauzere, C. (1912). Sur les changements qu'eprouvent les tour-billons cellulaires lorsque la temperature augmente, *C. R. Hebd. Seances Acad. Sci.* (Paris) **155**, 394-398.

[148] Davies, J. T. and Rideal, E. K. (1961). *Interfacial Phenomena* (Academic, N. Y.).

[149] Davis, S. H. (1969). Buoyancy-surface tension instability by the method of energy, *J. Fluid Mech.* **39**, 347-359.

[150] Davis, S. H. (1987). Thermocapillary instabilities, *Annu. Rev. Fluid Mech.* **19**, 403-435.

[151] Davis, S. H. and Homsy, G. M. (1980). Energy stability theory for free-surface problems: buoyancy-thermocapillary layers, *J. Fluid Mech.* **98**, 527-553.

[152] de Boer, P. C. T. (1984). Thermally-driven motion of strongly heated fluids, *Int. J .Heat Mass Transfer* **27**, 2239-2251.

[153] de Boer, P. C. T. (1986). Thermally-driven motion of highly viscous fluids, *Int. J .Heat Mass Transfer* **29**, 681-688.

[154] de Bruyn, J. R. (1992). Growth of fingers at a driven three-phase contact line, *Phys. Rev. A* **46**, R4500-R4503.

[155] de Gennes, P. G. (1985). Wetting: statics and dynamics, *Rev. Mod. Phys.* **57**, 827-863.

[156] Decker, W., Pesch, W. and Weber, A. (1994). Spiral defect chaos in Rayleigh-Benard convection, *Phys. Rev. Lett.* **73**, 648-651.

[157] Dee, G. T. and van Saarlos, W. (1988). Bistable systems with propagating fronts leading to pattern formation, *Phys. Rev. Lett.* **60**, 2641-2644.

[158] Defay, R., Prigogine, I., Bellemans, A. and Everett, D. H. (1966). *Surface Tension and Adsorption* (Longmans, London).

[159] Demekhin, E. A., Demekhin, I. A. and Shkadov, V. Y. (1983) Solitons in viscous films flowing down a vertical wall, *Izv. Akad. Nauk SSSR, Mekh. Zhidk. Gaza* **4**, 9-16 (in Russian).

[160] Demekhin, E. A., Kaplan, M. A. and Shkadov, V. Ya. (1987). Mathematical models of the theory of viscous liquid films, *Izv. Acad. Nauk SSSR, Mekh. Zhidk. Gaza* **6**, 73-81 (in Russian).

[161] Demekhin, E. A., Tokarev, G. Yu. and Shkadov, V. Ya. (1989), Instability and nonlinear waves on vertical liquid film flowing counter to a turbulent gas flow, *Teor. Osn. Khim. Tekhnol.* **23** (1), 54-64 (in Russian).

[162] Demekhin, E. A., Tokarev, G. Yu. and Shkadov, V. Ya. (1991). Hierarchy of bifurcations of space-periodic structures in a nonlinear model of active dissipative media, *Physica D* **52**, 338-361.

[163] Derjaguin, B. V., Muller, V. M. and Churaev, N. V. (1987). *Surface Forces* (Consultants Bureau, N. Y.).

[164] de Saedeleer, C. de, Garcimartin, A., Chavepeyer, G., Platten, J. K. and Lebon, G. (1996). The instability of a liquid layer heated from the side when the upper surface is open to the air, *Phys. Fluids* **8**, 670-676.

[165] Dionne, B., Silber, M. and Skeldon, A. (1997). Stability results for steady, spatially periodic planforms, *Nonlinearity* **10**, 321-353.

[166] di Pietro, N. D., Huh, C. and Cox, R. G. (1978). The hydrodynamics of the spreading of one liquid on the surface of another, *J. Fluid Mech.* **84**, 529-549.

[167] Doi, T. and Koster, J. N. (1993). Thermocapillary convection in two immiscible liquid layers with the free interface, *Phys. Fluids A* **5**, 1914-1927.

[168] Drazin, P. G. and Johnson, R. S. (1989). *Solitons. An Introduction* (University Press, Cambridge).

[169] Dupeyrat, M. and Michel, J. (1971). Instabillity observed at a liquid-liquid interface containing surface-active compounds, *J. Exp. Supp.* **18**, 269-273.

[170] Dupeyrat, M., Nakache, E. and Vignes-Adler, M. (1984). Chemically driven interfacial instabilities, in *Chemical Instabilities*, edited by G. Nicolis and F. Baras (Reidel, Dordrecht) 233-245.

[171] Dupre de Rennes, A. (1869), *Theorie Mecanique de la Chaleur* (Gauthier-Villars, Paris).

[172] Dussan V, E. B. (1976). The moving contact line: the slip boundary condition, *J. Fluid Mech.* **77**, 665-684.

[173] Dussan V, E. B. (1979). On the spreading of liquids on solid surfaces: static and dynamic contact lines, *Annu. Rev. Fluid Mech.* **11**, 371-400.

[174] Dussan V, E. B. (1985). On the ability of drops or bubbles to stick to non-horizontal surfaces of solids. II. Small drops or bubbles having contact angles of arbitrary size, *J. Fluid Mech.* **151**, 1-20.

[175] Dussan V, E. B. and Davis, S. H. (1974). On the motion of fluid-fluid interface along a solid surface, *J. Fluid Mech.* **65**, 71-95.

[176] Dussaud, A. D. and Troian, S. M. (1998). Dynamics of spontaneous spreading with evaporation on a deep fluid layer, *Phys. Fluids* **10**, 23-38.

[177] Echebarria, B. and Riecke, H. (2000). Sideband instabilities and defects of quasipatterns (preprint).

[178] Eckert, K., Bestehorn, M. and Thess, A. (1998). Square cells in surface-tension-driven Benard convection: experiment and theory, *J. Fluid Mech.* **356**, 155-197.

[179] Eckert, K. and Grahn, A. (1999). Plume and finger regimes driven by an exothermic interfacial reaction, *Phys. Rev. Lett.* **82**, 4436-4439.

[180] Eckert, K. and Thess, A. (1999). Nonbound dislocations in hexagonal patterns: pentagon lines in surface-tension-driven Benard convection, *Phys. Rev. E* **60**, 4117-4124.

[181] Eckert, K. and Thess, A. (2001) (private communication).

[182] Eckhaus, W. (1965). *Studies in Non-Linear Stability Theory* (Springer-Verlag, N. Y.).

[183] Edwards, D. A., Brenner, H. and Wasan D. T. (1991). *Interfacial Transport Processes and Rheology* (Butterworth-Heinemann, Boston).

[184] Ehrhard, P. and Davis, S. H. (1991). Non-isothermal spreading of liquid drops on horizontal plates, *J. Fluid Mech.* **229**, 365-388.

[185] Elbaum, M. and Lipson, S. G. (1994). How does a thin wetted film dry up, *Phys. Rev. Lett.* **72**, 3562-3565.

[186] Elbaum, M. and Lipson, S. G. (1995). Pattern formation in the evaporation of thin liquid films, *Israel J. Chem.* **35**, 27-32.

[187] Elder, K. R., Gunton, J. D. and Goldenfeld, N. (1997). Transition to spatiotemporal chaos in the damped Kuramoto-Shivashinsky equation, *Phys. Rev. E* **56**, 1613-1634.

[188] Elphick, C., Ierley, G. R., Regev, O. and Spiegel, E. A. (1991). Interacting localized structures with Galilean invariance, *Phys. Rev. A* **44**, 1110-1122.

[189] Esmail, N. M. and Shkadov, V. Ya. (1971). To nonlinear theory of waves in viscous liquid film, *Izv. Acad. Nauk SSSR, Mekh. Zhidk. Gaza* **4**, 54-59 (in Russian).

[190] Ezersky, A.B., Garcimartin, A., Burguete, J., Mancini, H. L. and Perez-Garcia, C. (1993). Hydrothermal waves in Marangoni convection in a cylindrical container, *Phys. Rev. E* **47**, 1126-1131.

[191] Falcke, M., Bar, M., Engel, H. and Eiswirth, M. (1992). Travelling waves in the CO oxidation on Pt(110): theory, *J. Chem. Phys.* **97**, 4555-4563.

[192] Faller, R. and Kramer, L. (1998). Phase chaos in the anisotropic complex Ginzburg-Landau equation, *Phys. Rev. E* **57**, R6249-R6252.

[193] Faller, R. and Kramer, L. (1999). Ordered defect chains in the 2D anisotropic complex Ginzburg-Landau equation, *Chaos, Solitons & Fractals* **10**, 745-752.

[194] Favre, E., Blumenfeld, L. and Daviaud, F. (1997). Instabilities of a liquid layer locally heated on its free surface, *Phys. Fluids* **9**, 1473-1475.

[195] Favre, H. (1935). *Etude Theorique et Experimentale des Ondes de Translation dans les Canaux Decouverts* (Dunod, Paris).

[196] Feudel, F., Feudel, U. and Brandenburg, A. (1993). On the bifurcation phenomena of the Kuramoto-Sivashinsky equation, *Int. J. Bifurcation Chaos* **3**, 1299-1303.

[197] Fraaije, J. G. E. M. and Cazabat, A. M. (1989). Dynamics of spreading on a liquid substrate, *J. Colloid Interface Sci.* **133**, 452-460.

[198] Franklin, B. (1774). Of the filling of waves by means of oil. Extracted from sundry letters between Benjamin Franklin, LL. D. F. R. S. William Brownrigg, M. D. F. R. S. and the Reverend Mr. Farifh, *Phil. Trans. Royal Soc. London* **64**, 445-460.

[199] Frenkel, A. L. (1992). Nonlinear theory of strongly undulating thin films flowing down a vertical cylinder, *Europhys. Lett.* **18**, 583-587.

[200] Frenkel, A. L. and Indireshkumar, K. (1996). Derivations and simulations of evolution equations of wavy film flows, in *Mathematical Modeling and Simulation in Hydrodynamic Stability*, edited by D. N. Riahi (World Scientific, Singapore) 35-81.

[201] Frisch, U., She, Z. S. and Thual, O. (1986). Viscoelastic behaviour of cellular solutions to the Kuramoto-Sivashinsky model, *J. Fluid Mech.* **168**, 221-240.

[202] Fulford, G. D. (1964). The flow of liquids in thin films, *Adv. Chem. Engng.* **5**, 151-236.

[203] Funada, T. (1987). Nonlinear surface waves driven by the Marangoni instability in a heat transfer system, *J. Phys. Soc. Jpn.* **56**, 2031-2038.

[204] Funada, T. and Kotani, M. (1986). A numerical study of nonlinear diffusion equation governing surface deformation in the Marangoni convection, *J. Phys. Soc. Jpn.* **55**, 3857-3862.

[205] Garazo, A. N. and Velarde, M. G. (1991). Dissipative Korteweg-de Vries description of Marangoni-Benard convection, *Phys. Fluids A* **3**, 2295-2300.

[206] Garazo, A. N. and Velarde, M. G. (1992). Marangoni-driven solitary waves, in *ESA SP-333* (European Space Agency, Paris) 711-715.

[207] Garcia-Ybarra, P. L. and Velarde, M. G. (1987). Oscillatory Marangoni-Benard interfacial instability and capillary-gravity waves in single- and two-component liquid layers with or without Soret thermal diffusion, *Phys. Fluids* **30**, 1649-1655.

[208] Garcia-Ybarra, P. L., Castillo, J. L. and Velarde, M. G. (1987a). A nonlinear evolution equation for Benard-Marangoni convection with deformable boundary, *Phys. Lett. A* **122**, 107-110.

[209] Garcia-Ybarra, P. L., Castillo, J. L. and Velarde, M. G. (1987b). Benard-Marangoni convection with a deformable interface and poorly conducting boundaries, *Phys. Fluids* **30**, 2655-2661.

[210] Gardner, P. L., Greene, J. M., Kruskal, M. D. and Miura, R. M. (1967) Method for solving Korteweg-de Vries equation, *Phys. Rev. Lett.* **19**, 1095-1097.

[211] Gaver, D. P. and Grotberg, J. B. (1990). The dynamics of a localized surfactant on a thin film, *J. Fluid Mech.* **213**, 127-148.

[212] Gaver, D. P. and Grotberg, J. B. (1992). Droplet spreading on a thin viscous film, *J. Fluid Mech.* **235**, 399-414.

[213] Georis, Ph., Hennenberg, M., Lebon, G. and Legros, J. C. (1999). Investigation of thermocapillary convection in a three-liquid-layer system, *J. Fluid Mech.* **389**, 209-228.

[214] Georis, Ph. and Legros, J. C. (1996). Pure thermocapillary convection in a multilayer system: first results from the IML-2 mission, in *Materials and Fluids Under Low Gravity*, edited by L. Ratke, H. Walter, and B. Feuerbacher (Springer, Berlin) 299-311.

[215] Gershuni, G. Z. and Zhukhovitsky, E. M. (1976). *Convective Stability of Incompressible Fluids* (Keter, Jerusalem).

[216] Gershuni, G. Z. and Zhukhovitsky, E. M. (1982). Monotonic and oscillatory instabilities of a two-layer system of immiscible liquids heated from below, *Sov. Phys. Dokl.* **27**, 531-533.

[217] Gershuni, G. Z., Zhukhovitsky, E. M. and Nepomnyashchy, A. A. (1989). *Stability of Convective Motions* (Nauka, Moscow) (in Russian).

[218] Gertsberg, V. L. and Sivashinsky, G. I. (1981). Large cells in Rayleigh-Benard convection, *Prog. Theor. Phys.* **66**, 1219-1229.

[219] Gilev, A. Yu., Nepomnyashchy, A. A. and Simanovskii, I. B. (1986). Onset of the thermogravitational convection in a two-layer system in the presence of a surfactant on the interface, *Zh. Prikl. Mekh. Tekhn. Fiz.* **5**, 76-81 (in Russian).

[220] Gilev, A. Yu., Nepomnyashchy, A. A. and Simanovskii, I. B. (1987a). Onset of convection in two-layer system under a combined action of Rayleigh and thermocapillary mechanisms of instability, *Fluid Dyn.* **22**, 142-145.

[221] Gilev, A. Yu., Nepomnyashchy, A. A. and Simanovskii, I. B. (1987b). Generation of thermocapillary and thermogravitational convection in an air-water system, *Fluid Mech.-Sov. Res.* **16**, 44-48.

[222] Gjevik, B. (1970), Occurence of finite-amplitude surface waves on falling liquid films, *Phys. Fluids* **13**, 1918-1925.

[223] Golovin, A. A. (1992). Mass transfer under interfacial turbulence-kinetic regularities, *Chem. Engng. Sci.* **47**, 2069-2080.

[224] Golovin, A. A., Gupalo, Yu. P. and Ryazantsev, Yu. S. (1986). On chemothermocapillary effect of the motion of drop in fluid, *Dokl. Akad. Nauk. SSSR* **290**, 35-40 (in Russian).

[225] Golovin, A. A., Nepomnyashchy, A. A. and Pismen, L. M. (1994). Interaction between short-scale Marangoni convection and long-scale deformational instability, *Phys. Fluids* **6**, 34-48.

[226] Golovin, A. A., Nepomnyashchy, A. A. and Pismen, L. M. (1995). Pattern formation in large-scale Marangoni convection with deformable interface, *Physica D* **81**, 117-147.

[227] Golovin, A. A., Nepomnyashchy, A. A. and Pismen, L. M. (1997). Nonlinear evolution and secondary instabilities of Marangoni convection in a liquid-gas system with deformable interface, *J. Fluid Mech.* **341**, 317-341.

[228] Golovin, A. A. and Nepomnyashchy, A. A. (1999). Nonlinear Marangoni waves in fluid systems with two interfaces (preprint).

[229] Gonzalez, S. and Castellanos, A. (1994). Korteweg-de Vries-Burgers equation for surface waves in nonideal conducting liquids, *Phys. Rev. E* **49**, 2935-2940.

[230] Goodrich, F. C. (1981). The theory of capillary excess viscosities, *Proc. Royal Soc. London A* **374**, 341-370.

[231] Goodwin, R. T. and Schowalter, W. R. (1995). Arbitrarily oriented capillary-viscous planar jets in the presence of gravity, *Phys. Fluids* **7**, 954-963.

[232] Goussis, D. A. and Kelly, R. E. (1990). On the thermocapillary instabilities in a liquid layer heated from below, *Int. J. Heat Mass Transfer* **33**, 2237-2245.

[233] Grimshaw, R. (1986). Theory of solitary waves in shallow fluids, in *Encyclopedia of Fluid Mechanics*, edited by N. P. Cheremisinoff (Gulf, Houston), vol. **2**, 3-25.

[234] Gumerman, R. J. and Homsy, G. M. (1974). Convective instabilities in concurrent two-phase flow, *AIChEJ.* **20**, 981-1160; Part II. Global stability, *ibidem* **20**, 1161-1172.

[235] Gunaratne, G. H., Ouyang, Q. and Swinney, H. L. (1994). Pattern formation in the presence of symmetries, *Phys. Rev. E* **50**, 2802-2820.

[236] Gupalo, Yu. P., Ryazantsev, Yu. S. and Chalyuk, A. T. (1972). Diffusion to a drop at high Peclet numbers and finite Reynolds numbers, *Izv. Akad. Nauk SSSR, Mekh. Zhidk. Gaza* **7** (2), 333-334 (in Russian).

[237] Gupalo, Yu. P., Rednikov, A. Ye. and Ryazantsev, Yu. S. (1989). Thermocapillary drift of a drop with non-linear surface tension dependence on temperature, *Izv. Akad. Nauk SSSR, Mekh. Zhidk. Gaza* **53** (3), 433-442 (in Russian).

[238] Guyon, E., Hulin, J. P., Petit, L. and Mitescu, C. (2001). *Physical Hydrodynamics* (University Press, Oxford).

[239] Hadamard, J. S. (1911). Mouvement permanent lent d'une sphere liquide et visquese dans un liquide visqueux, *C. R. Hebd. Seances Acad. Sci.* (Paris) **152**, 1735-1738.

[240] Hahnel, H., Delitzsch, V. and Eckelmann, H. (1989). The motion of droplets in a vertical temperature gradient, *Phys Fluids A* **1**, 1460-1466.

[241] Haj-Hariri, H., Nadim, A. and Borhan, A. (1990). Effect of inertia on the thermocapillary velocity of a drop, *J. Colloid Interface Sci.* **140**, 277-286.

[242] Haj-Hariri, H., Shi, Q. and Borhan, A. (1997). Thermocapillary motion of deformable drops at finite Reynolds and Marangoni numbers, *Phys. Fluids* **9**, 845-855.

[243] Haken, H. (1983a). *Synergetics* (Springer-Verlag, Berlin).

[244] Haken, H. (1983b). *Advanced Synergetics* (Springer-Verlag, Berlin).

[245] Hammerschmid, P. (1987). Title, *Stahl und Eisen* **107**, 61-80.

[246] Happel, J. and Brenner, H. (1983). *Low Reynolds Number Hydrodynamics* (Martinus Nijhoff, Boston).

[247] Hardy, S. C. (1979). The motion of bubbles in a vertical temperature gradient, *J. Colloid Interface Sci.* **69**, 157-162.

[248] Hari, A. and Nepomnyashchy, A. A. (1994). Dynamics of curved domain boundaries in convection patterns, *Phys. Rev. E* **50**, 1661-1664.

[249] Hari, A. and Nepomnyashchy, A. A. (2000). Nonpotential effects in dynamics of fronts between convection patterns, *Phys. Rev. E* **61**, 4835-4847.

[250] Harkins, W. D. (1952). *The Physical Chemistry of Surface Films* (Reinhold, N. Y.).

[251] Harkins, W. D. and Feldman, A. (1922). Films. The spreading of liquids and the spreading coefficient, *J. Am. Chem. Soc.* **44**, 2665-2675.

[252] Heiple, C. R., Roper, J. R., Stagner, R. T. and Aden, R. J. (1983). Surface active element effects on the shape of GTA, laser, and electron beam welds, *Welding Res. Suppl.* **62**, 72-77.

[253] Hennenberg, M., Sørensen, T. S., Steinchen-Sanfeld, A. and Sanfeld, A. (1975). Stabilite mecanique et chimique d'une interface plane, *J. Chim. Phys.* (Paris) **72**, 1202-1210.

[254] Hennenberg, M., Bisch, P. M., Vignes-Adler, M. and Sanfeld, A. (1979), Interfacial instability and longitudinal waves in liquid-liquid systems, in *Instability of Fluid Interfaces*, edited by T. S. Sørensen (Springer-Verlag, Berlin) 229-259.

[255] Hennenberg, M., Chu, X.-L., Sanfeld, A. and Velarde, M. G. (1992). Transverse and longitudinal waves at the air-liquid interface in the presence of an adsorption barrier, *J. Colloid Interface Sci.* **150**, 7-21.

[256] Hershey, A. V. (1939), Ridges in a liquid surface due to the temperature dependence of surface tension, *Phys. Rev.* **56**, 204.

[257] Hocking, L. M. (1977). A moving fluid interface. II. The removal of the force singularity by a slip flow, *J. Fluid Mech.* **79**, 209-229.

[258] Hocking, L. M. (1990). Spreading and instability of a viscosity fluid sheet, *J. Fluid Mech.* **211**, 373-392.

[259] Homsy, G. M. (1974). Model equations for wavy viscous film flow, *Lect. Appl. Math.* **15**, 191-201.

[260] Hondros, E. D., McLean, M. and Mills, K. C. (Editors) (1998). *Marangoni and Interfacial Phenomena in Materials Processing* (Royal Society, London).

[261] Hornung, H. (1988). Regular and Mach reflection of shock waves, *Annu. Rev. Fluid Mech.* **18**, 33-58.

[262] Hoult, D. P. (1972). Oil spreading on the sea, *Annu. Rev. Fluid Mech.* **4**, 341-368.

[263] Hu, X. Z. and Vignes-Adler, M. (1991). Effect Marangoni de solute-couplage entre transfert interfacial et dissolution, *Ann. Chim. France* **16**, 267-273.

[264] Hu, Y., Ecke, R. E. and Ahlers, G. (1995). Transition to spiral-defect chaos in low Prandtl number convection, *Phys. Rev. Lett.* **74**, 391-394.

[265] Huang, G.-X., Velarde, M. G. and Kurdyumov, V. N. (1998). Cylindrical solitary waves and their interaction in Benard-Marangoni layers, *Phys. Rev. E* **57**, 5473-5482.

[266] Huang, G.-X., Velarde, M. G. and Makarov, V. A. (2001). Dark solitons and their head-on collisions in Bose-Einstein condensates, *Phys. Rev. A* **64**, 13617-13627.

[267] Huh, C. and Mason, S. G. (1977). Effects of surface roughness on wetting (theoretical), *J. Colloid Interface Sci.* **60**, 11-38.

[268] Huh, C. and Scriven, L. E. (1971). Hydrodynamic model of a steady movement of a solid/liquid/fluid contact line, *J. Colloid Interface Sci.* **35**, 85-101.

[269] Huppert, H. E. (1982). Flow and instability of a viscous gravity current down a slope, *Nature* (London) **300**, 427-429.

[270] Hyman, J. M. and Nicolaenko, B. (1986). The Kuramoto-Sivashinsky equation: A bridge between PDE's and dynamical systems, *Physica D* **18**, 113-126.

[271] Hyman, J. M., Nicolaenko, B. and Zaleski, S. (1986). Order and complexity in the Kuramoto-Sivashinsky model of weakly turbulent interfaces, *Physica D* **23**, 265-292.

[272] Hyman, J. M. and Nicolaenko, B. (1987). Coherence and chaos in the Kuramoto-Velarde equation, in *Partial Differential Equations*, edited by M. Crandall (Academic, N. Y.) 200-220.

[273] Ibanez, J. L. and Velarde, M. G. (1977). Hydrochemical stability of an interface between two inmiscible liquids: the role of Langmuir-Hinshelwood saturation law, *J. Phys. Lett.* (Paris) **38**, 1479-1483.

[274] Inomoto, O., Kai, S., Ariyoshi, T. and Inanaqa, S. (1997). Hydrodynamical effects of chemical waves in quasi-two-dimensional solution in Belousov-Zhabotinsky reaction, *Int. J. Bifurcation Chaos* **7**, 989-996.

[275] Janiaud, B., Pumir, A., Bensimon, D., Croquette, V., Richter, H. and Kramer, L. (1992). The Eckhaus instability for traveling waves, *Physica D* **55**, 269-310.

[276] Jensen, O. E. (1994). Self-similar, surfactant-driven flows, *Phys Fluids* **6**, 1084-1094.

[277] Jensen, O. E. (1995). The spreading of insoluble surfactant at the free surface of a deep fluid layer, *J. Fluid Mech.* **293**, 349-378.

[278] Jensen, O. E. and Grotberg, J. B. (1992). Insoluble surfactant spreading on a thin viscous film: shock evolution and film rupture, *J. Fluid Mech.* **240**, 259-288.

[279] Jensen, O. E. and Grotberg, J. B. (1993). The spreading of heat or soluble surfactant along a thin liquid film, *Phys. Fluids A* **5**, 58-68.

[280] Ji, W. and Setterwall, F. (1994). On the instabilities of vertical falling liquid films in the presence of surface-active solute, *J. Fluid Mech.* **278**, 297-323.

[281] Joanny, J. F. and de Gennes, P. G. (1984). A model for contact angle hysteresis, *J. Chem. Phys.* **81**, 552-562.

[282] Johnson, E. S. (1975). Liquid encapsulated float zone melting for GaAs, *J. Crystal Growth* **30**, 249-256.

[283] Johnson, R. S. (1972). Shallow water waves on a viscous fluid-The undular bore, *Phys. Fluids* **15**, 1693-1699.

[284] Joo, S. W., Davis, S. H. and Bankoff, S. G. (1991). Long wave instabilities of heated falling films: two-dimensional theory of uniform layers, *J. Fluid Mech.* **320**, 117-146.

[285] Joo, S. W. and Davis, S. H. (1992). Instabilities of three-dimanesional viscous falling films, *J. Fluid Mech.* **242**, 529-538.

[286] Joos, P. and Pintens, J. (1976). Spreading kinetics of liquids on liquids, *J. Colloid Interface Sci.* **60**, 507-513.

[287] Joseph, D. D. (1976). *Stability of Fluid Motions*, 2 vols., (Springer-Verlag, N. Y.).

[288] Juel, A., Burgess, J. M., McCormick, W. D., Swift, J. B. and Swinney, H. L. (2000). Surface tension-driven convection patterns in two liquid layers, *Physica D* **143**, 169-186.

[289] Kai, S., Ooishi, E. and Imasaki, M. (1985). Experimental study of nonlinear waves on interface between two liquid phases with chemical reaction *J. Phys. Soc. Jpn.* **54**, 1274-1281.

[290] Kai, S., Muller, S. C., Mori, T. and Miki, M. (1991). Chemically driven nonlinear waves and oscillations at an oil-water interface, *Physica D* **50**, 412-428.

[291] Kai, S. and Miike, H. (1994). Hydrochemical soliton due to thermocapillary instability in Belousov-Zhabotinsky reaction, *Physica A* **204**, 346-358.

[292] Kalinin, V. V. and Starov, V. M. (1986). The viscous spread of drops over a wetted surface, *Colloid J. USSR* **48** (5), 767-771.

[293] Kapitza, P. L. (1948). Wave flow of thin layer of viscous fluid, *Zh. Exper. Teor. Fiz.* **18**, 3-28 (in Russian).

[294] Kapitza, P. L. and Kapitza, S. P. (1949). Wave flow of thin fluid layers of liquid, *Zh. Eksp. Teor. Fiz* **19**, 105-120 (in Russian).

[295] Kaplun, S. (1957). Low Reynolds number flow past a circular cylinder, *J. Math. Mech* **6**, 595-603.

[296] Kaplun, S. and Lagerstrom, P. A. (1957). Asymptotic expansions of Navier-Stokes solutions for small Reynolds numbers, *J. Math. Mech.* **6**, 585-593.

[297] Kardar, M., Parisi, G. and Zhang, Y. C. (1986). Dynamic scaling of growing interfaces, *Phys. Rev. Lett.* **56**, 889-892.

[298] Kats-Demianets, V., Oron, A. and Nepomnyashchy, A. A. (1997a). Linear stability of a tri-layer fluid system driven by the thermocapillary effect, *Acta Astronaut.* **40**, 655-661.

[299] Kats-Demianets, V., Oron, A. and Nepomnyashchy, A. A. (1997b). Marangoni instability in tri-layer liquid systems, *Eur. J. Mech. B/Fluids* **16**, 49-74.

[300] Kawahara, T. (1983). Formation of saturated solutions in a nonlinear dispersive system with instability and dissipation, *Phys. Rev. Lett.* **51**, 381-383.

[301] Kawahara, T. and Toh, S. (1985). Nonlinear dispersive periodic waves in the presence of instability and damping, *Phys. Fluids* **28**, 1636-1638.

[302] Kawahara, T. and Toh, S. (1988). Pulse interactions in an unstable dissipative-dispersive nonlinear system, *Phys. Fluids* **31**, 2103-2111.

[303] Kazhdan, D., Shtilman, L., Golovin, A. A. and Pismen, L. M. (1995). Nonlinear waves and turbulence in Marangoni convection, *Phys. Fluids* **7**, 2679-2685.

[304] Kirdyashkin, A. G. (1984). Thermogravitational and thermocapillary flows in a horizontal liquid layer under the conditions of a horizontal temperature gradient, *Int. J. Heat Mass Transfer* **27**, 1205-1218.

[305] Kiseleva, O. A., Sobolev, V. D., Starov, V. M. and Churaev, N. V. (1979). Changes in viscosity of water close to quartz surface, *Colloid J. USSR* **41**, 192-197.

[306] Kivshar, Yu. S. and Malomed, B. A. (1989). Dynamics of solitons in nearly integrable systems, *Rev. Mod. Phys.* **61**, 763-915.

[307] Kliakhandler, I. L., Nepomnyashchy, A. A., Simanovskii, I. B. and Zaks, M. A. (1998). Nonlinear Marangoni waves in multilayer systems, *Phys. Rev. E* **58**, 5765-5775.

[308] Kliakhandler, I. L., Porubov, A. V. and Velarde, M. G. (2000). Localized finite-amplitude disturbances and selection of solitary waves, *Phys. Rev. E* **62**, 4959-4962.

[309] Knobloch, E. (1990). Pattern selection in long-wavelength convection, *Physica D* **41**, 450-479.

[310] Kolmogorov, A. N., Petrovsky, I. G. and Piscounoff, N. S. (1937). Etude de l'equation de la diffusion avec croissance de la quantite de matiere et son application a un probleme biologique, *Bull. Math. Univ. Moscou* (Ser. Int.) A **1**, 1-25.

[311] Korteweg, D. J. and de Vries, G. (1895). On the change of form of long waves advancing in a rectangular channel, and on a new type of long stationary waves, *Phil. Mag.* **39**, 422-443.

[312] Koschmieder, E. L. (1967). On convection under an air surface, *J. Fluid Mech.* **30**, 9-15.

[313] Koschmieder, E. L. (1991). The wavelength of supercritical surface tension driven Benard convection, *Eur. J. Mech. B/Fluids* **10**, 233-237.

[314] Koschmieder, E. L. (1993). *Benard cells and Taylor vortices* (University Press, Cambridge).

[315] Koschmieder, E. L. and Biggerstaff, M. I. (1986). Onset of surface-tension-driven Benard convection, *J. Fluid Mech.* **167**, 49-64.

[316] Koschmieder, E. L. and Switzer, D. W. (1992). The wavenumbers of supercritical surface-tension-driven Benard convection, *J. Fluid Mech.* **240**, 533-548.

[317] Koulago, A. E. and Parseghian, D. (1996). A propos d'une equation de la dynamique ondulatoire dans les films liquides, *J. Phys. III* (Paris) **5**, 309-312.

[318] Kozhoukharova, Zh. and Slavchev, S. (1986). Computer simulation of the thermocapilary convection in a non-cylindrical floating zone, *J. Crystal Growth* **74**, 236-246.

[319] Krehl, P. and van der Geest, M. (1991). The discovery of the Mach reflection effect and its demonstration in an auditorium, *Shock Waves* **1**, 3-15.

[320] Krehl, P. (2001). History of shock waves, in *Handbook of Shock Waves* (Academic, N. Y.), vol. **1**, 1-142.

[321] Kuhlmann, H. C. (1999). *Thermocapillary Convection in Models of Crystal Growth* (Springer, Berlin).

[322] Kuhlmann, H. C. and Rath, H. J. (1993). Hydrodynamic instabilities in cylindrical thermocapillary liquid bridges, *J. Fluid Mech.* **247**, 247-274.

[323] Kuramoto, Y. and Tsuzuki, T. (1976). Persistent propagation of concentration waves in dissipative media for from thermal equilibrium, *Prog. Theor. Phys.* **55**, 356-369.

[324] Kurdyumov, V. N., Rednikov, A. Ye. and Ryazantsev, Yu. S. (1994). Thermocapillary motion of a bubble with heat generation at the surface, *Microg. Q.* **4**, 5-8.

[325] Kuske, R. and Milewski, P. (1999). Modulated two-dimensional patterns in reaction-diffusion systems, *Eur. J. Appl. Math.* **10**, 157-184.

[326] Kuznetsov, E. A., Nepomnyashchy, A. A. and Pismen, L. M. (1995). The new amplitude equation for Boussinesq convection and nonequilateral hexagonal patterns, *Phys. Lett. A* **205**, 261-265.

[327] Kuznetsov, E. A. and Spector, M. D. (1980). Weakly supercritical convection, *J. Appl. Mech. Tech. Phys.* **21**, 220-228.

[328] Lagerstrom, P. A. and Cole, J. D. (1955). Examples illustrating expansion procedures for the Navier-Stokes equations, *J. Rat. Mech. Anal.* **4**, 817-882.

[329] Lagerstrom, P. A. and Casten, R. G. (1972). Basic concepts underlying singular perturbation techniques, *SIAM Rev.* **14**, 63-120.

[330] Lamb, H. (1945). *Hydrodynamics*, 6th ed. (Dover, N. Y.).

[331] Lamb, G. L. (1980). *Elements of Soliton Theory* (Wiley, N. Y.).

[332] Landau, L. D. and Lifshitz, E. M. (1980). *Statistical Physics* (Pergamon, Oxford).

[333] Landau, L. D. and Lifshitz, E. M. (1987). *Fluid Mechanics* (Pergamon, Oxford).

[334] Leger, L. and Joanny, J. F. (1992). Liquid spreading, *Rep. Prog. Phys.* **55**, 431-486.

[335] Legros, J. C. (1986). Problems related to non-linear variations of surface tension, *Acta Astronaut.* **13**, 697-703.

[336] Legros, J. C. and Georis, Ph. (2001). Oscillatory thermocapillary instability, (preprint).

[337] Levchenko, E. B. and Chernyakov, A. L. (1981). Instability of surface waves in a nonuniformly heated liquid, *Sov. Phys. JETP* **54**, 102-105.

[338] Levich, V. G. (1962). *Physicochemical Hydrodynamics* (Prentice-Hall, Englewood Cliffs., N. J.).

[339] Levich, V. G. (1981). The influence of surface-active substances on the motion of liquids, *Physicochem. Hydrodyn.* **2**, 85-94.

[340] Levich, V. G. and Krylov, V. S. (1969). Surface-tension-driven phenomena, *Annu. Rev. Fluid Mech.* **1**, 293-316.

[341] Libchaber, A. and Maurer, J. (1978). Local probe in a Rayleigh-Benard experiment in liquid helium, *J. Phys. Lett.* (Paris) **39**, L369-L372.

[342] Lifshitz, E. M. and Pitaevskii, L. P. (1981). *Physical Kinetics* (Pergamon, Oxford).

[343] Liggieri, L., Ravera, F. and Passerone, A. (1996). A diffusion-based approach to mixed adsorption kinetics, *J. Colloid Interface Sci.* **114**, 351-359.

[344] Lin, C. C. and Segel, L. A. (1974). *Mathematics Applied to Deterministic Problems in the Natural Sciences* (Macmillan, N.Y.).

[345] Lin, S. P. and Krishna, M. V. G. (1977). Stability of a liquid film with respect to initially finite three-dimensional disturbances, *Phys. Fluids* **20**, 2005-2011.

[346] Lin, S. P. and Wang, C. Y. (1985). Modeling wavy film flows, in *Encyclopedia of Fluid Mechanics*, edited by N. P. Cheremisinoff (Gulf, Houston), vol. **1**, 931-951.

[347] Linde, H., X.-L. Chu and Velarde, M. G. (1993a). Oblique and head-on collisions of solitary waves in Marangoni-Benard convection, *Phys. Fluids A* **5**, 1068-1070.

[348] Linde, H., Chu, X.-L. and Velarde, M. G. (1993b). Solitary waves driven by Marangoni stresses, *Adv. Space Res.* **13**, 109-117.

[349] Linde, H., X.-L. Chu, Velarde, M. G. and Waldhelm, W. (1993c). Wall reflections of solitary waves in Marangoni-Benard convection, *Phys. Fluids A* **5**, 3162-3166.

[350] Linde, H. and Loeschcke, K. (1966). Rollzellen und oszillation beim warmeubergang zwischen gas und flussigkeit, *Chem. Ing. Tech.* **39**, 65-74.

[351] Linde, H., Schwarz, P. and Wilke, H. (1979). Dissipative structures and nonlinear kinetics of the Marangoni instability, in *Dynamics and Instability of Fluid Interfaces*, edited by T. S. Sørensen, (Springer-Verlag, Berlin) 75-119.

[352] Linde, H., Velarde, M. G., Wierschem, A., Waldhelm, W., Loeschcke, K. and Rednikov, A. Y. (1997). Interfacial wave motions due to Marangoni instability. I. Traveling periodic wave trains in square and annular containers, *J. Colloid Interface Sci.* **188**, 16-26.

[353] Linde, H., Velarde, M. G., Waldhelm, W. and Wierschem, A. (2001a). Interfacial wave motions due to Marangoni instability. III. Solitary waves and (periodic) wave trains and their collisions and reflections leading to dynamic network (cellular) patterns in large containers, *J. Colloid Interface Sci.* **236**, 214-224.

[354] Linde, H., Velarde, M. G., Waldhelm, W., Loeschcke, K. and Wierschem, A. (2001b). Interfacial wave motions due to Marangoni instability. IV. Interaction between surface and internal waves leading to a change from anomalous to normal wave dispersion, *J. Colloid Interface Sci.* (submitted).

[355] Liu, J., Paul, D. and Gollub, J. P. (1993). Measurements of the primary instabilities of film flows, *J. Fluid Mech.* **250**, 69-78.

[356] Liu, J., Schneider, J. B. and Gollub, J. P. (1995). Three-dimensional instabilities of film flows, *Phys. Fluids* **7**, 55-67.

[357] Liu, J. and Gollub, J. P. (1993). Onset of spatially chaotic waves on flowing films, *Phys. Rev. Lett.* **70**, 2289-2291.

[358] Liu, J. and Gollub, J. P. (1994). Solitary wave dynamics of film flows, *Phys. Fluids* **6**, 1702-1712.

[359] Liu, Q. S. and Roux, B. (1992). Instability of thermocapillary convection in multiple superposed immiscible liquid layers, in *ESA SP-333* (European Space Agency, Paris) 735-740.

[360] Liu, Q. S., Roux, B. and Velarde M. G. (1998). Thermocapillary convection in two-layer systems, *Int. J. Heat Mass Transfer* **41**, 1499-1511.

[361] Lopez, J. C., Bankoff, S. G. and Miksis, M. J. (1996). Non-isothermal spreading of a thin liquid film on an inclined plane, *J. Fluid Mech.* **324**, 261-286.

[362] Lopez, J. C., Miller, C. A. and Ruckenstein, E. (1976). Spreading kinetics of liquid drops on solids, *J. Colloid Interface Sci.* **56**, 460-468.

[363] Lucassen J. (1968). Longitudinal capillary waves, Part 1, Theory, *Trans. Faraday Soc.* **64**, 2221-2229; Part 2, Experiments, *ibidem* **64**, 2230-2235.

[364] Lucassen-Reynders, E. H. and Lucassen, J. (1969). Properties of capillary waves, *Adv. Colloid Interface Sci.* **2**, 347-395.

[365] Ludtge, F. H. R. (1869). Ueber die Ausbreitung der Flussigkeiten aufeinander, *Ann. Phys. Chem. (Poggendorff)* **137**, 362-370.

[366] Lugt, H. J. (1983). *Vortex Flow in Nature and Technology* (Wiley, N. Y.).

[367] Lyubimov, D. V., Lyubimova, T. P., Meradji, S. and Roux, B. (1997). Vibrational control of crystal growth from liquid phase, *J. Crystal Growth* **180**, 648-659.

[368] Lyubimov, D. V. and Zaks, M. A. (1983). Two mechanisms of the transition to chaos in finite-dimensional models of convection, *Physica D* **9**, 52-64.

[369] Mach, E. and Wosyka, J. (1875). Uber einige mechanische Wirkungen des elektrischen Funkens, *Sitzungsber. Akad. Wiss. Wien* **72** (II), 44-52.

[370] Malomed, B. A. (1997). Stability and grain boundaries in the dispersive Newell-Whitehead-Segel equation, *Phys. Scripta* **57**, 115-117.

[371] Malomed, B. A., Nepomnyashchy, A. A. and Nuz, A. E. (1994). Nonequilibrium hexagonal patterns, *Physica D* **70**, 357-369.

[372] Malomed, B. A., Nepomnyashchy, A. A. and Tribelskii, M. I. (1989). Two-dimensional quasiperiodic structures in nonequilibrium systems, *Sov. Phys. JETP* **69**, 388-396.

[373] Malomed, B. A., Nepomnyashchy, A. A. and Tribelskii, M. I. (1990). Domain boundaries in convective patterns, *Phys. Rev. A* **42**, 7244-7263.

[374] Malomed, B. A. and Tribelskii, M. I. (1987). Stability of stationary periodic structures for weakly supercritical convection and in related problems, *Sov. Phys. JETP* **65**, 305-310.

[375] Manneville, P. (1983a). A two-dimensional model for three-dimensional convective patterns in wide containers, *J. Phys.* (Paris) **44**, 759-765.

[376] Manneville, P. (1983b) Towards an understanding of weak turbulence close to the convection threshold in large aspect ratio systems, *J. Phys. Lett.* (Paris) **44**, 903-916.

[377] Manneville, P. (1990). *Dissipative Structures and Weak Turbulence* (Academic, San Diego).

[378] Marangoni, C. G. M. (1865). *Sull'expansione dell goccie di un liquido galleggianti sulla superficie di altro liquido* (Tipografia Fusi, Pavia).

[379] Marangoni, C. G. M. (1871a). Ueber die Ausbreitung der Tropfen einer Flussigkeit auf der Oberflache einer anderen, *Ann. Phys. Chem. (Poggendorff)* **143**, 337-354.

[380] Marangoni, C. G. M. (1871b). Title, *Nuovo Cimento* **5-6**, 239-245.

[381] Marra, J. and Huethorst, J. A. M. (1991). Physical principles of Marangoni drying, *Langmuir* **7**, 2748-2755.

[382] Maxwell, J. C. (1871). *Theory of Heat* (Longmans, Green, and Co., London).

[383] Maxwell, J. C. (1878). In *Encyclopaedia Britannica*, 9th ed. (Adam and Clark, Edinburgh), vol. **5**, 56.

[384] Maxworthy, T. (1976). Experiments on collisions between solitary waves, *J. Fluid Mech.* **76**, 177-185.

[385] Melville, W. K. (1980). On the Mach reflection of a solitary wave, *J. Fluid Mech.* **98**, 285-297.

[386] Mendes-Tatsis, M. A. and Perez de Ortiz, E. S. (1992). Spontaneous interfacial convection in liquid-liquid binary systems under microgravity, *Proc. Royal Soc. London A* **438**, 389-396.

[387] Mendes-Tatsis, M. A. and Perez de Ortiz, E. S. (1996). Marangoni instabilities in systems with an interfacial chemical reaction, *Chem. Eng. Sci.* **51**, 3755-3761.

[388] Merrit, R. M. and Subramanian, R. S. (1988). The migration of isolated gas bubbles in a vertical temperature gradient, *J. Colloid Interface Sci.* **125**, 333-339.

[389] Metzger, J. and Schwabe, D. (1988). Coupled buoyant thermocapillary convection, *Physicochem. Hydrodyn.* **10**, 263-282.

[390] Mihaljan, J. M. (1962). A rigorous exposition of the Boussinesq approximation applicable to a film layer of fluid, *Astrophys. J.* **136**, 1126-1133.

[391] Miike, H., Yamamoto, H., Kai, S. and Muller, S.C. (1993). Accelerating chemical waves accompanied by traveling hydrodynamic motion and surface deformation, *Phys. Rev. E* **48**, R1627-R1630.

[392] Miles, J. W. (1976). Korteweg-de Vries equation modified by viscosity, *Phys. Fluids* **19**, 1063.

[393] Miles, J. W. (1977). Obliquely interacting solitary waves, *J. Fluid Mech.* **79**, 157-169.

[394] Miles, J. W. (1977). Resonantly interacting solitary waves, *J. Fluid Mech.* **79**, 171-179.

[395] Miles, J. W. (1980). Solitary waves, *Annu. Rev. Fluid Mech.* **12**, 11-43.

[396] Miller, C. A. and Neogi, P. (1985). *Interfacial Phenomena* (Marcel Dekker, N. Y.).

[397] Miller, R., Wustneck, R., Kragel, J. and Kretschmar, G. (1996). Dilational and shear rheology of adsorption layers at liquid interface, *Colloids & Surfaces A: Physicochem. & Engng. Aspects* **11**, 75-118.

[398] Mills, K. C. (2001) The effect of interfacial phenomemna on Materials Processing, in *Interfacial Phenomena and the Marangoni Effect*, edited by M. G. Velarde and R. Kh. Zeytounian (CISM and Springer, Udine and Wien) (in the press).

[399] Mirie, R. M. and Su, C. H. (1982). Collisions between two solitary waves. II. A numerical study, *J. Fluid Mech.*, **115**, 475-492.

[400] Misbah, Ch. and Balance, A. (1994). Secondary instabilities in the stabilised Kuramoto-Sivashinsky equation, *Phys. Rev. E* **49**, 166-183.

[401] Mityushev, P. V. and Krylov, V. S. (1986). Hydrodynamic instabilities caused by electric forces acting upon electrolyte solution surfaces, *Sov. Electrochem.* **22**, 520-524.

[402] Moffatt, H. K. (1964). Viscous and resistive eddies near sharp corner, *J. Fluid Mech.* **18**, 1-18.

[403] Moore, D. W. (1963). The boundary layer on a spherical gas bubble, *J. Fluid Mech.* **16**, 161-176.

[404] Moran, F. (1960). *Los tensores cartesianos rectangulares* (Instituto Geografico Catastral, Madrid).

[405] Mori, H. and Kuramoto, Y. (1998). *Dissipative Structures and Chaos* (Springer-Verlag, Berlin).

[406] Morris, S. W., Bodenschatz, E., Cannell, D. S. and Ahlers, G. (1993). Spiral defect chaos in large aspect ratio Rayleigh-Benard convection, *Phys. Rev. Lett.* **71**, 2026-2029.

[407] Muller, S. C., Plesser, T. and Hess, B (1984). Coupling of gly-colitic oscillations and convective processes, in *Temporal Order*, edited by L. Rensing and N. I. Jaeger (Springer-Verlag, Berlin) 194-196.

[408] Muller, S. C., Plesser, T. and Hess, B (1985a). Surface tension driven convection in chemical and biochemical solution layers, *Ber. Bunsenges. Phys. Chem.* **89**, 654-658.

[409] Muller, S. C., Plesser, T., Boiteux, A. and Hess, B (1985b). Pattern formation and Marangooni convection during oscillating glycolisis, *Z. Naturforsch.* **40c**, 588-591.

[410] Myers, D. (1999) *Surface, Interfaces and Colloids. Principles and Applications* (Wiley-VCH. N. Y.).

[411] Nakache, E., Dupeyrat, M. and Vignes-Adler, M. (1983). Ex-perimental and theoretical study of an interfacial instability ay some oil-water interfaces involving a surface-active agent: I. Physicochemical description and outlines for a theoretical ap-proach, *J. Colloid Interface Sci.* **94**, 187-200.

[412] Nakache, E., Dupeyrat, M. and Vignes-Adler, M. (1984). The contribution of chemistry to new Marangoni mass-transfer in-stabilities at the oil/water interface, *Faraday Discuss. Chem. Soc.* **77**,189-196.

[413] Napolitano, L. G. (1978). Thermodynamics and dynamics of pure interfaces, *Acta Astronaut.* **5**, 655-670.

[414] Napolitano, L. G., Golia, C. and Viviani, A. (1984). Numerical simulation of unsteady thermal Marangoni flows, in *ESA-SP-222* (European Space Agency, Paris) 251-258.

[415] Nas, S. and Tryggvason, G. (1993). Computational investiga-tion of thermal migration of bubbles and drops, in *Fluid Me-chanics Phenomena in Microgravity* (American Society Mechan-ical Engineers, N.Y.).

[416] Nekorkin, V. I. and Velarde, M. G. (1994). Solitary waves of a dissipative Korteweg-de Vries equation describing Marangoni-Benard convection and other thermoconvective instabilities, *Int. J. Bifurcation Chaos* **4**, 1135-1146.

[417] Nepomnyashchy, A. A. (1974a). Stability of the wavy regimes in the film flowing down an inclined plane with respect to three-dimensional disturbances, in *Fluid Dynamics* (Perm State Univ., Perm) **316**, 91-104 (in Russian).

[418] Nepomnyashchy, A. A. (1974b). Stability of the wavy regimes in the film flowing down an inclined plane, *Fluid Dynam.* **9**, 354-359.

[419] Nepomnyashchy, A. A. (1976). Wavy motions in the layer of viscous fluid flowing down the inclined plane, in *Fluid Dynamics*, (Perm State Univ., Perm) **362**, 114-124 (in Russian).

[420] Nepomnyashchy, A. A. (1995). Order parameter equations for long-wavelength instabilities, *Physica D* **86**, 90-95.

[421] Nepomnyashchy, A. A. and Simanovskii, I. B. (1983a). Thermocapillary convection in a two-layer system, *Sov. Phys. Doklady* **28**, 838-839.

[422] Nepomnyashchy, A. A. and Simanovskii, I. B. (1983b). Thermocapillary convection in a two-layer system, *Fluid Dynam.* **18**, 629-633.

[423] Nepomnyashchy, A. A. and Simanovskii, I. B. (1984). Thermocapillary and thermogravitational convection in a two-layer system with a distorted interface, *Fluid Dynam.* **19**, 494-499.

[424] Nepomnyashchy, A. A. and Simanovskii, I. B. (1986). Thermocapillary convection in two-layer systems in the presence of surfactant on the interface, *Fluid Dynam.* **21**, 469-471.

[425] Nepomnyashchy, A. A. and Simanovskii, I. B. (1988). Generation of the thermocapillary convection in a two-layer system with the soluble surfactant, *Fluid Dynam.* **23**, 302-306.

[426] Nepomnyashchy, A. A. and Simanovskii, I. B. (1989a). Generation of the thermogravitational convection with the soluble surfactant on the interface, *Zh. Prikl. Mekh. Tekhn. Fiz.* **1**, 146-149 (in Russian).

[427] Nepomnyashchy, A. A. and Simanovskii, I. B. (1990). Generation of the thermocapillary convection in a two-layer system

with heat release at the interface, *Zh. Prikl. Mekh. Tekhn. Fiz.* **1**, 69-72 (in Russian).

[428] Nepomnyashchy, A. A. Simanovskii, I. B. and Braverman, L. M. (2001). Stability of thermocapillary flows with inclined temperature gradient (preprint).

[429] Nepomnyashchy, A. A. and Velarde, M. G. (1994). A three-dimensional description of solitary waves and their interaction in Marangoni-Benard layers, *Phys. Fluids* **6**, 187-198.

[430] Neuhaus, D., and Feuerbacher, B. (1987). Bubble motions induced by temperature gradient, in *ESA SP-256* (European Space Agency, Paris) 118-121.

[431] Newell, A. C. and Pomeau, Y. (1993). Turbulent crystals in macroscopic systems, *J. Phys. A* **26**, L429-L434.

[432] Newell, A. C. and Redekopp, L. G. (1977). Breakdown of Zakharov-Shabat theory and soliton creation, *Phys. Rev. Lett.* **38**, 377-380.

[433] Newell, A. C. and Whitehead, J. A. (1969). Finite bandwidth, finite amplitude convection, *J. Fluid Mech.* **38**, 279-303.

[434] Nield, D. A. (1964), Surface tension and buoyancy effects in cellular convection, *J. Fluid Mech.* **19**, 341-352.

[435] Nield, D. A. (1966). Streamlines in Benard convection cells induced by surface tension and buoyancy, *ZAMP* **17**, 226-232.

[436] Normand, C., Pomeau, Y. and Velarde, M.G. (1977). Convective instability: A Physicist's Approach, *Rev. Mod. Phys.* **49**, 581-624.

[437] Nuz, A. E., Nepomnyashchy, A. A., Golovin, A. A., Hari, A. A. and Pismen, L. M. (2000). Stability of rolls and hexagonal patterns in non-potential systems, *Physica D* **135**, 233-262.

[438] Oberbeck, A. (1879). Uber die Warmeleitung der Flussigkeiten bei Berucksichtigung der Stromungen infolge von Temperaturdifferenzen, *Ann. Phys. Chem.* **7**, 271-292.

[439] Ochiai, J., Kuwahara, K., Morioka, M., Enya, S., Sezaki, K., Maekawa, T. and Tanasawa, I. (1984), Experimental study on Marangoni convection, in *ESA-SP-222* (European Space Agency, Paris) 291-293.

[440] Ondarçuhu, T., Millan-Perez, J., Mancini, H. L., Garamartin, A. and Perez-Garcia, C. (1993a). Benard-Marangoni convective patterns in small cylindrical layers, *Phys. Rev. E* **48**, 1051-1057.

[441] Ondarçuhu, T., Mindlin, G., Mancini, H. L. and Perez-Garcia, C. (1993b). Dynamical patterns in Benard-Marangoni convection in a square container, *Phys. Rev. Lett.* **70**, 3892-3895.

[442] Or, A. C., Kelly, R. E., Cortelezzi, L. and Speyer, J. L. (1999). Control of long-wavelength Marangoni-Benard convection, *J. Fluid Mech.* **387**, 321-341.

[443] Orell, A. and Westwater, J. W. (1962). Spontaneous interfacial cellular convection accompanying mass transfer: ethylene glycol - acetic acid - ethyl acetate, *AIChE J.* **8**, 350-356.

[444] Oron, A. Davis, S. H. and Bankoff, G. (1997). Long-scale evolution of thin liquid films, *Rev. Mod. Phys.* **69**, 931-980.

[445] Oron, A. and Rosenau, Ph. (1997). Evolution and formation of dispersive-dissipative patterns, *Phys. Rev. E* **55**, R1267-R1270.

[446] Oseen, C. W. (1910). Uber die Stokes'sche Formel, und uber eine verwendte Aufgabe in der Hydrodynamik, *Ark. Math. Astronom. Fys.* **6**, No. 29, 1-20.

[447] Oseen, C. W. (1913). Uber den Gultigkeitsbereich der stokesschen widerstandsformel, *Ark. Math. Astronom. Fys.* **9**, No. 16, 1-15.

[448] Ostrach, S. (1977). Motion induced by capillarity, in *Physicochemical Hydrodynamics*, edited by B. Spalding (Advance Publ., London) vol. 2, 571-589.

[449] Ostrach, S. (1982). Low-gravity fluid flows, *Annu. Rev. Fluid Mech.* **14**, 313-345.

[450] Ostrach, S. (1983). Fluid Mechanics in Crystal Growth. The 1982 Freeman Scholar Lecture, *J. Fluids Engng.* **105**, 5-30.

[451] Palacios, J. (1964). *Analisis Dimensional* (Espasa-Calpe, Madrid).

[452] Palm, E. (1960). On the tendency towards hexagonal cells in steady convection, *J. Fluid Mech.* **19**, 183-192.

[453] Palmer, H. J. and Berg, J. C. (1971). Convective instability in liquid pools heated from below, *J. Fluid Mech.* **47**, 779-787.

[454] Palmer, H. J. and Berg, J. C. (1972). Hydrodynamic stability of surfactant solutions heated from below, *J. Fluid Mech.* **51**, 385-402.

[455] Palmer, H. J. and Maheshri, J. C. (1981). Enhanced interfacial heat transfer by differential vapor recoil instabilities, *Int. J. Heat Mass Transfer* **24**, 117-123.

[456] Paniconi, M. and Elder, K. R. (1997). Stationary, dynamical and chaotic states of the two-dimensional damped Kuramoto-Sivashinsky equation, *Phys. Rev. E* **56**, 2713-2721.

[457] Parmentier, P. M., Regnier, V. C. and Lebon, G. (1993). Buoyant-thermocapillary instabilities in medium-Prandtl-number fluid layers subject to a horizontal temperature gradient, *Int. J. Heat Mass Transfer* **36**, 2417-2427.

[458] Parmentier, P. M., Regnier, V. C. Lebon, G. and Legros, J. C. (1996). Nonlinear analysis of coupled gravitational and capillary thermoconvection in thin fluid layers, *Phys. Rev. E* **54**, 411-423.

[459] Pearson, J. R. A. (1958). On convection cells induced by surface tension, *J. Fluid Mech.* **4**, 489-500.

[460] Perez-Cordon, R. and Velarde, M. G. (1975). On the (nonlinear) foundations of Boussinesq approximation applicable to a thin layer of fluid, *J. Phys.* (Paris) **36**, 591-601.

[461] Perez de Ortiz, E. S. (1992). Marangoni phenomena, in *Science and Practice of Liquid-Liquid Extraction*, edited by J. D. Thornton (Clarendon, Oxford), vol. **1**, 157-209.

[462] Perry R. H. and Chelton, C. H. (1993). *Chemical Engineers' Handbook* (McGraw Hill, N. Y.).

[463] Pesch, W. (1996). Complex spatio-temporal convection patterns, *Chaos* **6**, 348-357.

[464] Petre, G. and Azouni, M. (1984). Experimental Evidence for the Minimum of Surface Tension with Temperature at Aqueous Solutions/Air Interfaces, *J. Colloid Interface Sci.* **98**, 261-263.

[465] Petre, G., Azouni, M. A. and Tshinyama, K. (1993) Marangoni Convection at Alcohol Aqueous Solutions-Air Interfaces. *Appl. Sci. Res.* **50**, 97-106.

[466] Petviashvili, V. I. and Tsvelodub, O. Y. (1978). Horse-shoe shaped solitons on a flowing viscous film of fluid, *Sov. Phys. Dokl.* **23**, 117-118.

[467] Pismen, L. M. (1980). Pattern selection at the bifurcation point, *J. Chem Phys.* **72**, 1900-1907.

[468] Pismen, L. M. (1997). Interaction of reaction-diffusion fronts and Marangoni flow on the interface of a deep fluid, *Phys. Rev. Lett.* **78**, 382-385.

[469] Pismen, L. M. (1999). *Vortices in Nonlinear Fields* (Clarendon, Oxford).

[470] Pismen, L. M. and Nepomnyashchy, A. A. (1994). Propagation of hexagonal patterns, *Europhys. Lett.* **27**, 433-436.

[471] Pismen, L. M. and Pomeau, Y. (2000). Disjoining potential and spreading of thin liquid layers in the diffuse-interface model coupled to hydrodynamics, *Phys. Rev. E* **62**, 2480-2492.

[472] Plateau, J. (1873) *Statique Experimentale et Theorique des Liquides Soumis aux Seules Forces Moleculaires*, vol. **1**, (Gauthier-Villars, Paris).

[473] Pocheau, A. Croquette, V. and Le Gal, P. (1985). Turbulence in a cylindrical container of argon near threshold of convection, *Phys. Rev. Lett.* **55**, 1094-1097.

[474] Polezhaev, V. I., Dubovik, K. G., Nikitin, S. A., Prostomolotov, A. I. and Fedyushkin, A. I. (1981). Convection during crystal growth on earth and in space, *J. Crystal Growth* **52**, 465-470.

[475] Pomeau, Y. (1986). Front motion, metastability and subcritical bifurcations in hydrodynamics, *Physica D* **23**, 3-11.

[476] Pomeau, Y. and Vannimenus, J. (1985). Contact angle on heterogeneous surfaces: weak heterogeneities, *J. Colloid Interface Sci.* **104**, 477-488.

[477] Pomeau, Y., Zaleski, S. and Manneville, P. (1983). Dislocation motion in cellular structures, *Phys. Rev. A.* **27**, 2710-2726.

[478] Pontes, J., Christov, C. I. and Velarde, M. G. (1996). Numerical study of patterns and their evolution in finite geometries, *Int. J. Bifurcation Chaos* **6**, 1883-1890.

[479] Pontes, J., Christov, C. I. and Velarde, M. G. (1999). Numerical approach to pattern selection in a model problem for Benard convection in finite fluid layer, *Ann. Univ. Sofia* **93**, 157-175.

[480] Powell, J. A., Newell, A. C. and Jones, C. K. R. T. (1991). Competition between generic and nongeneric fronts in envelope equations, *Phys. Rev. A* **44**, 3636-3652.

[481] Prakash, A. and Koster, J. N. (1993). Natural and thermocapillary convection in three layers, *Eur. J. Mech. B/Fluids* **12**, 635-655.

[482] Priede, J. and Gerbeth, G. (1997). Convective, absolute and global instabilities of thermocapillary-buoyancy convection in extended layers, *Phys. Rev. E* **56**, 4187-4199.

[483] Priede, J. and Gerbeth, G. (2000). Hydrothermal wave instability of thermocapillary-driven convection, in a transverse magnetic field, *J. Fluid Mech.* **404**, 211-250.

[484] Proudman, L. and Pearson, J. R. A. (1957). Expansion at small Reynolds numbers for flow past a sphere and a circular cylinder, *J. Fluid Mech.* **2**, 237-262.

[485] Pshenichnikov, A. F. and Tokmenina, G. A. (1983). Deformation of the free surface of a liquid by thermocapillary motion, *Fluid Dynam.* **18**, 463-465.

[486] Pukhnachev, V. V. (1987). *Thermocapillary Convection at Reduced Gravitation* (Institute of Hydrodynamics SB AS USSR, Novosibirsk).

[487] Pumir, A. Manneville, P. and Pomeau, Y. (1983). On solitary waves running down an inclined plane, *J. Fluid Mech.* **135**, 27-34.

[488] Rabinovich, L. M., Vyazmin, A. V. and Buyevich, Yu. A. (1995). Chemo-Marangoni convection. III. Pattern parameters: interface mass transfer, *J. Colloid Interface Sci.* **173**, 1-7.

[489] Rabinovich, M. I. and Tsimring, L. S. (1994). Dynamics of dislocations in hexagonal patterns, *Phys. Rev. E* **49**, R35-38.

[490] Raphaël, E. and de Gennes, P. G. (1989). Dynamics of wetting with non ideal surfaces. The single defect problem, *J. Chem Phys.* **90**, 7577-7584.

[491] Rasenat, S., Busse, F. H. and Rehberg, I. (1989). A theoretical and experimental study of double-layer convection, *J. Fluid Mech.* **199**, 519-540.

[492] Ratke, L., Walter, H. and Feuerbacher, B. (Editors) (1989). *Materials and Fluids under Low Gravity* (Springer-Verlag, Berlin).

[493] Ravera, F., Liggieri, L. and Steinchen, A. (1993). Sorption kinetics considered as a renormalized diffusion process, *J. Colloid Interface Sci.* **156**, 109-116.

[494] Ravera, F., Liggieri, L., Passerone, A. and Steinchen, A. (1994) Sorption kinetics at liquid-liquid interfaces with the surface-active component soluble in both phases, *J. Colloid Interface Sci.* **163**, 309-314.

[495] Rayleigh, Lord. (1876). On waves, *Phil. Mag.* **1**, 257-279.

[496] Rayleigh, Lord. (1916). On convection currents in a horizontal layer of fluid, when the higher temperature is on the under side, *Phil. Mag.* **32**, 529-536.

[497] Ready, J. F. (1965). Effects due to absorption of laser radiation, *J. Appl. Phys.* **36**, 462-468.

[498] Rednikov, A. Ye., Ryazantsev, Yu. S. and Velarde, M. G. (1994a). Drop motion with surfactant transfer in a homogeneous surrounding, *Phys. Fluids* **6**, 451-468.

[499] Rednikov, A. Ye., Ryazantsev, Yu. S. and Velarde, M. G. (1994b). On the development of translational subcritical Marangoni instability for a drop with uniform internal heat generation, *J. Colloid Interface Sci.* **164**, 168-180.

[500] Rednikov, A. Ye., Ryazantsev, Yu. S. and Velarde, M. G. (1994c). Drop motion with surfactant transfer in an inhomogeneous medium, *Int. J. Heat Mass Transfer* **37**, Suppl. 1, 361-374.

[501] Rednikov, A. Ye., Ryazantsev, Yu. S. and Velarde, M. G. (1994d). Active drops and drop motions due to non-equilibrium phenomena, *J. Non-Equilib. Thermodyn.* **19**, 95-113.

[502] Rednikov, A. Ye., Ryazantsev, Yu. S. and Velarde, M. G. (1994e). Drop motion and the Marangoni effect. Interaction of modes, *Phys. Scripta* T **55**, 115-118.

[503] Rednikov, A. Ye., Kurdyumov, V. N., Ryazantsev, Yu. S. and Velarde, M. G. (1995a). The role of time-varying gravity on the motion of a drop induced by Marangoni instability, *Phys. Fluids* **7**, 2670-2678.

[504] Rednikov, A. Ye., Colinet, P., Velarde, M. G. and Legros, J. C. (1998). Two-layer Benard-Marangoni instability and the limit of transverse and longitudinal waves, *Phys. Rev. E* **57**, 2872-2884.

[505] Rednikov, A. Ye., Colinet, P., Velarde, M. G. and Legros, J. C. (2000a). Oscillatory instability and high-frequency wave modes in a Marangoni-Benard layer with deformable free surface, *J. Non-Equilib. Thermodyn.* **25**, 381-405.

[506] Rednikov, A. Ye., Colinet, P., Velarde, M. G. and Legros, J. C. (2000b). Rayleigh-Marangoni oscillatory instability in a horizontal liquid layer heated from above: coupling and mode-mixing of internal and surface dilational waves, *J. Fluid Mech.* **405**, 57-77.

[507] Rednikov, A. Ye., Velarde, M. G., Ryazantsev, Yu. S., Nepomnyashchy, A. A. and Kurdyumov, V. N. (1995b). Cnoidal wave trains and solitary waves in a dissipation-modified Korteweg-de Vries equation, *Acta Appl. Math.* **39**, 457-475.

[508] Redon, C., Brochard-Wyart, F., Hervet, H. and Rondelez, F (1992). Spreading of Heavy Droplets. II. Experiments, *J. Colloid Interface Sci.* **149**, 580-591.

[509] Regel, L. A. (1987). *Materials Science in Space* (Halsted, N.Y.).

[510] Reichenbach, J. and Linde, H. (1981). Linear perturbation analysis of surface-tension-driven convection at a plane interface (Marangoni instability), *J. Colloid Interface Sci.* **84**, 433-443.

[511] Renouard, D. P., Seabra-Santos, F. J. and Temperville, A. M. (1985). Experimental study of the generation, damping, and reflection of a solitary wave, *Dyn. Atmosph. Oceans* **9**, 341-358.

[512] Riley, R. J. and Neitzel, G. P. (1998). Instability of thermocapillary-buoyancy convection in shallow layers, Part 1, Characterization of steady and oscillatory instabilities, *J. Fluid Mech.* **359**, 143-164.

[513] Rodriguez-Bernal, A. (1992). Initial value problem and asymptotic low dimensional behavior in the Kuramoto-Velarde equation, *Nonl. Anal. Theory Meth. Appl.* **19**, 643-685.

[514] Rosner, D. E. (1986). *Transport Processes in Chemically Reacting Flow Systems* (Butterworths, Boston).

[515] Ross, S. and Becher, P. (1992). The history of the spreading coefficient, *J. Colloid Interface Sci.* **149**, 575-579.

[516] Rotermund, H. H., Jakubitch, S., von Oertzen, A. and Ertl, G. (1991). Solitons in a surface reaction, *Phys. Rev. Lett.* **66**, 3083-3086.

[517] Ruckenstein, E. and Berbente, C. (1964). The occurrence of interfacial turbulence in the case of diffusion accompanied by chemical reaction, *Chem. Engng. Sci.* **19**, 329-347.

[518] Ruckenstein, E., Smigelschi, O. and Suciu, D. G. (1970). A steady dissolving drop method for studying the pure Marangoni effect, *Chem. Engng. Sci.* **25**, 1249-1254.

[519] Russell, J. S. (1844). Report on waves, in *Rep. 14th Meet. British Ass. Adv. Sci.*, York, 311-390 (J. Murray, London). Reprinted as *Appendix* in Russell (1885).

[520] Russell, J. S. (1885). *The Wave of Translation in the Oceans of Water, Air and Ether* (Trubner, London).

[521] Ruyer-Quil, C. and Manneville, P. (2000). Transition to turbulence of fluid flowing down on inclined plane: modeling and simulation, in *Advances in Turbulence*, edited by U. Frisch (Kluwer, Dordrecht) 93-96.

[522] Ryazantsev, Yu. S. (1985). Thermocapillary motion of a reacting droplet in a chemically active medium, *Izv. Akad. Nauk SSSR, Mekh. Zhidk. Gaza* **3**, 180-183 (in Russian). English translation: *Fluid Dynam.* **20**, 491-495.

[523] Rybczynski, W. (1911). Uber die fortschreitende Bewegung einer Flussingen Kugel in einem zahen Medium, *Bull. Int. Acad. Pol. Lett. Cl. Sci. Math. Nat. Sci. Cracovie* **A**, 40-46.

[524] Rybicki, A. and Florian, J. M. (1987). Thermocapillary effects in liquid bridges, *Phys. Fluids* **30**, 1956-1962.

[525] Sadhal, S. S. and Johnson, R. E. (1983). Stokes flow past drops and bubbles coated with thin films. Part 1: Stagnant cap of surfactant film - exact solution, *J. Fluid Mech.* **126**, 237-250.

[526] Sadhal, S. S., Ayyaswamy, P. S. and Chung, J. N. (1997). *Transport Phenomena with Drops and Bubbles* (Springer-Verlag, N. Y.).

[527] Salamon, T. R., Armstrong, R. C. and Brown, R. A. (1994). Traveling waves on vertical films: Numerical analysis using the finite element method, *Phys. Fluids* **6**, 2202-2220.

[528] Sander, J. and Hutter, K. (1991). On the development of the theory of the solitary wave. A historical essay, *Acta Mech.* **86**, 111-152.

[529] Sanfeld, A. and Steinchen, A. (1975). Coupling between a transconformation surface reaction and hydrodynamic motion, *Biophys. Chem.* **3**, 99-106.

[530] Sanfeld, A., Steinchen, A., Hennenberg, M., Bisch, P. M., Van Lamswerde-Gallez, D. and Dalle-Vedove, W. (1979). Mechanical, chemical, and electrical constraints and hydrodynamic interfacial instability, in *Dynamics and Instability of Fluid Interfaces*, edited by T. S. Sørensen (Springer-Verlag, Berlin) 168-204.

[531] Sanfeld, A. and Steinchen, A. (1984). Motion induced by surface-chemical and electrochemical kinetics, *Faraday Discuss. Chem. Soc.* **77**, 1-11.

[532] Santiago-Rosanne, M., Vignes-Adler, M. and Velarde, M. G. (1997). Dissolution of a drop on a liquid surface leading to surface waves and interfacial turbulence, *J. Colloid Interface Sci.* **191**, 65-80.

[533] Santiago-Rosanne, M., Vignes-Adler, M. and Velarde, M. G. (2001). On the spreading of partially miscible liquids, *J. Colloid Interface Sci.* **234**, 375-383.

[534] Sawistowski, H. (1971). Interfacial phenomena, in *Recent Advances in Liquid-Liquid Extraction*, edited by C. Hanson (Pergamon, Oxford) 293-365.

[535] Sawistowski, H. (1981). Interfacial convection, *Ber. Bunsenges. Phys. Chem.* **85**, 905-909.

[536] Scanlon, J. W. and Segel, L. A. (1967). Finite amplitude cellular convection induced by surface tension, *J. Fluid Mech.* **29**, 149-162.

[537] Schatz, M. F. and Neitzel, G. P. (2001). Experiments on thermocapillary instabilities, *Annu. Rev. Fluid Mech.* **33**, 93-127.

[538] Schatz, M. F., VanHook, S. J., McCormick, W. D., Swift, J. B. and Swinney, H. L. (1995). Onset of surface-tension-driven Benard convection, *Phys. Rev. Lett.* **75**, 1938-1941.

[539] Schatz, M. F., VanHook, S. J., McCormick, W. D., Swift, J. B. and Swinney, H. L. (1999). Time-independent square patterns in surface-tension-driven Benard convection, *Phys. Fluids* **11**, 2577-2582.

[540] Schrader, M. E. and Loeb, G. I. (Editors) (1992). *Modern Approaches to Wettability* (Plenum, N. Y.).

[541] Schwabe, D., Moller, U., Schneider, J. and Scharmann, A. (1992). Instabilities of shallow dynamic thermocapillary liquid layers, *Phys. Fluids A* **4**, 2368-2381.

[542] Schwarz, E. (1970). On the occurrence of Marangoni instability, *Warme - und Stoffubertragung* **3**, 131-133.

[543] Scott, A. C., Chu, F. Y. F. and McLaughlin, D. W. (1973). The soliton: A new concept in applied science, *Proc. IEEE* **61**, 1443-1483.

[544] Scriven, L. E. (1960). Dynamics of a fluid interface. Equation of motion for Newtonian surface fluids, *Chem. Eng. Sci.* **12**, 98-108.

[545] Scriven, L. E. and Sternling, C. V. (1960). The Marangoni effects, *Nature* (London) **187**, 186-188.

[546] Scriven, L. E. and Sternling, C. V. (1964). On cellular convection driven by surface-tension gradients: effects of mean surface tension and surface viscosity, *J. Fluid Mech.* **19**, 321-340.

[547] Segel, L. A. (1969). Distant sidewalls cause slow amplitude modulation of cellular convection, *J. Fluid Mech.* **38**, 203-224.

[548] Segel, L. A. (1972). Simplification and scaling, *SIAM Rev.* **14**, 547-570.

[549] Segel, L. A. (1977). *Mathematics Applied to Continuum Mechanics* (Macmillan, N. Y.).

[550] Sen, A. K. and Davis, S. H. (1982). Steady thermocapillary flows in two-dimensional slots, *J. Fluid Mech.* **121**, 163-186.

[551] Seppecher, P. (1996). Moving contact lines in the Cahn-Hilliard theory, *Int. J. Engng. Sci.* **34**, 977-992.

[552] Series, R. W. and Hurle, D. T. J. (1991). The use of magnetic fields in semiconductor crystal growth, *J. Crystal Growth* **113**, 305-328.

[553] Shankar, N. and Subramanian, R. S. (1988). The Stokes motion of a gas bubble due to interfacial tension gradients at low to moderate Marangoni numbers, *J. Colloid Interface Sci.* **123**, 512-522.

[554] Shen, S. S. (1993). *Course on Nonlinear Waves* (Kluwer, Dordrecht).

[555] Sherwood, T. S. and Wei, J. C. (1957). Interfacial phenomena in liquid extraction, *Ind. Engng. Chem.* **49**, 1030-1034.

[556] Shevtsova, V. M., Kuhlmann, H. C. and Rath, H. J. (1996). Thermocapillary convection in liquid bridges with a deformed free surface, in *Materials and Fluids under Low Gravity*, edited by L. Ratke, H. Walter and B. Feuerbacher (Springer, Berlin) 323-330.

[557] Shikhmurzaev, Yu. D. (1994). Mathematical modeling of wetting hydrodynamics, *Fluid Dyn. Res.* **13**, 45-64.

[558] Shikhmurzaev, Yu. D. (1997). Moving contact lines in liquid/liquid/solid systems, *J. Fluid Mech.* **334**, 211-249.

[559] Shkadov, V. Ya. (1967). Wave conditions in the flow on thin layer of a viscous liquid under the action of gravity, *Izv. Akad. Nauk SSSR, Mekh. Zhidk. Gaza* **1**, 43-51 (in Russian).

[560] Shkadov, V. Ya. (1968). Theory of wave flows of a thin film of viscous liquid, *Izv. Akad. Nauk SSSR, Mekh. Zhidk. Gaza* **2**, 20-25 (in Russian).

[561] Shkadov, V. Ya. (1977). Solitary waves in layer of viscous liquid. *Izv. Akad. Nauk SSSR, Mech. Zhidk. Gaza* **1**, 63-66 (in Russian).

[562] Shkadov, V. Ya. and Sisoev, G. M. (2000). Wavy falling liquid films; theory and computation instead of physical experiment, in *Nonlinear Waves in Multi-Phase Flow*, edited by H.-C. Chang (Kluwer, Dordrecht) 1-10.

[563] Shkadov, V. Ya., Velarde, M. G. and Shkadova, V. P. (2001). Falling films and the Marangoni effect, (preprint).

[564] Shklyaev O. (2001) (private communication).

[565] Shtilman, L. and Sivashinsky, G. (1991). Hexagonal structure of large-scale Marangoni convection, *Physica D* **52**, 477-488.

[566] Siggia, E. D. and Zippelius, A. (1981). Dynamics of defects in Rayleigh-Benard convection, *Phys. Rev. A* **24**, 1036-1049.

[567] Simanovskii, I., Georis, Ph., Hennenberg, M., Van Vaerenberg, S., Wertgeim, I. and Legros, J. C. (1992). Numerical investigation of Marangoni-Benard instability in multilayer systems, in *ESA SP-333* (European Space Agency, Paris) 729-740.

[568] Simanovskii, I. B. and Nepomnyashchy, A. A. (1993). *Convective Instabilities in Systems with Interface* (Gordon and Breach, N. Y.).

[569] Sivashinsky, G. I. (1977). Nonlinear analysis of hydrodynamic instability of laminar flames, Part 1, Derivation of basic equations, *Acta Astronaut.* **4**, 1177-1206.

[570] Skeldon, A. C. and Silber, M. (1998). New stability results for patterns in a model of long-wavelength convection, *Physica D* **122**, 117-123.

[571] Smith, K. A. (1966). On convective instability induced by surface-tension gradients, *J. Fluid Mech.* **24**, 401-414.

[572] Smith, M. K. and Davis, S. H. (1983a), Instabilities of dynamic thermocapillary liquid layers, Part 1, Convective instabilities, *J. Fluid Mech.* **132**, 119-144.

[573] Smith, M. K. and Davis, S. H. (1983b). Instabilities of dynamic thermocapillary liquid layers, Part 2, Surface-wave instabilities, *J. Fluid Mech.* **132**, 145-162.

[574] Smith, M. K. (1986). Instability mechanisms in dynamic thermocapillary liquid layers, *Phys. Fluids* **29**, 3182-3186.

[575] Sobolev, V. D., Churaev, N. V., Velarde, M. G. and Zorin, Z. M. (2000). Surface tension and dynamic contact angle of water in thin quartz capillaries, *J. Colloid Interface Sci.* **222**, 51-54.

[576] Sreenivasan, S. and Lin, S. P. (1978). Surface tension driven instability of a liquid film flow down a heated incline, *Int. J. Heat Mass Transfer* **21**, 1517-1526.

[577] Stainthorp, F. P. and Allen, J. M. (1965). The development of repples on the surface of a liquid film flowing inside a vertical tube, *Trans. Inst. Chem. Eng.* **43**, 85-91.

[578] Starov, V. M. (1992). Equilibrium and Hysteresis Contact Angles, *Adv. Colloid Interface Sci.* **39**, 147-173.

[579] Starov, V. M. and Churaev, N. V. (1976). Thickness of wetting films on rough substrates, *Kolloid Zh.* **39**, 1112-1117 (in Russian).

[580] Starov, V. M., Kalinin, V. V. and Chen, J.-D. (1994). Spreading of liquid drops over solid substrata, *Adv. Colloid Interface Sci.* **50**, 187-222.

[581] Starov, V. M., de Ryck, A. and Velarde, M. G. (1997). On the spreading of an insoluble surfactant over a thin viscous liquid layer, *J. Colloid Interface Sci.* **190**, 104-113.

[582] Starov, V. M., Kosvintsev, S. R., Sobolev, V. D. and Velarde, M. G. (2001a). Spreading of liquid drops over saturated porous layers, *J. Colloid Interface Sci.* (submitted).

[583] Starov, V. M., Kosvintsev, S. R., Sobolev, V. D. and Velarde, M. G. (2001b). Spreading of liquid drops over dry porous layers, *J. Colloid Interface Sci.* (submitted).

[584] Stebe, K. J. and Barthes-Biesel, D. (1995). Marangoni effects of adsorption-desorption controlled surfactants on the leading end of an infinitely long bubble in a capillary, *J. Fluid Mech.* **286**, 25-48.

[585] Steinchen, A. and Sanfeld, A. (1973). Chemical and hydrodynamic stability of an interface with an autocatalytic reaction, *Chem. Phys.* **1**, 156-160.

[586] Sternling, C. V. and Scriven, L. E. (1959). Interfacial turbulence: Hydrodynamic instability and the Marangoni effect, *AIChE J.* **5**, 514-523.

[587] Stewartson, K. and Stuart, J. P. (1971). A non-linear instability theory for a wave system in plane Poiseuille flow, *J. Fluid Mech.* **48**, 529-545.

[588] Stokes, G. G. (1851). On the effect of internal friction of fluids on the motion of pendulums, *Trans. Camb. Phil. Soc.* **9**, 8-106.

[589] Strani, M., Riva, R. and Graziani, G. (1983). Thermocapillary convection in a rectangular cavity: Asymptotic theory and numerical simulation, *J. Fluid Mech.* **130**, 347-376.

[590] Strehlo, R., A. (1970). *Combustion Fundamentals* (McGraw-Hill, N. Y.) 310-311.

[591] Su, C. H. and Mirie, R. M. (1980). On head-on collisions between two solitary waves, *J. Fluid Mech.* **98**, 509-525.

[592] Subramanian, R. S. (1981). Slow migration of a gas bubble in a thermal gradient, *AIChE J.* **27**, 646-654.

[593] Subramanian, R. S. (1992). The motion of bubbles and drops in reduced gravity, in *Transport Processes in Bubbles, Drops, and Particles*, edited by R. P. Chahabra and D. De Kee (Hemisphere, N. Y.) 1-40.

[594] Suciu, D. G., Smigelschi, O. and Ruckenstein, E. (1967). Some experiments on the Marangoni effect, *AIChE J.* **13**, 1120-1124.

[595] Suciu, D. G., Smigelschi, O. and Ruckenstein, E. (1969). On the structure of dissolving thin liquid films, *AIChE J.* **15**, 686-689.

[596] Suciu, D. G., Smigelschi, O. and Ruckenstein, E. (1970). The spreading of liquids on liquids, *J. Colloid Interface Sci.* **88**, 520-528.

[597] Sushchik, M. M. and Tsimring, L. S. (1994). The Eckhaus instability in hexagonal patterns, *Physica D* **74**, 90-106.

[598] Swift, J. B. and Hohenberg, P. C. (1977). Hydrodynamic fluctuations at the convective instability, *Phys. Rev. A* **15**, 319-328.

[599] Szekely, J. (1979). *Fluid Flow Phenomena in Metals Processing* (Academic, N. Y.).

[600] Takashima, M. (1981a). Surface tension driven instability in a horizontal liquid layer with a deformable free surface, I. Stationary convection, *J. Phys. Soc. Jpn.* **50**, 2745-2750.

[601] Takashima, M. (1981b). Surface tension driven instability in a horizontal liquid layer with a deformable free surface, II. Overstability, *J. Phys. Soc. Jpn.* **50**, 2751-2756.

[602] Tan M. J., Bankoff, S. G. and Davis, S. H. (1990). Steady thermocapilary flows of thin liquid layers, *Phys. Fluids A* **2**, 313-321.

[603] Tanford, Ch. (1989). *Ben Franklin Stilled the Waves. An Informal History of Pouring Oil on Water with reflections on the ups and downs of scientific life in general* (Duke University Press, London).

[604] Tang, J. and Bau, H. H. (1994). Stabilization of the no-motion state in the Rayleigh-Benard problem, *Proc. Royal Soc. London A* **447**, 587-607.

[605] Tanner, L. H. (1979). The spreading of silicone oil drops on horizontal surfaces, *J. Phys. D* **12**, 1473-1484.

[606] Tanny, J., Chen, C. C. and Chen, C. F. (1995). Effects of interaction between Marangoni and double-diffusive instabilities, *J. Fluid Mech.* **303**, 1-21.

[607] Taylor, T. D. and Acrivos, A. (1964). On the deformation and drag of a falling viscous drop at low Reynolds number, *J. Fluid Mech.* **18**, 466 -476.

[608] Terada, T. and Second-year Students of Physics (1928). Some experiments, *Tokyo Imp. Univ., Aero. Res. Inst. Rept.* **3**, 1-47.

[609] Thess, A. and Bestehorn, M. (1995), Planform selection in Benard-Marangoni convection: l-hexagons versus g-hexagons, *Phys. Rev. E* **52**, 6358-6367.

[610] Thess, A. and Orszag, S. A. (1994). Temperature spectrum in surface tension driven Benard convection, *Phys. Rev. Lett.* **73**, 541-544.

[611] Thess, A. and Orszag, S. A. (1995). Surface tension driven Benard convection at infinite Prandtl number, *J. Fluid Mech.* **283**, 201-230.

[612] Thiele, U., Velarde, M. G. and Neuffer, K. (2001a). Dewetting: Film rupture by nucleation in the spinodal regime, *Phys. Rev. Lett.* **87**, 16104-16107.

[613] Thiele, U., Neuffer, K., Pomeau, Y. and Velarde, M. G. (2001b). Film rupture in the diffuse interface model coupled to hydrodynamics, *Phys. Rev. E* (to appear).

[614] Thiele, U., Neuffer, K., Bestehorn, M., Pomeau, Y. and Velarde, M. G. (2001c). Sliding drops in the diffuse interface model coupled to hydrodynamics, *Phys. Rev. E* (to appear).

[615] Thompson, R. L., De Witt, K. J. and Labus, T. L. (1980). Marangoni bubble motion phenomenon in zero gravity, *Chem. Engng. Commun.* **5**, 299-314.

[616] Thomson, J. (1855). On certain curious motions observable at the surfaces of wine and other alcoholic liquors, *Philos. Mag.* **10**, 330-333.

[617] Thomson, J. (1882). On a changing tesselated structure in certain liquids, *Procs. Royal Phil. Soc.* (Glasgow), **13**, 464-468.

[618] Thornton, J. (1987). Interfacial phenomena and mass transfer in liquid-liquid extraction, *Chem. and Ind.* **6**, 193-200.

[619] Tolstoi, D. M. (1952a). Molecular theory of fluid sliding on solid surfaces, *Dokl. Akad. Nauk. SSSR* **85**, 1089-1093 (in Russian).

[620] Tolstoi, D. M. (1952b). Mercury sliding on glass, *Dokl. Akad. Nauk. SSSR* **85**, 1329-1335 (in Russian).

[621] Tomita, H. and Abe, K. (2000). Numerical simulation of pattern formation in the Benard-Marangoni convection, *Phys. Fluids* **12**, 1389-1400.

[622] Toh, S. and Kawahara, T. (1985). On the stability of soliton-like pulses in a nolinear dispersive system with instability and dissipation, *J. Phys. Soc. Jpn.* **54**, 1257-1269.

[623] Tomlison, C. (1864). On the motions of eugenic acid on the surface of water, *Phil. Mag.* S4, No. 185, Suppl. Vol. **27**, 528-537.

[624] Tomlison, C. (1869). On the motions of camphor on the surface of water, *Phil. Mag.* **38**, N 257, 409-424.

[625] Topper, J. and Kawahara, T. (1978). Approximate equations for long nonlinear waves on a viscous film, *J. Phys. Soc. Jpn.* **44**, 663-666.

[626] Triantafyllou, M. S. and Triantafyllou, G. S. (1991). Frequency coalescence and mode localization phenomena: a geometric theory, *J. Sound Vibr.* **150**, 485-500.

[627] Trifonov, Y. Y. (1989). Bifurcations of two-dimensional into three-dimensional wave regimes for a verticallly flowing liquid film, *Izv. Akad. Nauk SSSR, Mekh. Zhidk. Gaza* **5**, 109-114 (in Russian).

[628] Trifonov, Y. Y. (1990). Bifurcations of two-dimensional into three-dimensional wave regimes for a vertically flowing liquid film, *Fluid Dyn.* **25**, 741-745.

[629] Trifonov, Y. Y. (1991). Stability of two-dimensional traveling waves on a vertically draining liquid film to three-dimensional perturbations, *J. Appl. Mech. Tech. Phys.* **32**, 210-214 (in Russian).

[630] Trifonov, Y. Y. and Tsvelodub, O. Y. (1986). Three-dimensional steady state traveling waves on a vertically flowing liquid film, *J. Appl. Mech. Tech. Phys.* **27**, 821-828.

[631] Trifonov, Y. Y. and Tsvelodub, O. Y. (1991). Nonlinear waves on the surface of a falling liquid film. Part 1. Waves of the first family and their stability, *J. Fluid. Mech.* **229**, 531-554.

[632] Troian, S. M., Herzbolheimer, E., Safran, S. A. and Joanny, J. F. (1989a). Fingering instabilities of driven spreading films, *Europhys. Lett.* **10**, 25-30.

[633] Troian, S. M., Wu, X. L. and Safran, S. A. (1989b). Fingering instability in thin wetting films, *Phys. Rev. Lett.* **62**, 1496-1499.

[634] Tsimring, L. S. (1996). Dynamics of penta-hepta defects in hexagonal patterns, *Physica D* **89**, 368-380.

[635] Tsvelodub, O. Y. (1980a). Solitons on a falling film at intermediate flow rates, *Zh. Prikl. Mekh. Tekh. Fiz.* **3**, 64-70 (in Russian).

[636] Tsvelodub, O. Y. (1980b). Stationary traveling waves on a film flowing down an inclined plane, *Izv. Akad. Nauk SSSR, Mekh. Zhidk. Gaza* **4**, 142-146 (in Russian).

[637] Tsvelodub, O. Y. and Trifonov, Y. Y. (1992). Nonlinear waves on the surface of a falling liquid film. Part 2. Bifurcations of the first-family waves and other types of nonlinear waves, *J. Fluid Mech.* **244**, 146-169.

[638] Ursell, F. (1983). The long-wave paradox in the theory of gravity waves, *Proc. Cambridge Phil. Soc.* **49**, 685-694.

[639] van der Blij, F. (1978). Some details of the history of the Korteweg-de Vries equation, *Niew Arch. Wisk.* **26**, 54-64.

[640] Van der Mensbrugghe, G. (1870). Sur la tension superficielle des liquides au point de vue de certains mouvements observes a leur surface, *Mem. couronnees & Mem. Savants etrangers, Acad. Royale Belgique* (Brussels) **34**, No. 1, 1-67.

[641] Van der Mensbrugghe, G. L. (1873). Sur la tension superficielle des liquides, consideree au point de vue de certains mouvements observes a leur surface (second memoire), *Mem. couronnees & Mem. Savants etrangers, Acad. Royale Belgique* (Brussels) **37**, No. 4, 1-32.

[642] Van Dyke, M. (1975). *Perturbation Methods in Fluid Mechanics* (Parabolic, Stanford, CA).

[643] Van Dyke, M. (1982). *An Album of Fluid Motion* (Parabolic, Stanford).

[644] van Gurp, G. J., Eggermont, G. E. J., Tamminga, Y., Stacy, W. T. and Gijsbers, J. R. M. (1979). Cellular structure and silicide formation in laser-irradiated metal-silicon, *Appl. Phys. Lett.* **35**, 273-275.

[645] VanHook, S. J., Schatz, M. F., Swift, J. B., McCormick, W. D. and Swinney, H. L. (1997). Long-wavelength surface-tension-driven Benard convection: experiment and theory, *J. Fluid Mech.* **345**, 45-78.

[646] Vanden-Broeck, J. M. (1991). Elevation solitary waves with surface tension, *Phys. Fluids A* **3**, 2659-2663.

[647] van Saarlos, W. (1989). Front propagation into unstable states. II. Linear versus nonlinear marginal stability and rate of convergence, *Phys. Rev. A* **39**, 6367-6390.

[648] Varley, C. (1836). Circulation in oil of turpentine, spirit of wine &c, *Trans. Roy. Soc. Arts Sc. Mauritius* **50**, 190-191.

[649] Velarde, M. G. and Castillo, J. L. (1982). Transport and reactive phenomena leading to interfacial instability, in *Convective Transport and Instability Phenomena*, edited by J. Zierep and H. Oertel (Braun-Verlag, Karlsruhe) 235-264.

[650] Velarde, M. G. and Christov, C. I. (Editors) (1995). *Fluid Physics* (World Scientific, Singapore).

[651] Velarde, M. G., Nekorkin, V. I. and Maksimov, A. G. (1995). Further results on the evolution of solitary waves and their bound states of a dissipative Korteweg-de Vries equation, *Int. J. Bifurcation Chaos* **5**, 831-839.

[652] Velarde, M. G., Nepomnyashchy, A. A. and Hennenberg, M. (2000). Onset of oscillatory interfacial instability and wave motions in Benard layers, *Adv. Appl. Mech.* **37**, 167-238.

[653] Velarde, M. G. and Normand, Ch. (1980). Convection, *Sci. Am.* **243** (1), 92-108.

[654] Velarde, M. G. and Perez-Cordon, R. (1976). On the (nonlinear) foundations of Boussinesq approximation applicable to a thin layer of fluid (II) Viscous dissipation and large cell gap effects, *J. Phys.* (Paris) **37**, 177-182.

[655] Velarde, M. G., Rednikov, A. Ye. and Linde, H. (1999). Waves generated by surface tension gradients and instability, in *Fluid Dynamics at Interfaces*, edited by W. Shyy and R. Narayanan (University Press, Cambridge) 43-56.

[656] Velarde, M. G. and Rednikov, A. Ye. (2000). Nonlinear waves and (dissipative) solitons in thin liquid layers subjected to surfactant gradients, in *Nonlinear Waves in Multi-Phase Flow*, edited by H.-C. Chang (Kluwer, Dordrecht) 57-67.

[657] Villers, D. and Platten, J. K. (1987). Separation of Marangoni convection from gravitational convection in earth experiments, *Physicochem. Hydrodyn.* **8**, 173-183.

[658] Vinogradova, O. I. (1999). Slippage of water over hydrophobic surfaces, *Int. J. Mineral Process.* **56**, 31-60.

[659] Vochten, R. and Petre, G. (1973). Study of the Heat of Reversible Adsorption at the Air-Solution Interface. Experimental Determination of the Heat of Reversible Adsorption of Some Alcohols, *J. Colloid Interface Sci.* **42**, 320-327.

[660] Walter, H. U. (Editor) (1987). *Fluid Sciences and Materials Science in Space* (Springer-Verlag, Berlin).

[661] Warmuzinski, K. and Tanczyk, M. (1991). Oscillatory Marangoni instability during absorption accompanied by chemical reaction, *Chem. Eng. Sci.* **46** (8), 2031-2039.

[662] Weber, E. H. (1854). Mikroskopische Beobachtungen sehr gesetzmassiger Bewegungen, welche die Bildung von Niederschlagen harziger Korper aus Weingeist begleiten, *Ber. Verhandl. Kon. Sachs. Ges. Wiss. Leipzig* **II**, 57-69.

[663] Weh, L. and Linde, H. (1997). Marangoni-stress-driven "solitonic" (periodic) wave trains rotating in an annular container during heat transfer, *J. Colloid Interface Sci.* **187**, 159-165.

[664] Wei, H.-L. and Subramanian, R. S. (1994). Interactions between two bubbles under isothermal conditions and in a downward temperature gradient, *Phys. Fluids* **6**, 2971-2978.

[665] Weidman, P. D., Linde, H. and Velarde, M. G. (1992). Evidence for solitary wave behavior in Marangoni-Benard convection, *Phys. Fluids A* **4**, 921-926.

[666] Weidman, P. D. and Maxworthy, T. (1978). Experiments on strong interactions between solitary waves, *J. Fluid Mech.* **85**, 417-431.

[667] Whitham, G. B. (1974). *Linear and Nonlinear Waves* (Wiley, N.Y.).

[668] Wiegel, R. L. (1964). *Oceanographical Engineering* (Prentice-Hall, Englewood Clifts).

[669] Wiegel, R. L. (1964). Water wave equivalent of Mach-reflection, in *Procs. 9th. Conf. Coastal Engineering* (ASCE, Lisbon) 82-102.

[670] Wierschem, A., Cerisier, P., Gallet, P. and Velarde, M. G. (1997). Creation and extinction of cells in Benard convection, *J. Non-Equilib. Thermodyn.* **22**, 162-168.

[671] Wierschem, A., Linde, H. and Velarde, M. G. (2000). Internal waves excited by the Marangoni effect, *Phys. Rev. E* **62**, 6522-6530.

[672] Wierschem, A., Velarde, M. G., Linde, H. and Waldhelm, W. (1999). Interfacial wave motions due to Marangoni instability. II. Three-dimensional characteristics of surface waves in annular containers, *J. Colloid Interface Sci.* **212**, 365-383.

[673] Wierschem, A., Linde, H. and Velarde, M. G. (2001). Properties of surface wave trains excited by mass transfer through a liquid surface, *Phys. Rev. E* (to appear).

[674] Wilke, H. and Loser, W. (1983). Numerical calculation of Marangoni convection in a rectangular open boat, *Cryst. Res. Tech.* **18**, 825-833.

[675] Williams, R. (1977). The advancing front of a spreading liquid, *Nature* (London) **266**, 153.

[676] Wilson, S. K. (1993a). The effect of a uniform magnetic field on the onset of steady Benard-Marangoni convection in a layer of conducting fluid, *J. Engng. Maths.* **27**, 161-188.

[677] Wilson, S. K. (1993b). The effect of a uniform magnetic field on the onset of Marangoni convection in a layer of conducting fluid, *Q. J. Mech. Appl. Maths.* **46**, 211-248.

[678] Wozniak, G., Siekmann, J. and Srulijes, J. (1988). Thermocapillary bubble and drop dynamics under reduced gravity-survey and prospects, *Z. Flugwiss. Weltraumforsch.* **12**, 137-144.

[679] Wu, T. Y. (1998). Nonlinear waves and solitons in water, *Physica D* **123**, 48-63.

[680] Xu, J. J. and Davis, S. H. (1983). Liquid bridges with thermocapillarity, *Phys. Fluids* **26**, 2880-2886.

[681] Yamada, T. and Kuramoto, Y. (1976). A reduced model showing chemical turbulence, *Prog. Theor. Phys.* **56**, 681-683.

[682] Yih, C.-S. (1963). Stability of liquid flow down an inclined plane, *Phys. Fluids* **6**, 321-334.

[683] Yih, C.-S. (1963). Stability of liquid flow down an inclined plane, *Phys. Fluids* **6**, 321-334.

[684] Young, T. (1805). An essay on the cohesion of fluids, *Philos. Trans. Royal Soc. London* **95**, 65-87.

[685] Young, N. O., Goldstein, J. S. and Block, M. J. (1959). The motion of bubbles in a vertical temperature gradient, *J. Fluid Mech.* **6**, 350-356.

[686] Zabusky, N. J. and Galvin, C. J. (1971). Shallow-water waves, the Korteweg-de Vries equation and solitons, *J. Fluid Mech.* **47**, 811-824.

[687] Zabusky, N. J. and Kruskal, M. D. (1965). Interaction of solitons in a collisionless plasma and the recurrence of initial states, *Phys. Rev. Lett.* **15**, 240-243.

[688] Zebib, A., Homsy, G. N. and Meiburg, E. (1985). High Marangoni number convection in a square cavity, *Phys. Fluids* **28**, 3467-3476.

[689] Zeren, R. V. and Reynolds, W. C. (1972). Thermal instabilities in two-fluid horizontal layers, *J. Fluid Mech.* **53**, 305-327.

[690] Zeytounian, R. Kh. (1995). Nonlinear long waves on water and solitons, *Uspekhi Fiz. Nauk* **165**, 1403-1456 (in Russian). English translation: *Physics-Uspekhi* **38**, 1333-1381.

[691] Zeytounian, R. Kh. (1998). The Benard-Marangoni thermocapillary instability problem, *Uspekhi Fiz. Nauk* **168**, 259-286 (in Russian). English translation: *Physics-Uspekhi* **41**, 241-267.

[692] Zhao, A. X., Wagner, C., Narayanan, R. and Friedrich, R. (1995). Bilayer Rayleigh-Marangoni convection: transitions in flow structures at the interface, *Proc. Royal Soc. Lond. A* **451**, 487-502.

[693] Zharikov, E. V., Prikhod'ko, L. V. and Storozhev, N. R. (1990). Fluid-flow formation resulting from forced vibration of a growing crystal, *J. Crystal Growth* **99**, 910-914.

[694] Zimmerman, W. B. (2001). Excitation of surface waves due to thermocapillary effects on a stable stratified fluid layer, (preprint).

[695] Zuev, A. L. and Pshenichnikov, A. F. (1987). Deformation and fluid film rupture caused by thermocapillary convection, *J. Appl. Mech. Techn. Phys.* **28**, 399-403.

# Index

active drops, 49, 57, 58, 59, 61, 63, 67, 69
adsorption, 1, 51
adsorption-desorption kinetics, 4, 16, 274
aging of a solitary wave, 174
anomalous thermo/soluto capillarity, 3, 194, 272
autonomous motion, 62, 64, 66, 68
axisymmetric flow, 36

Benard cells, 92
Benard-Marangoni convection, 91, 98, 108, 109, 110, 115, 125, 127, 130, 131, 132, 134, 139, 142, 150, 156
Biot number, 12, 93, 153, 263
Bond number, 12, 183
  dynamic, 85, 94, 189
Boussinesq approximation, 7, 12, 13
Boussinesq-Korteweg-de Vries equation, 165
  dissipation-modified, 164, 171, 178, 181
Brunt-Väisälä frequency, 189, 227
buoyancy force, 7, 45
buoyancy–driven convection, 7, 21, 106

buoyancy-driven (Rayleigh) instability, 205, 253

capillary length, 94, 182
capillary or crispation number, 12, 44, 94, 171, 184, 206, 265, 279
capillary-gravity wave, 94, 165, 166, 168, 183, 225, 227
cnoidal waves, 172, 177
cold corner, 36
contact angle
  advancing, 72
  dynamic, 72
  receding/recessing, 72
  static, 71
creeping flow, 40

desorption, 1, 275
dilational (longitudinal) wave, 165, 168, 187, 224, 226, 227
dilational viscosity, 9, 47, 168
disjoining pressure, 74
dissipative solitons, 175, 176, 228, 229, 230, 231, 232, 234, 239, 240, 241, 242, 243, 245

Eckhaus stability, 126

fixed drop, 60
Froude number, 263